# Springer Geography

The Springer Geography series seeks to publish a broad portfolio of scientific books, aiming at researchers, students, and everyone interested in geographical research.

The series includes peer-reviewed monographs, edited volumes, textbooks, and conference proceedings. It covers the major topics in geography and geographical sciences including, but not limited to; Economic Geography, Landscape and Urban Planning, Urban Geography, Physical Geography and Environmental Geography.

**Springer Geography—now indexed in Scopus**

Maria Korneykova · Viacheslav Vasenev ·
Elvira Dovletyarova · Riccardo Valentini ·
Sergey Gorbov · Denis Vinnikov · Diana Dushkova
Editors

# Smart and Sustainable Urban Ecosystems: Challenges and Solutions

## Proceedings of Smart and Sustainable Cities 2022

Springer

*Editors*
Maria Korneykova
Smart Urban Nature Centre
Agrarian-Technological Institute
Peoples' Friendship University of Russia
(RUDN University)
Moscow, Russia

Elvira Dovletyarova
Department of Landscape Design
and Sustainable Ecosystems
Agrarian-Technological Institute
Peoples' Friendship University of Russia
(RUDN University)
Moscow, Russia

Sergey Gorbov
Southern Federal University
Rostov-on-Don, Russia

Diana Dushkova
Helmholtz Centre for Environmental
Research
Leipzig, Germany

Viacheslav Vasenev
Soil Geography and Landscape Group
Wageningen University and Research
Wageningen, The Netherlands

Riccardo Valentini
Department for Innovation in Biological,
Agro-Food and Forest Systems
Euro-Mediterranean Centre for Climate
Change
Tuscia University
Viterbo, Italy

Denis Vinnikov
Medical Institute
Peoples' Friendship University of Russia
(RUDN University)
Moscow, Russia

ISSN 2194-315X ISSN 2194-3168 (electronic)
Springer Geography
ISBN 978-3-031-37215-5 ISBN 978-3-031-37216-2 (eBook)
https://doi.org/10.1007/978-3-031-37216-2



This Springer imprint is published by the registered company Springer Nature Switzerland AG
The registered company address is: Gewerbestrasse 11, 6330 Cham, Switzerland

# Organization

## Committee

## Chief Patron

Vladimir M. Philippov, Peoples Friendship University of Russia (RUDN University), Russia

## Scientific Program Chair

Riccardo Valentini, Tuscia University, Italy/Peoples Friendship University of Russia (RUDN University), Russia

## Scientific Program Committee

Jean Louis Morel, Université de Lorraine, France
Rattan Lal, Ohio State University, USA
Samuilov Konstantin, Peoples Friendship University of Russia (RUDN University), Russia
Jetse Stoorvogel, Wageningen University, Netherlands
Zhongqi Cheng, City University of New York, USA
Nikolay S. Kasimov, Lomonosov Moscow State University, Russia
Ivan I. Vasenev, Peoples Friendship University of Russia (RUDN University), Russia
Christophe Schwartz, Université de Lorraine, France
Dovletyarova Elvira, Peoples Friendship University of Russia (RUDN University), Russia
Gerd Wessolek, Technical University of Berlin, Germany

## Organizing Chair



## Organizing Co-chair



## Organizing Committee

Olga Romzaykina, Peoples Friendship University of Russia (RUDN University), Russia
Tatiana Zhukova, Peoples Friendship University of Russia (RUDN University), Russia
Vlad Pluyshchikov, Peoples Friendship University of Russia (RUDN University), Russia
Leonid Voskresensky, Peoples Friendship University of Russia (RUDN University), Russia
Ramilla Hajiaghaeva, Peoples Friendship University of Russia (RUDN University), Russia
Morgan Williams, The University of California, USA
Konstantin Samuilov, Peoples Friendship University of Russia (RUDN University), Russia
Polina Petrovskaya, Peoples Friendship University of Russia (RUDN University), Russia
Pavel Konstantinov, Lomonosov Moscow State University, Russia
Tatiana Morin, Urban Soils Institute, New York
Irina Veretelnikova, Peoples Friendship University of Russia (RUDN University), Russia

## Organized by

International Union of Soil Sciences

RUDN
university

# Preface

This manuscript contains a selection of edited, refereed and revised papers, which were presented at the International Conference "Smart and Sustainable Cities-2022 (SSC-2022)" entitled "Sustainable urban ecosystems: challenges and solutions". This cross- and multi-disciplinary conference was hosted by RUDN University, Moscow, Russia, on 20–22 July 2022. The SSC-2022 continued the series of international events focused on the environmental consequences of urbanization and possible solutions for sustainable development of urban ecosystems: Megacities 2050 conference (2016) and SUITMA 9 congress (2017), Smart and Sustainable Cities conferences (2018 and 2022). Compared to the previous events, the SSC-2022 aimed to strengthen connection between challenges and solutions for sustainable urban development. Recreational load, soil pollution, wildfires and other anthropogenic and environmental risks urban ecosystems are exposed to be discussed in the context of corresponding solutions, such as bioremediation, blue-green infrastructures, landscape planning and nature-based solutions. A particular attention was given to technological solutions, express and non-destructive soil survey methods, remote and proximal sensing, real-time monitoring and Internet of things.

SSC-2022 conference provided a panel for an interdisciplinary discussion between scientific and research community, municipal services, landscape architects and engineers, environmental protection agencies and practitioners in urban management and greenery. We would like to thank more than 150 participants and 80 speakers who contributed with plenary, oral and poster presentations, round tables and excursions. We wish to express our special gratitude to the authors who contributed to these proceedings. The volume contains 23 research papers, which were selected by the scientific committee with the additional help of external expert reviewers from over 50 submissions. The authors were asked to consider the reviewers' comments and have made all necessary edits to improve the quality of the papers. The papers addresses the following main topics: 1) environmental challenges in urban ecosystems: recreational load and over-compaction, wildfires, soil pollution by hydrocarbon and heavy metals in different bioclimatic conditions (from tundra to steppe zones); 2) technologies and solutions, including advances techniques in monitoring biodiversity, carbon and water balance, and other ecosystem services at different spatial and temporal scales, nature-based solutions and landscape planning projects.

The conference was organized under the umbrella of the International Union of Soil Sciences. The organizational and financial support to the SSC-2022 Conference was provided by "RUDN University Strategic Academic Leadership Program". We would like to express our gratitude to the many people who put essential efforts to ensure this successful conference: keynote speakers, members of organizing and scientific committees, conveners of sessions and round tables, reviewers and technical editors.

We believe these proceedings will be valuable and informative for researchers, practitioners and policy-makers, involved in sustainable urban development.

Moscow, Russia — Maria Korneykova
Wageningen, The Netherlands — Viacheslav Vasenev
Moscow, Russia — Elvira Dovletyarova
Viterbo, Italy — Riccardo Valentini
Rostov-on-Don, Russia — Sergey Gorbov
Moscow, Russia — Denis Vinnikov
Leipzig, Germany — Diana Dushkova

# Contents

1 **Hydrocarbon Content and Enzymatic Activity of Urban Soils in Kola Subarctic** ..... 1
Vladimir Myazin

2 **Acute and Chronic Phytotoxicity of Subarctic Urban Soils and Industrial Wastes** ..... 11
Irina Mosendz, Marina Slukovskaya, Anna Shirokaya, Svetlana Drogobuzhskaya, Liubov' Ivanova, and Irina Kremenetskaya

3 **The Ecotoxicological State of Urban Soils of the Saint Petersburg City** ..... 27
Vyacheslav Polyakov, Azamat Suleymanov, Alexander Kozlov, Ivan Kushnov, Timur Nizamutdinov, Ekaterina Kozlova, and Evgeny Abakumov

4 **Assessment of Soil Quality in Urban Green Areas of Two Russian Cities by Means of Chemical and Biological Methods** ..... 43
A. Kamalova, A. Gordeev, P. Galitskaya, and S. Selivanovskaya

5 **Assessment of Urban Soil Pollution by Heavy Metals (Russian Federation, Republic of Bashkortostan)** ..... 67
R. F. Khasanova, I. N. Semenova, Ya. T. Suyundukov, and G. R. Ilbulova

6 **Distribution of Microelements in the Mineral Part of Chernozems Under Different Types of Plant Communities in the Rostov Agglomeration** ..... 77
N. V. Salnik, S. N. Gorbov, E. N. Minaeva, A. K. Sherstnev, I. P. Melnikova, and S. S. Tagiverdiev

**7 Seasonal Dynamics of Mobile Phosphorus and Potassium in Podzol Soils Within Slope Catena of Different Anthropogenic Load at the RSAU-MTAA Forest Experimental Station** ..... 87
Solomon Melaku Melese, Ivan Ivanovich Vasenev, and Dmitry Vladimirovich Morev

**8 Grazing Effect on Carbon Stocks and Fluxes in Soils of the Mountainous Pastures** ..... 101
R. Bekmurzaeva, Z. Iriskhanova, K. Ivashchenko, M. Korneykova, D. Sarzhanov, K. Makhinya, D. Gosse, and M. Nakhaev

**9 Ecosystem Services and Ecological State of Postpyrogenic Soils in Urban Forest Environments** ..... 111
Ekaterina Chebykina, Evgeny Abakumov, and Anna Shchepeleva

**10 The Dynamics Peculiarities of the Xylem Sap of Coniferous Woody Plants on the Example of Picea Abies in Rostov-on-Don** ..... 123
R. N. Ospischev, S. N. Gorbov, and A. M. Yaroslavtsev

**11 Daily and Seasonal Dynamics of Mixed Forest Biodiversity in the Moscow Region According to Acoustic Monitoring Data** ..... 131
M. Goretskaya, I. Beme, J. Betancourt, M. Perov, and V. Matasov

**12 A Tremendous Green Roof or Biodiversity Museum? First Outcomes from Soil Survey in Zaryadye Park** ..... 143
V. Vasenev, M. Korneykova, I. Krokhmal, E. Kozlova, A. Robert, A. Losev, D. Sarzhanov, Y. Sotnikova, K. Makhinya, D. Gosse, E. Dovletyarova, and M. Nakhaev

**13 Carbon Stocks and Fluxes in Soils of the Urban Park in Grozny City** ..... 159
K. Ivashchenko, R. Bekmurzaeva, Z. Iriskhanova, V. Vasenev, D. Sarzhanov, M. Korneykova, and M. Nakhaev

**14 Water Table and Dissolved Organic Carbon Seasonal Dynamic at the Different Ecosystems of the Ombrotrophic Bog (Mukhrino, West Siberia)** ..... 169
Evgeny A. Zarov, Anastasiia V. Meshcheryakova, Viktoria S. Shanyova, and Vitaliy Yu Khoroshavin

**15 Specifics of Accumulation and Profile Distribution of Organic Carbon in Soils of Park and Recreational Areas of Rostov Agglomeration** ..... 181
P. N. Skripnikov, S. N. Gorbov, A. Y. Matetskaya, and V. A. Ivolgina

**16 Microbial Activity of Technosols Based on Peat-Sand Mixtures in Different Climatic Zones** ..... 195
Maria N. Vasileva, Maria V. Korneykova, Ekaterina V. Kozlova, Viacheslav I. Vasenev, and Yulia L. Sotnikova

**17 Strategies for Phytoremediation of Polluted Sites in the Subarctic** ..... 207
M. A. Pistsova, A. V. Gromova, M. V. Slukovskaya, Yu. A. Dvornikov, and A. L. Gurinov

**18 Remediation of Urban Soils with Nitrogen and Microbiological Fertilizers when Contaminated by Oil, Gasoline and Fuel Oil** ..... 225
Tatiana Minnikova, Anna Ruseva, and Sergey Kolesnikov

**19 Tree Health of *Betula Pubescens* L. for Sustainable Urban Ecosystems (in the Railway Impact Zone on Kola Peninsula)** ..... 237
Natalya V. Saltan and Ekaterina A. Sviatkovskaya

**20 Ecological Revitalization Master Plan of Lipetsk City Based on the HBV Hydrological Modelling of a Small Ungauged Lipovka River (Russia)** ..... 251
Pavel Terskii, Anatoly Tsyplenkov, Artem Gurinov, Anna Antoniuk, Igor Shchukin, and Aleksey Sayanov

**21 Development of Environmentally Safe Concentration of Silver Content in Soils of Different Resistance to Chemical Pollution** ..... 269
Natalia Tsepina, Sergey Kolesnikov, and Tatiana Minnikova

**22 Cost–Benefit Assessment for Maintenance of Urban Green Infrastructure at the University Campus in Moscow: Application of GreenSpaces and TreeTalker Technologies to Regulating Ecosystem Services** ..... 277
V. Matasov, V. Grigoreva, K. Makhinya, M. Kozyreva, O. Romzaikina, O. Maximova, A. Pakina, A. Filyushkina, and V. Vasenev

**23 Urban Livability: A Place of Environmental Compounds in a Mindset** ..... 295
Vitaly A. Kryukov and Elena I. Golubeva

**Index** ..... 315

# Chapter 1
# Hydrocarbon Content and Enzymatic Activity of Urban Soils in Kola Subarctic

**Vladimir Myazin**

**Abstract** The soils of the urbanized territories of the Kola Subarctic have been studied. The content of hydrocarbons in soil samples in various functional zones of Murmansk ranged from 57 to 2287 mg/kg. The maximum content of hydrocarbons was noted near roads with intensive traffic. The average value of hydrocarbon content in the residential zone of Murmansk was 538 ± 95 mg/kg, and in the recreational zone—185 ± 50 mg/kg. The biological activity of the urban soils of Murmansk, determined by the activity of soil enzymes, is at medium and high level, but there is a decrease in the activity of hydrolytic enzymes and an increase in the activity of redox enzymes. As a criterion for assessing the level of pollution, it can use the percentage ratio of hydrocarbons to soil organic carbon. In accordance with the method, 60% of the studied samples from the urban soils of Murmansk can be classified as uncontaminated or lightly polluted, and 40%—polluted and highly polluted.

**Keywords** Urban soils · Pollution · Hydrocarbons · Enzymatic activity · Subarctic zone

## 1.1 Introduction

The natural components of the urban environment experience an extremely high degree of anthropogenic and technogenic pressure. One of the priority pollutants in urban environments are oil and products of its processing. Therefore, the urban soils require special attention in the constant influence of transport, industry, construction and combustion of fossil fuels. As a result, in the soils of urban areas, a certain complex of hydrocarbons of various classes is formed, which are in different states of aggregation.

V. Myazin (✉)
Institute of North Industrial Ecology Problems—Subdivision of the Federal Research Centre "Kola Science Centre of Russian Academy of Science", Apatity, Russia
e-mail: v.myazin@ksc.ru
URL: http://inep.ksc.ru



M. Korneykova et al. (eds.), *Smart and Sustainable Urban Ecosystems: Challenges and Solutions*, Springer Geography, https://doi.org/10.1007/978-3-031-37216-2_1

The constant load on the soil leads to a change in all its components, from agrochemical and physical properties to microbiological and biochemical parameters. Under the anthropogenic impact, microbial diversity, biochemical parameters and biological activity of soils change first.

Many studies have been carried out in various countries aimed at assessing the urban soils, including the determination of the content of hydrocarbons in the soils of urban landscapes (Adeniyi and Owoade 2010; Erofeeva and Anikina 2021; Garcia-Flores et al. 2009; Khalilzade et al. 2020; Nikolaeva et al. 2017; Quiñonez-Plaza et al. 2017) and the activity of soil enzymes (Baranova et al. 2010; Kazimov and Ali 2012; Trifonova and Zabelina 2017). At the same time, there are few similar works carried out in the northern regions (Alekseev et al. 2016; Korneykova et al. 2021; Vishnevaya et al. 2016).

The study of the urban environment of the Kola Subarctic requires special attention, since the northern ecosystems are considered the most vulnerable to technogenic stress and have a low rate of natural self-recovery after damage. The health of people, the well-being and safety of the urban population depends on the state of the biotic components of the soil and soil cover.

The aim of this study was to assess the content of hydrocarbons in the urban soils of the Kola Subarctic (Murmansk) and the enzymatic activity of urban soils.

## 1.2 Materials and Methods

*Sites Description.* Soil samples were taken in the residential area of the Murmansk city at 35 sites (in yards and near roads with different traffic intensity). The site № 26 on the western coast of the Kola Bay in the forest-tundra, located at a distance from pollution sources and experiencing minimal anthropogenic load, was the background site (Fig. 1.1).

*Sampling.* For the study, the top layer of soil was sampling to a depth of 5–7 cm. The organogenic horizons have the maximum sorption capacity and accumulate the largest amount of hydrocarbons. The samples were placed in bags and transported to the laboratory, dried to air-dry state and analyzed.

*Analysis Methods.* The total petroleum hydrocarbons (TPH) in the soil were determined by IR spectrometry (PND F 16.1:2.2.22-98 1998), based on the extraction of hydrocarbons with carbon tetrachloride, purification of the extract on an aluminum oxide column and subsequent determination of the concentration of hydrocarbons in the eluate using an AN-2 analyzer "Neftekhimavtomatika".

The content of organic carbon in the soil (SOC) was determined by the Nikitin's method with a colorimetric ending according to Orlov-Grindel (Mineev 2001).

The activity of soil dehydrogenase was determined by a colorimetric method based on the reduction of the colorless salt of 2,3,5-triphenyltetrazolium chloride to red triphenylformazan; phosphatase activity—colorimetrically by the Dubovenko

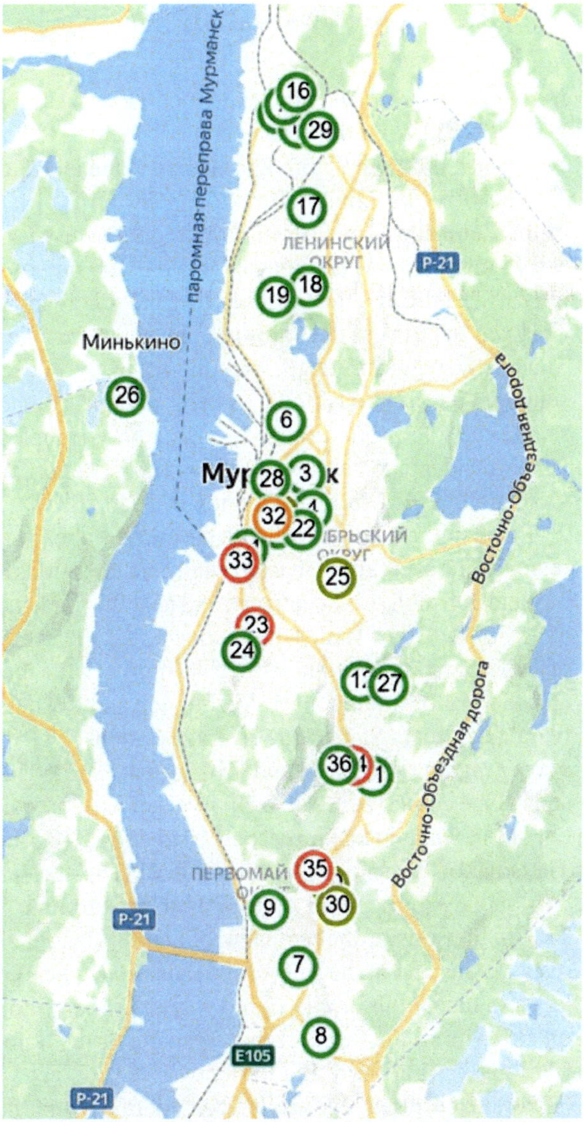


**Fig. 1.1** Map of sampling sites in Murmansk

method, using sodium phenolphthalein phosphate as a substrate; urease activity—colorimetrically with Nessler's reagent according to the Shcherbakova method (Mineev 2001).

*Statistical Processing.* Statistical analysis of the data was carried out using the applied programs Statistica 6.0 and Microsoft Excel 2007. Student's t-test was used to determine the reliability of the coefficients. Pearson's method (significance level 0.05) was used to calculate the correlation coefficient ($r$).

## 1.3 Results and Discussion

*Hydrocarbon Content.* The content of TPH in soil samples at different sites ranged from 57 to 2287 mg/kg. The average value of TPH content in the residential area of Murmansk is 538 ± 95 mg/kg (median value is 302 mg/kg), in the recreation area—185 ± 50 mg/kg. The minimum content of TPH is typical for the residential area of the Leninsky district (256 ± 32 mg/kg), the maximum for the Oktyabrsky (661 ± 195 mg/kg) and Pervomaisky (605 ± 200 mg/kg) districts, which is due to more intensive traffic and the presence thermal power facilities. The maximum content of TPH was noted near the roads with heavy traffic. According to Murmanskstat in the Murmansk region in 2019, the amount of hydrocarbon emissions from stationary sources was 8.7 thous. tons, volatile organic compounds—3.2 thous. tons. At the same time, the proportion of unsatisfactory samples exceeding the MPC on roads in the residential area in 2020 was 0.93 (Report on the state and environmental protection of the Murmansk region in 2020 2021).

The results are consistent with the other studies of urban soils carried out in various regions. The average and background concentrations of TPH in Moscow in 2007 were 689 and 308 mg/kg, respectively, and near the road, it reached 7281 mg/kg (Nikolaeva et al. 2017). Other studies have shown that the average content of TPH in the soils of Moscow ranged from 40 to 588 mg/kg with increases up to 876–1144 mg/kg (Erofeeva and Anikina 2021; Yuzefovich and Kosheleva 2009). In the administrative building sector in Samara, the average content of TPH was 732.9 mg/kg, maximum—1605.7 mg/kg (Vorobieva and Prokhorova 2017). The average content of TPH in the soils of the coastal zones of small rivers in Khabarovsk ranged from 11.3 to 366.6 mg/kg, but in some areas, it reached 1925 mg/kg (Koshelkov and Matyushkina 2018). The soils of Krasnoyarsk content of TPH in the range of 100–500 mg/kg, which corresponds to the level of increased background. The average concentrations of TPH in the soils of the recreational and residential zones were 110 and 125 mg/kg, respectively, and in the soils of the industrial zone, they reached moderate (1180 mg/kg) and dangerous levels (2300 mg/kg) (Sharafutdinov et al. 2018). The content of TPH within the largest urban systems of the Saratov region ranged from 248 to 1127 mg/kg (Larionov and Larionov 2009).

For the northern regions, the content of TPH in the soil is within the same limits. In the soil samples of most settlements of the Yamalo-Nenets Autonomous Okrug, the content of TPH was up to 70 mg/kg. However, higher concentrations (up to 1100 mg/kg) were also noted (Alekseev et al. 2016). The average content of TPH in the soils of Arkhangelsk ranges from 466.20 to 1342.2 mg/kg, reaching 4788 mg/kg (Vishnevaya et al. 2016).

The higher TPH content in the roadside areas of Murmansk is also consistent with results obtained in other studies. The degree of contamination of the soil with TPH in the city of Sumgayit (Azerbaijan) depended on the functional zone: the content of TPH in the roadside soils is on average 2.8 times higher compared to the park soils (Khalilzade et al. 2020). The concentration of TPH in the roadside soil of the city of Tijuana (Mexico) was 608 mg/kg, in road dust, it was 7 times higher—4105 mg/

kg (from 1186 to 9982 mg/kg). The highest concentrations of TPH were found in residential and commercial areas with heavy bus, taxi and truck traffic. Nearly all asphalt roads showed higher TPH concentrations compared to concrete roads. At the same time, the ANOVA test showed that there is no significant difference between TPH concentrations and traffic intensity (Garcia-Flores et al. 2009). The content of TPH in the roadside soils of various functional areas of Ibadan City (Nigeria) ranges from 167 to 923 mg/kg (Onianwa 1995). The content of TPH in the roadside soil in Lagos, Nigeria, was 40 times higher than the average values in the background soil of this region, and in the area of the bus stop, the TPH content reached 16,110 mg/kg (Adeniyi and Owoade 2010).

*Soil Enzyme Activity.* The urban soils of Murmansk are characterized by a decrease in the activity of hydrolytic enzymes (urease, phosphatase) and an increase in the activity of redox enzymes (dehydrogenase). The activity of urease involved in nitrogen metabolism was 29.39 ± 12.06 mg $NH_4$/10 g in urban soils and 132.45 ± 72.26 mg $NH_4$/10 g in the background soil. The activity of phosphatase, which carries out the biological mobilization of organic phosphorus, was 6.16 ± 0.52 mg $P_2O_5$/10 g and 17.40 ± 4.01 mg $P_2O_5$/10 g in urban and background soils, respectively. The activity of dehydrogenase, which is an indicator of the intensity of microbiological decomposition of substances, was almost 2 times higher in urban soils (0.72 ± 0.28 mg TPP/10 g) than in the soil of the background plot (0.40 ± 0.29 mg TPP/10 g) however, this difference statistically not reliable.

A decrease in the activity of hydrolytic enzymes in urban soils was shown earlier in the study of soils in the city of Apatity in the Murmansk region (Korneykova et al. 2021), as well as in other regions (Baranova et al. 2010; Kazimov and Ali 2012). An increase in dehydrogenase activity can be caused by a rearrangement of the soil microbiota community, an increase in the number and proportion of bacteria in soil microorganisms (Korneykova et al. 2021). This can be considered as an adaptive response of the soil to anthropogenic influence, pollution neutralization and self-purification (Trifonova and Zabelina 2017).

Correlation analysis did not show any relationship between the activity of the studied enzymes and the content of TPH in the soil. At the same time, significant correlations were established between the content of organic carbon and the activity of urease ($r = 0.64$, $t = 5.03$; d$f = 34$, $p = 0.05$) and phosphatase ($r = 0.84$, $t = 9$, 30; d$f = 34$, $p = 0.05$).

Thus, in the urban soils of Murmansk, there was a tendency to increase the activity of redox enzymes and a significant decrease in the activity of hydrolytic enzymes, which is a soil response to anthropogenic impact in general. The activity of enzymes is largely determined by the content of organic carbon, pH value, the number and composition of microorganisms and not by the detected content of hydrocarbons. According to the scale for the comparative assessment of the biological activity of soils based on the activity of enzymes, the background soils are classified as very high biological activity soils (according to the activity of hydrolytic enzymes), and urban soils are classified as medium and high biological activity soils.

*Assessing the Degree of Hydrocarbon Pollution.* The assessment of the soil pollution level with oil products is very difficult due to the lack of approved standards. At present, the standards for the TPH content have been developed for individual regions, and the Murmansk region is not one of them. For example, the regional standards for the TPH content in soils of residential areas in the Samara region and Saint Petersburg are no more than 180 mg/kg of soil (Regional standard for soil protection in St. Petersburg 1994; Vorobieva and Prokhorova 2017). In accordance with the scale developed for the city of Arkhangelsk, soils with the TPH content of less than 5 mg/kg are uncontaminated, from 5 to 100 mg/kg are slightly polluted, from 101 to 500 mg/kg are moderately polluted and more than 500 mg/kg are heavily polluted (Vishnevaya and Popova 2016; Vishnevaya et al. 2016). These are rather low MPC standards in comparison with other regions and countries. For example, the MPCs for hydrocarbons in soils set by Mexican law are 3000 mg/kg for residential areas and 6000 mg/kg for industrial and commercial areas; with mixed land use MPC is 3000 mg/kg (Quiñonez-Plaza et al. 2017).

Using these regulatory documents, it is difficult to assess the level of hydrocarbon pollution, since any soil contains compounds defined as of hydrocarbon, which are its natural components. For example, the content of hydrocarbons at the background site near Murmansk is 540 mg/kg, which is higher than these standards. Therefore, the content of TPH in the soil must have a regional standard, which takes into account the characteristics of soil types and the presence of specific and non-specific organic compounds in it. Determination of TPH in uncontaminated soil, along with the studied one, should be mandatory (Okolelova et al. 2019).

To assess the effect of biogenic hydrocarbons on the content of TPH and to more objectively interpret the results, it is necessary to take into account the content of organic carbon in the soil. As a criterion for assessing the level of pollution, it can use the percentage ratio of the amount of TPH to the amount of soil organic carbon:

$$(\mathrm{TPH}/\mathrm{SOC}) \times 100, \tag{1.1}$$

where TPH is the content of hydrocarbons in the soil, mg/kg, SOC is the content of organic carbon in the soil, mg/kg.

An analysis of the TPH content and organic carbon in the soils of Murmansk showed that in the background soil in the forest-tundra area, this ratio is 0.12–0.27, and in urban soils, it reaches 6.62. Based on this criterion, it can evaluate the level of soil pollution in the city of Murmansk with hydrocarbons on the following scale: uncontaminated—0–0.3; lightly polluted—0.3–1; polluted—1–3; highly polluted—more than 3.

According to this method, 60% of the studied samples from the urban soils of Murmansk can be classified as uncontaminated or lightly polluted, and 40%—as polluted and highly polluted (Table 1.1).

**Table 1.1** Content of TPH and the level of soil pollution in various functional zones of Murmansk

| Functional zone | Content of organic carbon, % | Content of TPH, mg/kg | Ratio (TPH/SOC), % | Pollution level |
|---|---|---|---|---|
| Recreation zone | 12.5–16.8 | 135–235 | 0.11–0.14 | Uncontaminated |
| Residential zone | 0.6–29.3 | 57–1037 | 0.18–2.95 | Uncontaminated, lightly polluted, polluted |
| Roadside zone | 2.6–8.1 | 723–2287 | 1.08–6.62 | Polluted and highly polluted |
| Background in the forest-tundra | 27.5 | 540 | 0.20 | Uncontaminated |

## 1.4 Conclusion

This study made it possible to determine the amount of hydrocarbons in urban soils in various areas of Murmansk and to assess their degree of pollution. A method for assessing the degree of contamination of urban soils with hydrocarbons is proposed, taking into account the content of organic carbon in the soil, which makes it possible to more objectively assess the degree of contamination. In accordance with the method, 60% of the studied samples from the urban soils of Murmansk can be classified as uncontaminated or lightly polluted, and 40%—polluted and highly polluted. The most polluted soils are located near roads and in areas with a high concentration of industrial enterprises, including thermal power facilities. At the same time, there are highly polluted areas in the yards of residential buildings also. A high degree of pollution is determined by hydrocarbon emissions as a result of fuel combustion, leakage of oil products in case of vehicle malfunctions, settling of solid particles of aerotechnogenic emissions from industrial enterprises, asphalt and rubber particles from the roadway.

The biological activity of the urban soils of Murmansk, determined by the activity of soil enzymes, is at a medium and high level, but there is a decrease in the activity of hydrolytic enzymes and an increase in the activity of redox enzymes. This indicates the processes of restructuring the community of soil microorganisms as a response to anthropogenic impact. In general, urban soils in the residential and recreational areas of Murmansk are characterized by a low level of hydrocarbon pollution and a relatively high biological activity.

An assessment of the degree of pollution of urban soils with hydrocarbons, taking into account the content of organic carbon and the biological activity, provides more objective results about the state of urban ecosystems. It will make possible to determine the direction and rate of transformation of disturbed ecosystems and the possibility of their restoration.

**Acknowledgements** This research was funded by State scientific program №122022400109-7 (FMEZ-2022-0011).

## References

Adeniyi AA, Owoade OJ (2010) Total petroleum hydrocarbons and trace heavy metals in roadside soils along the Lagos-Badagry expressway, Nigeria. Environ Monit Assess 167:625–630

Alekseev II, Abakumov EV, Shamilishvili GA, Lodygin ED (2016) The content of heavy metals, arsenic and hydrocarbons in the soils of the settlements of the Yamalo-Nenets Autonomous Okrug. Hyg Sanitation 95(9):818–821. https://doi.org/10.18821/0016-9900-2016-95-9-818-821 (in Russian)

Baranova EV, Ilyshkina LN, Polyakova AV (2010) The change of enzymatic activity of soils under city conditions. Anthropogenic Transform Nat Environ 1:144–149 (in Russian)

Erofeeva VV, Anikina EV (2021) Assessment of soil pollution in urban ecosystems (on the example of Moscow). Int Sci Res J 4(106):53–57. Part 2. https://doi.org/10.23670/IRJ.2021.106.4.034 (in Russian)

Garcia-Flores E, Herrera-Castanon D, Vega-Cruz E, Lopez-Laureano G, Rodríguez-Ventura G, Espinoza-Gomez H, Wakida FT (2009) Sources of oil and grease in the stormwater of Tijuana, Mexico. IWA DIPCON B Abstr 13:388–389

Kazimov MA, Ali FM (2012) The hygienic importance of the research of enzymatic activity of soils from roadside territories. Siberian Med J 110(3):118–122 (in Russian)

Khalilzade VJ, Bayram KKh, Bagirova ChZ (2020) Assessment of pollution of urban soils using biomonitoring methods (on the example of Sumgayit). Geogr Environ Living Syst 3:142–153. https://doi.org/10.18384/2712-7621-2020-3-142-153. (in Russian)

Korneykova MV, Redkina VV, Fokina NV, Myazin VA, Soshina AS (2021) Soil microorganisms in the urban ecosystems of the Russian Subarctic (Murmansk region, Apatity). Czech Polar Rep 11(2):333–351. https://doi.org/10.5817/CPR2021-2-23

Koshelkov AM, Matyushkina LA (2018) Assessment of chemical pollution of soils in water protection zones of small rivers in the city of Khabarovsk. Reg Prob 21(2):76–85. https://doi.org/10.31433/1605-220X-2018-21-2-76-85. (in Russian)

Larionov MV, Larionov NV (2009) On the state of urban soils in the Saratov region. Vestnik KrasGAU. Soil Sci 10:35–38 (in Russian)

Mineev VG (2001) Workshop on agricultural chemistry: Proc. Allowance, 2nd ed., revised and additional. Publishing House of Moscow State University, Moscow, p 689 (in Russian)

Nikolaeva O, Rozanova M, Karpukhin M (2017) Distribution of traffic-related contaminants in urban topsoils across a highway in Moscow. J Soils Sediments 17:1045–1053. https://doi.org/10.1007/s11368-016-1587-y

Okolelova AA, Kaplya VN, Lapchenkov AG (2019) Evaluation of the content of petroleum products in soils. Sci Bull Belgorod State Univ Ser Nat Sci 43(1):76–86. https://doi.org/10.18413/2075-4671-2019-43-1-76-86 (in Russian)

Onianwa PC (1995) Petroleum hydrocarbon pollution of urban topsoil in Ibadan City, Nigeria. Environ Int 21(3):341–343

PND F 16.1:2.2.22-98 (1998) Methodology for measuring the mass fraction of hydrocarbons in soils and bottom sediments by IR spectrometry. Moscow, p 17 (in Russian)

Quiñonez-Plaza A, Toyohiko Wakida F, Temores-Peña J, Rodriguez-Mendivi DD, Garcia-Flores E, Pastrana-Corral MA, Melendez-Lopez SG (2017) Total petroleum hydrocarbons and heavy metals in road-deposited sediments in Tijuana, Mexico. J Soils Sediments 17:2873–2886. https://doi.org/10.1007/s11368-017-1778-1

Regional standard for soil protection in St. Petersburg (1994) Order N 891-r. https://docs.cntd.ru/document/9102762 (in Russian)

Report on the state and environmental protection of the Murmansk region in 2020 (2021). Murmansk. https://gov-murman.ru/region/environmentstate (in Russian)

Sharafutdinov RA, Mitev AR, Romanov AA, Borisova IV (2018) Content of oil products in the soil cover of Krasnoyarsk. Vestnik KrasGAU. Biol Sci 6:289–293 (in Russian)

Trifonova TA, Zabelina ON (2017) Changes in the biological activity of heavy metal- and oil-polluted soils in urban recreation territories. Eurasian Soil Sci 50(4):483–490. https://doi.org/10.1134/S1064229317040147

Vishnevaya YuS, Popova LF (2016) Influence of transport on the content of hydrocarbons and oil products in the soils of the residential landscape of the city of Arkhangelsk. Universum Chem Biol Electron Sci Mag 4(22). http://7universum.com/ru/nature/archive/item/3025 (in Russian)

Vishnevaya YuS, Popova LF, Popov SS (2016) Spatial-temporal dynamics of the content of hydrocarbons and oil products in the soils of the residential landscape of the city of Arkhangelsk. Izv. universities. Chem Chem Technol 59(10):88–94 (in Russian)

Vorobieva KYu, Prokhorova NV (2017) Benz(a)pyrene and oil products in the soil cover of settlements of different status in the Samara region. Proc Samara Sci Center Russ Acad Sci 19(2(2)):239–243 (in Russian)

Yuzefovich AM, Kosheleva NE (2009) Pollution of soils in the residential area of Moscow and its relationship with natural and anthropogenic factors. Theor Appl Ecol 3:35–42 (in Russian)

# Chapter 2
# Acute and Chronic Phytotoxicity of Subarctic Urban Soils and Industrial Wastes

**Irina Mosendz, Marina Slukovskaya, Anna Shirokaya, Svetlana Drogobuzhskaya, Liubov' Ivanova, and Irina Kremenetskaya**

**Abstract** Soil degradation is a combination of natural and anthropogenic processes that lead to changes in soil functions, quantitative and/or qualitative deterioration of composition and properties, and a decrease in the natural and economic significance of lands. The state of the soil cover in Russia is far from satisfactory, and in some areas even critical. The Murmansk region is highly industrialized and urbanized region in the Arctic zone. We have tested the acute and chronic phytotoxicity of various industrial materials (expanded vermiculite, nepheline waste and quartz waste), urban and suburban soils of industrial, traffic and recreational zones, soil of urban farm, and conditionally background soil. The water-soluble, plant available, and acid-soluble metal fractions were analyzed. To assess the acute toxicity of industrial wastes, pre-grown hydroponic lawns from *Lolium* spp. were used, and the duration of the experiment was 7 days. The chronic toxicity was assessed using marigolds (*Tagetes* spp.), after 28 and 60 days. The highest level of metals was found in soils of the suburban zone near the plant of nonferrous metallurgy and in the traffic zones, but it was revealed that the presence of nutrient elements in high concentrations can partly neutralize the toxic effect of these soils. Phytotesting showed that the weak

I. Mosendz · M. Slukovskaya (✉)
Laboratory of Nature-Inspired Technologies and Environmental Safety of the Arctic region, Kola Science Centre, Russian Academy of Sciences, Apatity, Russia
e-mail: m.slukovskaya@ksc.ru

I. Mosendz · M. Slukovskaya · A. Shirokaya · S. Drogobuzhskaya · I. Kremenetskaya
I.V. Tananaev Institute of Chemistry and Technology of Rare Elements and Mineral Raw Materials, Kola Science Centre, Russian Academy of Sciences, Apatity, Russia

A. Shirokaya
Department of Landscape Design and Sustainable Ecosystems, RUDN University, Moscow, Russia

L. Ivanova
N.A. Avrorin Polar-Alpine Botanical Garden-Institute, Russian Academy of Sciences, Apatity, Russia

Kola Science Centre, Institute of North Industrial Ecology Problems, Russian Academy of Sciences, Apatity, Russia

 
M. Korneykova et al. (eds.), *Smart and Sustainable Urban Ecosystems: Challenges and Solutions*, Springer Geography, https://doi.org/10.1007/978-3-031-37216-2_2

acceleration of roots to the substrate after 3–7 days of exposition is the express indicator of the presence of high concentrations of toxic metals in the substrates and an unfavorable pH level. Thus, we propose the following low-cost algorithm for the rapid field assessment of soil phytotoxicity in the Subarctic: testing of acute phytotoxicity using hydroponic lawns; testing of chronic phytotoxicity using marigolds in case when acute phytotoxicity is absent; the chemical analysis of mobile fractions of heavy metals and nutrients only for soils which revealed the toxic effect to develop the recommendations for improving soil quality or to make decision about their replacing.

**Keywords** Lawn · Marigold · Phytotoxicity · Urban soil · Industrial waste · Subarctic

## 2.1 Introduction

Various pollutants are released into the environment in large quantities. Industrial activity, poor agricultural practices and urbanization accelerate environmental pollution. Soil pollution is currently attracting considerable public attention as the problem is growing rapidly, especially in highly industrialized regions of the World. Heavy metals are among the most hazardous substances in the soil environment due to their high persistence and toxicity to biota. In soil, heavy metals tend to be strongly adsorbed to the soil matrix and, unlike organic pollutants, are not degraded by microorganisms or by chemical oxidation. The accumulation of heavy metals in the soil can reduce crop yields, cause disease, affect food security and impede sustainable development (Minisha et al. 2021).

Remediation of sites contaminated with heavy metals using chemical or physical methods is a complex task. Phytoremediation is a "low-tech, high-efficiency" purification method that can be used as an alternative to expensive methods. Numerous studies have been carried out aimed at developing an effective and economical way to restore soils contaminated with heavy metals (Alkorta et al. 2004; Erakhrumen 2007). However, in the conditions of polar latitudes, the accumulation of biomass during the year is slow and extremely uneven, which significantly affects the efficiency of such technologies. There are environmental-friendly technologies for the remediation of soil objects using mining waste, the products of which have an acid-neutralizing ability and sorption properties (Slukovskaya et al. 2018). This technology has been tested on the territory of the Subarctic in the impact zone of a copper–nickel plant, but mining wastes have various composition and should be tested before the use in remediation.

Environmental hazard assessment of soils and industrial substrates could be carried out by quantitative methods based on comparing the physicochemical composition with the permissible content of pollutants, and by qualitative assessment accounting the reaction of organisms in biotesting. Physicochemical approach cannot provide a comprehensive assessment of soil toxicity due to the antagonistic or synergistic action of different chemical elements and also has such limitations as costs and labor.

Biotesting methods are based on the organism response to the negative impact of pollutants, and they can provide reliable information about the quality of the substrate (Timofeeva 2011). One of the biotesting methods determines the acute and chronic toxicity of controlled objects for plant organisms. The toxicity criterion is a reliable quantitative change in parameters of testing organisms, based on which a conclusion about the soil toxicity is made.

Phytotesting in an express laboratory version (seed germination in contact with an aqueous soil extract or soil in a humid chamber), carried out in a short period (4–5 days), is known as acute phytotoxicity (Blok et al. 2008), when a longer vegetation experiment (3–6 weeks) in known as chronic phytotoxicity (GOST R ISO 2010).

One suitable plant organism for assessing the degree of toxicity may be the genus *Tagetes* L. (*Asteraceae*) or marigold which is an annual plant 20–30 cm high and widely distributed throughout the World (Politi et al. 2016). The reproduction rate of marigolds is high, the life cycle is short, growth is fast and the root system is well developed (Chatterjee et al. 2012; Goswami and Das 2017; Chintakovid et al. 2008; Sathya et al. 2020). *Tagetes* L. is very popular due to its ease of cultivation and wide adaptability, including poor soils (Coelho et al. 2017). Marigolds show rapid phytoextraction of heavy metals from the soil during their initial growth phase (Choudhury et al. 2016). *Lolium multiflorum* L. was used to assess acute phytotoxicity. The plant is one of the valuable grass species for fast creation of plant communities due to its rapid growth and high biomass, which is necessary for phytoremediation of contaminated areas in harsh climatic conditions (Slukovskaya et al. 2018). Biological reclamation of disturbed areas and industrial dumps, as well as greening of urban spaces in the Subarctic climatic conditions, often requires quick decision-making on the choice of an action strategy and, in particular, such parameters as the composition of soil constructions and plant assortment. In this regard, the purpose of the work was to substantiate the method of field phytotesting of acute and chronic toxicity of urban soils and mining wastes with varying degrees of pollution, applicable in the Subarctic zone.

## 2.2 Materials and Methods

### *2.2.1 Characteristics of Industrial Wastes and Urban Soils*

Samples were collected from the depth 0–15 cm from different land-use zones in the Murmansk region, Russia (NE Europe).

#### 2.2.1.1 Industrial Zone

Tested substrates were sampled from the depositions of different mining wastes.

*Expanded vermiculite* (**IV**) (LLC Kovdorlyuda, Kovdor) with a particle size of 1–2 mm was obtained from the vermiculite concentrate of the Kovdor deposit by firing at 500–550 °C in an electric modular trigger furnace designed by Nizhegorodov (Kremenetskaya et al. 2020). The initial sample was a typical vermiculite for the Kovdor deposit with a noticeable admixture of phlogopite. Vermiculite chemical composition (wt%): $SiO_2$ (30.9), MgO (27.0), $Al_2O_3$ (9.6), $Fe_2O_3$ (5.3), CaO (4.0), $Na_2O$ (3.3), $K_2O$ (0.9), C (0.5), $H_2O$ (7.7). The acidity of the water extract was 8.97.

*Quartz waste* (**IQ**) (JSC Olkon, Olenegorsk) was formed during the processing of ferruginous quartzites, and the production of iron ore concentrate (Suvorova et al. 2012). After the enrichment of iron ores, fine-grained waste was formed, millions of tons of which were annually deposited into the tailings. The annual mass of waste disposal is 53936401 tons/year. Chemical composition (wt%): $SiO_2$ (66.8), $Fe_2O_3$ (19.7), CaO (4.8), $Al_2O_3$ (4.0), MgO (2.9), $K_2O$ (0.9), Cl (0.25), MnO (0.24), $TiO_2$ (0.2), S (0.14), $P_2O_5$ (0.07). The acidity of the water extract was 8.07.

*Nepheline waste* (**IN**) (LLC Lovozersky Mining and Processing Plant, Revda) was formed after the extraction and processing of loparite ores, followed by the production of a concentrate of rare earth metals, as well as niobium and tantalum. The annual mass of waste disposal is 397413 tons/year. The content of elements in the sample (wt.%): Na (9.08), K (2.98), Ca (0.89), Mg (0.25), Mn (0.18), Al (10.57), Fe (4.4), Sr (0.19). The acidity of the water extract was 8.89.

#### 2.2.1.2 Urban and Suburban Zone

Soils were collected from three industrial cities in Murmansk region, Russia: Monchegorsk, Apatity and Kirovsk.

*Soil of industrially polluted zone* (degraded podzol) (**PD**) was sampled from the suburban territory near JSC Kola MMC, Monchegorsk. This area is located on the industrial wasteland and is highly polluted. The proportion of element content in the mobile fraction for potentially toxic metals was (%): Cu (58.1), Ni (44.7), Pb (36.4), Fe (26.5), Al (20.2). The acidity of the water extract of the soil was 4.2–4.8, the content of organic carbon-1.3%, the carbon content of humic acids-0.35%, the content of total nitrogen-0.07%, of which 2.5–4.0 mg/kg of nitrate nitrogen and 0.8 mg/kg of ammonium nitrogen (Ivanova et al. 2021).

*Soils of traffic zone* were sampled from the territory near the administration building of Monchegorsk (**TM**) and a roadside site in Apatity (**TA**) (Saltan et al. 2021). The total composition (wt.%) of the soil in Monchegorsk: Na (4.02), Mg (1.3), Al (8.35), Si (71.3), P (0.4), K (2.13), Ca (2.4), Mn (0.16), Fe (9.7), Co (0.008), Ni (0.08), Cu (0.06), Zn (0.02), Pb (0.004). The acidity of the water extract was 6.78. The composition of the soil sample in Apatity (wt.%)—Na (4.2), Mg (1.5), Al (8.7), Si (68.9), P (0.1), K (3.7), Ca (3.06), Mn (0.13), Fe (8.97), Co (0.008), Ni (0.13), Cu (0.42), Zn (0.1), Pb (0.004). The acidity of the water extract is 6.23.

*Soils of recreational zone*: Soil sample was taken in the city park of Kirovsk (**RK**). The total composition of the sample (wt.%): Na (4.75), Mg (0.81), Al (13.33), Si (62.1), P (1.1), K (3 0.04), Ca (3.4), Mn (0.14), Fe (11.27), Co (0.003), Ni (0.02), Cu (0.03), Zn (0.02), Pb (0.008). The acidity of the water extract is 5.79.

*Soils of urban farming zone*: The soil sample was taken from the farm field of the Polar Experimental Station of the Vavilov All-Russian Institute of Plant Genetic Resources (**UF**). The total composition of the sample (wt.%): Na (5.06), Mg (1.14), Al (13.88), Si (60.44), P (0.5), K (2.33), Ca (3.89), Mn (0.15), Fe (12.51), Co (0.003), Ni (0.017), Cu (0.03), Zn (0.014), Pb (0.002). The acidity of the water extract was 6.15.

*Soils of background zone*: Soils were sampled at the experimental site of the Polar Alpine Botanical Garden-institute (**BG**). The total chemical composition of the sample (wt.%): Na (5.3), Mg (0.97), Al (14.4), Si (61.06), P (0.15), K (4.73), Ca (2.48), Mn (0.17), Fe (10.6), Co (0.003), Ni (0.013), Cu (0.028), Zn (0.02), Pb (0.001). The acidity of the water extract was 5.79.

### 2.2.2 Phytotoxicity Testing

To test the acute toxicity of samples of industrial wastes, seeds of annual ryegrass (*Lolium multiflorum* L.), manufacturer: Agroprogress, class Izorskii RS-1 were used. Hydroponic carpet lawns were previously grown during 7 days in the 1 cm layer of expanded vermiculite on the plastic tray. After this, turfs were layered on the moistened industrial wastes. Test parameters—assessment of root development and growth to the substrate were evaluated after 7 days of the experiment. The acute phytotoxicity of samples of soils and wastes from industrial zone was assessed using *Lolium multiflorum* L. since this plant was used in the technology of remediation of industrially disturbed areas using mining waste as the dominant species of grass cover and was proven to be resistant to harsh climatic conditions of Arctic and Subarctic (Slukovskaya et al. 2018).

To assess the chronic phytotoxicity of samples, we used erect marigolds (*Tagetes patula* L.), manufacturer: AELITA (GOST 12260-81). *Tagetes* L. is the most common plants used in urban landscaping and they are able to grow and blossom in soils of good quality in the Subarctic. Marigolds were preliminarily germinated on filter paper in Petri dishes for three days before planting in tested samples. The duration of the experiment for the samples of the industrial zone was 28 days of vegetation, for the samples of the urban zone—60 days. The test parameters are the length of aboveground plant organs, water balance and the content of chlorophylls a and b in plant leaves. The experiments were carried out in triplicate, and the results were processed statistically ($p < 0.95$). The content of chlorophyll a, b and water band index (WBI) was determined using a SpectraVue Leaf Spectrometer CI-710 s. WBI is the index used for the assessing of water stress and related to them indicators, i.e., productivity determination, etc.

Phytotesting was carried out in laboratory conditions at air temperature of 20–22 °C. Plants were grown in a well-ventilated room using both daylight and artificial lighting.

### *2.2.3 Methods of Soil and Waste Analysis*

To determine the pH, 5 g of soil and mineral samples was placed in 50 ml of distilled water, and the pH of the suspensions was measured after 1 h using an Expert-001 pH-meter-ionomer liquid analyzer with an ESL-63-07SR glass laboratory electrode and auxiliary laboratory silver chloride electrode EVL-1M3.1.

The total content of heavy metals (HM) and content of their acid-soluble and water-soluble fractions were analyzed. Total content of metals was determined after autoclave microwave decomposition in the SW4 system in the DAK 100 autoclaves (Berghof, Germany). A mixture of HF and $HNO_3$ concentrated acids was used to decompose the soil samples. Content of water-soluble fraction (WSF) was determined after treatment of fresh soil (10 g) with deionized water (50 ml), shaking (1 h) and filtering using membrane filter with pore size 0.45 μm. For samples of urban and suburban soils, the content of acid-soluble metal fraction (ASF) was determined after treatment of air-dried soil (1 g) with 0.2 N HCl (50 ml), mixing (1 min), shaking (15 min) and filtering through 'blue ribbon' filter paper with pore size 2–3 μm. For industrial zone samples, a stronger 1N $HNO_3$ acid was used to obtain acid-soluble metal fraction. Solutions were analyzed using a Perkin Elmer ELAN 9000 DRCe inductively coupled plasma mass spectrometer. Multielement calibration solutions Inorganic Ventures (USA) (IV-STOCK-29, IV-STOCK-21, IV-STOCK-28) were used for the instrumental calibration. For quality control, standard reference materials were used: CRM-SOIL-A (High purity standards, USA), GSO 7126-84 Bil-1, GSO 902-76 SP-2, GSO 903-76 SP-3, SDPS, SKR -1 (Russia).

## 2.3 Results

### *2.3.1 Characteristics of Industrial Wastes*

The assessment of the chemical properties of mineral substrates was carried out based on the results of determining the water-soluble, available to plants and acid-soluble fractions of chemical elements (Table 2.1).

**Table 2.1** Results of chemical fractionation of mineral substrates

| Concentration, mg•kg$^{-1}$ | | | | | | | | | | | | |
|---|---|---|---|---|---|---|---|---|---|---|---|---|
| Sample | Na | Ca | Mg | Al | Fe | P | K | Mn | Cu | Ni | Zn | Si | S |
| *Water-soluble fraction* | | | | | | | | | | | | | |
| PD | 18 | *63* | *18* | 7.75 | 7 | 1.375 | 9 | 0.65 | **11** | **3.1** | 0.12 | 20 | **85** |
| IQ | 18 | *58* | *16* | 3.3 | 7.25 | 1.15 | 11 | 0.32 | 0.22 | 0.07 | 0.1 | 13 | 18 |
| IN | 88 | 0.2 | 0.7 | 2.52 | 1.4 | 12.25 | 5 | 0.03 | 0.24 | 0.024 | 0.17 | 9 | 0 |
| *Plant available* (1 M $CH_3COOH + NH_4OH$, *single extraction) fraction* | | | | | | | | | | | | | |
| PD | 10 | 22 | 6 | **427** | **110** | 4.75 | 17.75 | 1.1 | **175** | **6.75** | 0 | 32 | **300** |
| IQ | 9 | *1100* | 11 | 4 | 37 | 0 | 16.5 | 12.75 | 0.142 | 0.217 | 4.75 | 2 | 32 |
| IN | **1100** | 202 | 8 | **470** | 22 | *52.5* | *150* | *32.25* | 0.825 | 0.177 | *10* | *515* | 0 |
| *Plant available* (1 M $CH_3COOH + NH_4OH$, *second + third extractions) fraction* | | | | | | | | | | | | | |
| PD | 12 | 35 | 6 | **700** | **187** | 13.25 | 15.5 | 5.75 | **102.5** | **4.35** | 0.15 | 77 | **42** |
| IQ | 11 | *390* | 11 | 2.4 | 28 | 0 | 15.25 | 6.25 | 0.375 | 0.065 | 1.65 | 5 | 0 |
| IN | 925 | 77 | 3 | **625** | 57 | *28.5* | *77.5* | *14.75* | 0 | 0.105 | *4.25* | *430* | 0 |
| *Acid-soluble* (1N $HNO_3$) *fraction* | | | | | | | | | | | | | |
| PD | 26 | 392 | 72 | **4450** | **825** | 117.5 | 20.5 | 14.75 | **200** | **13.75** | 11.5 | 1675 | 45 |
| IQ | 24 | 725 | 67 | 60 | 190 | 222.5 | 20.5 | 4.07 | 0 | 0.127 | 7 | 80 | 57 |
| IN | **17,575** | *2450* | 57 | **32,450** | 590 | *1300* | *5050* | *102.5* | 0 | 0 | 31 | *44,850* | **1000** |

*Note* Elevated levels of elements that may be significant for the substrate toxicity are highlighted in color, a negative effect is bold, a positive effect is italic

Extremely high concentrations of plant available fraction of such elements as Na, Al, P, K and Mn were obtained for IN substrate. High concentrations of the first two elements can have a negative effect on plant growth, and the last three are nutrients that can be beneficial for plants. The IQ substrate contains high concentrations of Ca and is depleted in both nutrients and toxic metals. Naturally, the highest level of metals is characteristic of PD, in which, there are no elements in elevated concentrations that can neutralize the negative impact, primarily of Cu, Ni and sulfates.

## 2.3.2 *Study of Phytotoxicity of Industrial Wastes*

### 2.3.2.1 Acute Phytotoxicity

The appearance of plants and roots after exposure for 14 days is presented in Table 2.2. As a result of the express test, it was revealed that the complete growth of the roots of *L. multiflorum* occurred in experiments with expanded vermiculite and agricultural soil, which were characterized by the most optimal pH values. Growth of mini lawns to other mineral substrates was not noted. The degree of development of the root system was assessed by visual inspection in points from 1 to 10, where 1 is the absence of roots growth, 3 is very weak growth, 8–9 is good growth and 10 is excellent growth (Table 2.2). In the variant with nepheline waste (IN), an active development of the root system occurred (10 points), for IV, this indicator was 9 points, for UF and IQ—8 points. The most toxic was the contaminated podzol soil (3 points). The results obtained correspond to the results of the chemical analysis of the substrates. The lack of growth of mini lawns reflects the presence of high concentrations of toxic metals (PD) in the substrates and unfavorable pH levels (IN and IQ), and the development of the root system correlates with the availability of nutrients (IN).

Thus, the mini lawns test makes it possible to compare the acute toxicity of different substrates and assess the possibility of their turfing and root development.

**Table 2.2** Root appearance and development of the root system of *Lolium multiflorum* L

| Index | IQ | IN | PD | UF | IV |
|---|---|---|---|---|---|
| Appearance |  |  |  | 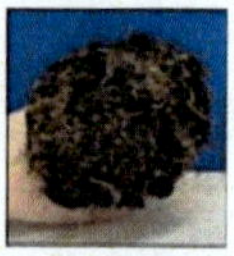 |  |
| Growth to the substrate | No | No | No | Yes | Yes |
| Roots development | 8 points | 10 points | 3 points | 8 points | 9 points |

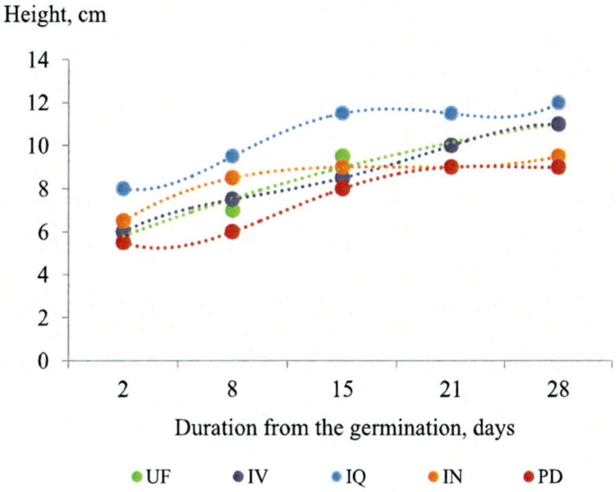


**Fig. 2.1** Height of *Tagetes patula* L. in the experiment with industrial wastes

#### 2.3.2.2 Chronic Phytotoxicity

Phytotesting with marigolds showed that the least favorable substrate for plant growth and development is degraded podzol, which is characterized by the highest content of Cu and Ni (Fig. 2.1). The maximum height of marigolds on the degraded podzol reached $9 \pm 0.5$ cm at the 21st day of vegetation. The greatest height of marigolds was present in the experiment with quartz tails (IQ)—$12 \pm 0.5$ cm. In comparison with the control soil sample, experiments with mineral substrates were characterized by the best results due to the high content of macronutrients such as Mg and Ca.

The results of the study of water balance index (WBI) are presented in Fig. 2.2. This index is designed to assess the moisture content in vegetation. The water content is an important indicator; high humidity is characteristic of healthy vegetation. WBI is used for water stress analysis, productivity determination, fire hazard analysis, irrigated land management, etc. (Penuelas et al. 1995; Champagne et al. 2003).

The lowest WBI values were in the experiment with nepheline waste (IN), and the rest of substrates had almost the same optimal level.

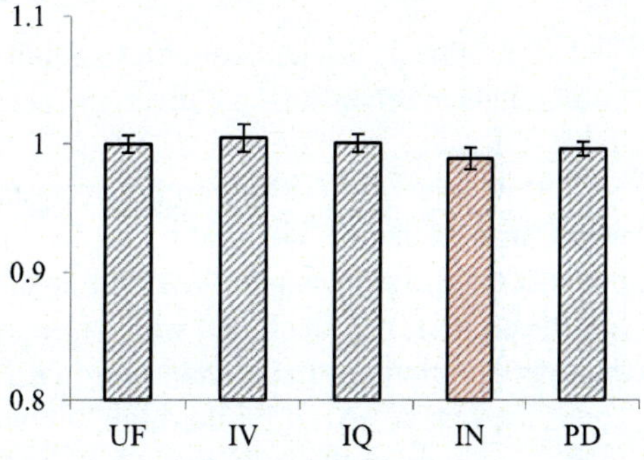


**Fig. 2.2** Water band index of *Tagetes patula* L. in the experiment with industrial wastes

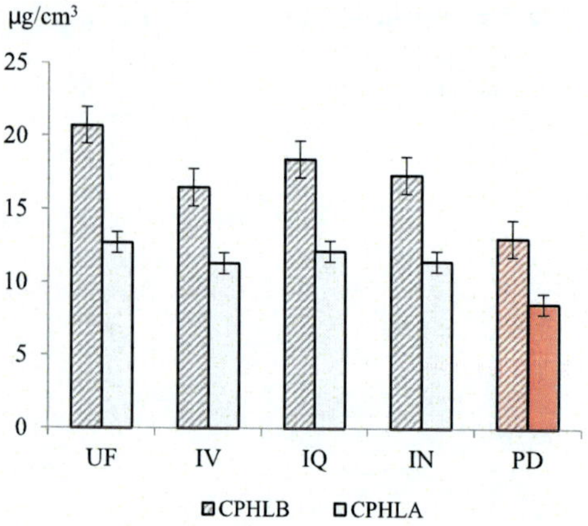


**Fig. 2.3** Chlorophyll content in the leaf blade of *Tagetes patula* L. in the experiment with industrial wastes

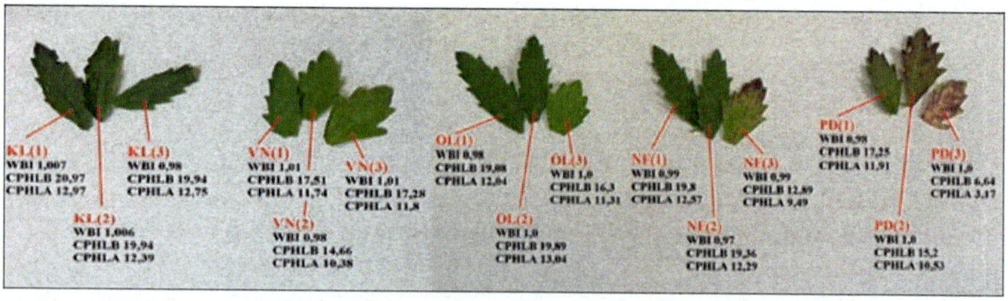


**Fig. 2.4** Image of leaf blades of *Tagetes patula* L. analyzed on a leaf spectrometer

The maximum value of chlorophyll content in the leaf blade belonged to the marigolds grown on the control soil (Fig. 2.3). The minimum content of chlorophylls was noted in the experiment with degraded podzol, which was also reflected in the visual inspection of the leaves (Fig. 2.4).

### 2.3.3 *Characteristics of Urban Soils*

The hydrophysical characteristics, acidity and content of carbon and nitrogen compounds of urban soils are given in Table 2.3, the results of chemical analysis—in Table 2.4.

Soil samples differ in the content of organic carbon, availability of mobile nitrogen, actual and exchangeable acidity. According to the availability of phosphorus and potassium, all samples are classified as highly endowed. The TM and UF samples had the highest content of $NO_3$ which is likely related to the input of mineral fertilizers as a part of planting maintenance activities. The TM and TA samples from traffic zones are characterized by the high content of mobile Cu and Ni, which is related to the proximity to the nonferrous factory (TM) or logistic factor—transportation of

**Table 2.3** Characteristics of urban soil samples

| Land-use zone | Urban farm | Background | Recreational | Traffic, city center | Traffic, divine lane |
|---|---|---|---|---|---|
| Index | UF | BG | RK | TM | TA |
| Bulk density, kg/dm$^3$ | 0.88 | 1.05 | 0.41 | 1.01 | 0.63 |
| Humidity of field samples, % | 21.45 | 23.23 | 50.94 | 18.98 | 36.42 |
| Humidity of air-dried samples, % | 2.82 | 2.67 | 3.61 | 1.57 | 2.29 |
| Loss of ignition, 900 °C | 14.01 | 5.52 | 20.58 | 8.51 | 17.98 |
| Field water capacity, % | 67.49 | 39.83 | 135.29 | 51.14 | 96.71 |
| $NH_4^+$, mg/kg | 3.9 | 4.8 | 2.1 | 2.1 | 0.9 |
| $NO_3^-$, mg/kg | 145.9 | 34.7 | 18.1 | 253.6 | 21.9 |
| C, % | 6.2 | 3.4 | 9.3 | 3.5 | 8.3 |
| N, % | 0.3 | 0.1 | 0.5 | 0.2 | 0.3 |
| H, % | 1.1 | 0.9 | 1.4 | 0.7 | 1.2 |
| pH ($H_2O$) | 6.15 | 6.42 | 5.79 | 6.78 | 6.23 |
| pH (KCl) | 4.81 | 5.1 | 4.33 | 5.94 | 4.6 |
| Eh | 216 | 266 | 209 | 208 | 242 |

contaminated substrate from sand career to other towns of the region (TA). The BG sample is characterized by a high content of Al due to the natural soil properties of Podzols.

### 2.3.4 Phytotoxicity of Urban Soils

The results of the assessment of chronic toxicity of urban soils are presented in Fig. 2.5.

The most toxic was the soil with a high content of Cu and Ni (TA, traffic zone in Apatity), where the plants died on the first day of vegetation. High levels of Cu and Ni were also found in the TM soil, where however the presence of calcium and phosphorus partly neutralized the metal toxicity. The background (BG) soil was more toxic than the TM soil due to the high content of mobile Al, about 2 wt.%; plant height on the last growing day was 6.3 ± 0.5 cm.

**Table 2.4** Chemical composition of urban soil samples

| Sample | Na | Mg | Al | Si | P | K | Ca | Mn | Fe | Co | Ni | Cu | Zn |
|---|---|---|---|---|---|---|---|---|---|---|---|---|---|
| *Total content*, wt.% | | | | | | | | | | | | | |
| UF | 2.26 | 0.51 | 6.21 | 27.03 | 0.22 | 1.04 | 1.74 | 0.07 | 5.60 | 15 | 80 | 134 | 66 |
| BG | 2.47 | 0.45 | 6.76 | 28.53 | 0.07 | 2.21 | 1.16 | 0.08 | 4.96 | 14 | 61 | 132 | 97 |
| RK | 1.81 | 0.31 | 5.08 | 23.66 | 0.42 | 1.16 | 1.28 | 0.05 | 4.29 | 12 | 78 | 132 | 83 |
| TM | 1.89 | 0.61 | 3.92 | 33.43 | 0.17 | 1.00 | 1.13 | 0.07 | 4.58 | 38 | **378** | **305** | 113 |
| TA | 1.57 | 0.58 | 3.33 | 26.34 | 0.04 | 1.45 | 1.17 | 0.05 | 3.43 | 29 | **513** | **1609** | **390** |
| *Acid-soluble fraction* (0.2 N HCl), mg/kg | | | | | | | | | | | | | |
| UF | 195 | 445 | 5850 | 1860 | *2130* | *820* | *4800* | 133 | 1105 | 1.7 | 6 | 19 | 25 |
| BG | 390 | 295 | **19,500** | 7100 | *785* | *375* | *2650* | 160 | 645 | 1.4 | 4 | 12 | 13 |
| RK | 355 | 340 | 8050 | 1105 | *5250* | *540* | *13,650* | 131 | 1145 | 1.2 | 11 | 18 | 56 |
| TM | 142 | 430 | 3350 | 2030 | *1985* | *230* | *6450* | 78 | 1080 | 4.7 | **105** | **200** | 41 |
| TA | 1200 | 585 | 4900 | 3850 | *350* | *700* | *4050* | 43 | 840 | 9.3 | **265** | **1500** | 32 |
| *Water-soluble fraction*, mg/kg | | | | | | | | | | | | | |
| UF | 10 | 1.4 | 0.6 | 4.3 | 2.9 | 41 | 5.9 | 0.04 | 0.24 | 0.001 | 0.008 | 0.030 | 0.005 |
| BG | 10 | 0.7 | **9.3** | 13.7 | 0.7 | 16 | 3.6 | 0.15 | 1.38 | 0.008 | 0.050 | 0.245 | 0.208 |
| RK | 7 | 0.4 | 1.1 | 9.4 | 0.7 | 14 | 2.3 | 0.02 | 0.57 | 0.000 | 0.009 | 0.034 | 0.010 |
| TM | 10 | 2.5 | 0.7 | 2.9 | 2.5 | 14 | 15.3 | 0.02 | 0.57 | 0.001 | 0.119 | 0.132 | 0.009 |
| TA | 15 | 0.5 | 1.7 | 30.0 | 0.3 | 7 | 1.3 | 0.02 | 0.87 | 0.003 | 0.070 | 0.796 | 0.075 |

*Note* Elevated levels of elements that may be significant for the substrate toxicity are highlighted in color, a negative effect is bold, a positive effect is italic

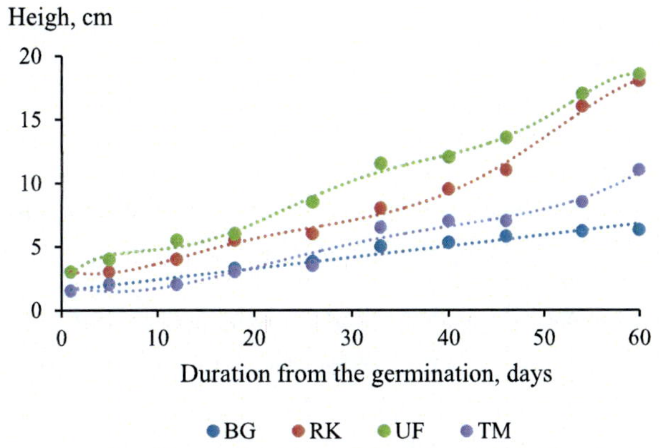


**Fig. 2.5** Height of *Tagetes patula* L. in the experiment with urban soils

The chronic toxicity of RK and UF soils was the same; the height of the marigolds was 18.5 ± 0.5 cm at the end of the experiment. The RK soil sample contained mobile Ni and Al, but also contained mobile Ca, which is known to reduce the toxic effect of metals.

## 2.4 Conclusions

Existing enterprises and an increased regional background for the content of heavy metals in several regions of the Russian Arctic zone require the preservation of natural vegetation, regular planting of greenery and maintaining a high quality of vegetation cover. In the Murmansk region, there are mining enterprises that store thousands of tons of waste, which can be used in the technologies of restoration of industrially polluted and poor urban soils. These products contain layered materials that have high sorption activity with respect to metals, a high specific surface area and the ability to retain moisture and are available in quantities sufficient to the remediation of large areas.

Complex study of the chemical properties of urban soils and industrial wastes and phytotesting of their acute and chronic toxicity using test plants *Lolium multiflorum* L. and *Tagetes patula* L.) shown that the use of these species provide an adequate assessment of the complex phytotoxicity effect of the analyzed materials. At the same time, the data on the chemical composition mainly on the content of mobile forms of the components is needed for the development of recommendations for improving soil quality.

The assessment of acute and chronic phytotoxicity showed the high toxicity of podzol soil from the suburban area of the city Monchegorsk located near the nonferrous plant. This soil contains high concentrations of metals and is depleted in nutrients for essential plants, which are also able to neutralize the toxic effects of metals. The most active development of the root system was observed in variants with expanded vermiculite, nepheline waste and quartz waste. Therefore, these wastes can be easy remediated and also used as components of soil constructions on the areas with lost soil cover.

Phytotesting of urban soils from different land-use zones of three subarctic cities showed that the most toxic was soil with a high content of mobile copper and nickel from the traffic zone of Apatity town, where all plants died. Another tested soil from traffic zone of Monchegorsk town also contained a high concentration of copper and nickel, but this soil was highly provided with phosphorus and calcium, which levels the toxicity of metals.

Thus, we propose the following algorithm for the rapid field assessment of soil phytotoxicity in the Subarctic:

(1) The testing of the acute phytotoxicity (5–7 days) by the root attachment to soil of pre-grown hydroponic lawns from *Lolium* spp.
(2) The testing of chronic phytotoxicity (21 days) using marigolds (*Tagetes* spp.) in case when roots of grasses demonstrated the growth in tested substrate.
(3) The chemical analysis of mobile fractions of heavy metals and nutrients for soils which revealed the toxic effect to take measures to improve the quality of such soils or decide to replace them.

**Acknowledgements** The study of industrial wastes and soils was supported by the Russian Science Foundation Project #21-77-10111 and Council of Grants of the President of Russian Federation MK-2697.2021.1.5, the study of urban soils—in the framework of the state assignment No. FMEZ-2022-0022 of the Russian Academy of Sciences.

## References

Alkorta I, Hernandez-Allica J, Becerril J, Amezaga I, Albizu I, Garbisu C (2004) Recent findings on the phytoremediation of soils contaminated with environmentally toxicheavy metals and metalloids such as zinc, cadmium, lead, and arsenic. Rev Environ Sci Biotechnol 3:71–90. https://doi.org/10.1023/B:RESB.0000040059.70899.3d

Blok C, Persoone G, Wever G (2008) A practical and low cost microbiotest to assess the phytotoxic potential of growing media and soil. ISHS Acta Horticulturae 779:367–374. https://doi.org/10.17660/actahortic.2008.779.46

Champagne CM, Staenz K, Bannari A, McNairin H, Deguise JC (2003) Validation of a hyperspectral curve fitting model for the estimation of plant water content of agricultural canopies. Remote Sens Environ 87:148–160. https://doi.org/10.1016/S0034-4257(03)00137-8

Chatterjee S, Singh L, Chattopadhyay B, Datta S, Mukhopadhyay SK (2012) A study on the waste metal remediation using floriculture at East Calcutta Wetlands, a Ramsar site in India. Environ Monit Assess 184:5139–5150. https://doi.org/10.1007/s10661-011-2328-8

Chintakovid W, Visoottiviseth P, Khokiattiwong S, Lauengsuchonkul S (2008) Potential of the hybrid marigolds for arsenic phytoremediation and income generation of remediators in Ron Phibun District, Thailand. Chemosphere 70:1532–1537. https://doi.org/10.1016/j.chemosphere.2007.08.031

Choudhury MR, Islam MS, Ahmed ZU, Nayar F (2016) Phytoremediation of heavy metal contaminated buriganga riverbed sediment by Indian mustard and marigold plants. Environ Prog Sustain Energy 35:117–124. https://doi.org/10.1002/ep.12213

Coelho LC, Bastos ARR, Pinho PJ, Souza GA, Carvalho JG, Coelho VAT, Oliveira LCA, Domingues RR, Faquin V (2017) Marigold (Tagetes erecta): the potential value in the phytoremediation of chromium. Pedosphere 27:559–568. https://doi.org/10.1016/S1002-0160(17)60351-5

Erakhrumen AA (2007) Phytoremediation: an environmentally sound technology for pollutionprevention, control and remediation in developing countries. Educ Res Rev 2:151–156

GOST R ISO 22030-2009 (2010) Soil quality. Biological methods. Chronic phytotoxicity in relation to higher plants. M.: Standartinform

Goswami S, Das S (2017) Screening of cadmium and copper phytoremediation ability of Tagetes erecta using biochemical parameters and scanning electron microscopyenergy-dispersive X-ray microanalysis. Environ Toxicol Chem 36:2533–2542. https://doi.org/10.1002/etc.3768

Ivanova TK, Slukovskaya MV, Mosendz IA, Krasavtseva EA, Maksimova VV, Kanareykina IP, Shirokaya AA, Kremenetskaya IP (2021) Modified materials based on layered minerals as ameliorants for the remediation of podzol in the industrial barren. RUDN J Agronomy Animal Ind 16(4):370–388. https://doi.org/10.22363/2312-797X-2021-16-4-370-388

Kremenetskaya IP, Alekseeva SA, Slukovskaya MV, Mosendz IA, Drogobuzhskaya SV, Ivanova LA (2020) Expanded vermiculite-reached product obtained from mining waste: the effect of roasting temperature on the agronomic properties. Physicochem Probl Miner Process 56(1):103–113. https://doi.org/10.5277/ppmp19086

Minisha MT, Shah IK, Varghese GK, Kaushal RK (2021) Application of Aztec Marigold (Tagetes erecta L.) for phytoremediation of heavy metal polluted lateritic soil. Environ Chem Ecotoxicol 3:17–22. https://doi.org/10.1016/j.enceco.2020.10.007

Penuelas J, Filella I, Biel C, Serrano L, Save R (1995) The reflectance at the 950–970 region as an indicator of plant water status. Int J Remote Sens 14:1887–1905

Politi FA, Queiroz-Fernandes G, Rodrigues ER, Freitas JA, Pietro RC (2016) Antifungal, antiradical and cytotoxic activities of extractives obtained from Tagetes patula L. (Asteraceae), a potential acaricide plant species. Microb Pathog 95:15–20. https://doi.org/10.1016/j.micpath.2016.02.016

Saltan N, Slukovskaya M, Mikhaylova I, Zarov E, Skripnikov P, Gorbov S, Khvostova A, Drogobuzhskaya S, Shirokaya A, Kremenetskaya I (2021) Assessment of soil heavy metal pollution by land use zones in small towns of the industrialized Arctic Region, Russia. In: Vasenev V, et al (eds) Advanced technologies for sustainable development of urban green infrastructure. SSC 2020. Springer Geography. Springer, Cham. https://doi.org/10.1007/978-3-030-75285-9_10

Sathya V, Mahimairaja S, Bharani A, Krishnaveni A (2020) Influence of soil bioamendments on the availabilty of nickel and phytoextraction capability of Marigold from the contaminated soil. Int J Plant Soil Sci 31:1–12. https://doi.org/10.9734/ijpss/2019/v31i530221

Slukovskaya MV, Kremenetskaya IP, Drogobuzhskaya SV, Ivanova LA, Mosendz IA, Novikov AI (2018) Serpentine mining wastes - materials for soil rehabilitation in Cu-Ni polluted wastelands. Soil Sci 183:141–149. https://doi.org/10.1097/SS.0000000000000236

Suvorova OV, Laschuk VV, Makarov DV, Bokareva VA, Kozhina IS (2012) Investigation of waste of enrichment of ironic quartzites as a raw material for obtaining building ceramics. Proceedings of the Fersman Scientific Session of the Geological Institute, KSC RAS 9. K & M, Apatity, pp 366–369

Timofeeva SS (2011) Modern methods of environmental diagnostics of soils. Bull ISTU 11(58):88–94

# Chapter 3
# The Ecotoxicological State of Urban Soils of the Saint Petersburg City

**Vyacheslav Polyakov, Azamat Suleymanov, Alexander Kozlov, Ivan Kushnov, Timur Nizamutdinov, Ekaterina Kozlova, and Evgeny Abakumov**

**Abstract** The intensive urbanization of environments and technogenic activity has an effect on the accumulation of trace elements in the soil and increases the toxicological risk to the terrestrial ecosystems and human health. We studied the distribution of seven priority trace elements (As, Cd, Pb, Zn, Ni, Cu, Hg) in the soil of the northernmost city in the world with a population more than 1 million people. To identify the spatial distribution of the trace elements, the GIS technologies have been used. Based on the data obtained, interactive maps of soil pollution were made. It was found that the content of the studied trace elements exceeds the background values for the region and has a "hot spots" character. Four of the studied elements (Cu, Zn, Pb, As) on average exceed values of maximum permissible concentrations in soil. The highest levels of pollution are concentrated in the central part of the city, and this is associated with the location of major transport roads, as well as railway stations located in the city center. Accumulation of trace elements occurs in bottom sediments in the rivers of St. Petersburg, and this is associated with lateral runoff of dust from roads and adjacent areas. The level of contamination of bottom sediments varied with the degree of proximity to major transportation hubs. According to the analysis of the index, Zc was found that 57.35% of all the points studied have extremely high levels of pollution. The data obtained indicates that the accumulation of extremely high concentrations of trace elements can lead to a deterioration in the quality of life of the population.

**Keywords** Soil pollution · Zc index · Heavy metals · Risk elements

V. Polyakov (✉) · A. Kozlov · I. Kushnov · T. Nizamutdinov · E. Abakumov
Department of Applied Ecology, Faculty of Biology, St. Petersburg State University, St. Petersburg, Russian Federation
e-mail: v.polyakov@spbu.ru

A. Suleymanov
Department of Soil Science, Ufa Federal Research Centre, Ufa Institute of Biology, Russian Academy of Sciences, Ufa, Russian Federation

E. Kozlova
RUDN University, Moscow, Russian Federation



M. Korneykova et al. (eds.), *Smart and Sustainable Urban Ecosystems: Challenges and Solutions*, Springer Geography, https://doi.org/10.1007/978-3-031-37216-2_3

## 3.1 Introduction

Nowadays, 76% of the people of developed countries (Canada, USA, Germany, France, Japan, South Korea and, etc.) live in cities due to the rapid rates of urbanization (Vasenev et al. 2012). The process of urbanization includes both the formation of new urban areas and the expansion of existing cities (Bullock 1991). Rapidly progressing urbanization and an increase in the area occupied by cities and settlements lead to the formation of urban ecosystems (Blum 2005). An urban ecosystem is a natural urban system consisting of fragments of natural ecosystems surrounded by houses, industrial zones and highways (Polyakov et al. 2020). Urban ecosystem is characterized by the creation of new types of artificially created systems as a result of degradation, destruction and (or) replacement of natural systems and is characterized by less recreational value, disruption of biocycling, reduction of biodiversity both in composition and in structural and functional characteristics (Shamilishvili et al. 2016). In the process of urbanization, natural landscapes undergo a key transformation—the soil and vegetation cover change, and the resulting urban ecosystems are radically different from natural ecosystems (Polyakov et al. 2021). The specificity of soil formation within the urban environment is mainly influenced by the anthropogenic factor, as well as the specific heterogeneity and heterochronicity of soil composition and properties (Ji et al. 2019). Natural landscapes are transformed into anthropogenic ones, and there are local changes in biochemical cycles, volumes of accumulation and removal of chemical elements (Polyakov et al. 2021). Among the main factors that affect the processes of soil formation in the urban environment, it is necessary to distinguish: formation of soils on bulk, reclaimed or mixed soils and cultural layer; presence of inclusions of construction and household garbage, both in the upper horizons and in the whole profile change in acid-alkaline balance with a trend toward alkalization; high pollution with heavy metals and petroleum products; change in physical and chemical properties of soils (reduced moisture capacity, increased compaction, stoniness and other features), profile growth due to intensive aerial spraying (Kapelkina 2010).

Urban soils, as a depositing component of the urban landscape, accumulate many trace elements, including heavy metals and metalloids, as well as organic compounds toxic to living organisms (Lado et al. 2008). Regressive-accumulative distribution of trace elements over the soil profile is characteristic of technogenic territories regardless of the soil type: metals accumulate in the upper humus horizon of the soil and their concentration in the lower horizons decreases sharply. The character of redistribution of heavy metals and metalloids in the profile is influenced by a complex of soil factors such as: the particle size distribution of soils, pH reaction, soil organic matter content, cation exchange capacity, the presence of geochemical barriers as well as drainage (Lu et al. 2014).

Atmosphere and hydrosphere—the most dynamic systems, and changes in their geochemical appearance occur over short periods of time, in contrast to the pedosphere (Maksimova and Abakumov 2015). The study of the geochemical parameters of the soil cover gives an idea of the centers of stable pollution by chemical elements, formed over a certain period of exposure to pollution sources (Ji et al. 2019). In this regard, the environmental assessment of the territory of cities is rational to use the geochemical method, which is based not on the study of the emission of pollutants from man-made sources (which are focused on departmental monitoring systems), but on the study of the accumulation of pollutants in the depositing natural environments, soils in particular (Korchagina et al. 2014). Despite the obvious importance of information on the ecological condition and functioning of urban soils, at the moment there are actually no objective maps of urban soils or a unified approach to their classification (Polyakov et al. 2021). The main reason is the unique spatial variability of urban soils and their properties (Vasenev et al. 2012). The leading environmental control organization in Europe is the European Environment Agency (EEA), and EIONET-CSI system is used to collect data, which is controlled by the European Soil Data Centre, where data on soil contamination in European countries is accumulated and processed (ESDAC: The European Soil Data Centre 2022). The datasets obtained from both governmental organizations and individual researchers are aimed at studying the pathways of human exposure which is a key issue in contaminated regions, and the integration of soil contamination datasets with epidemiological data is a key step for the development of effective environmental biomonitoring and soil cover studies of urban areas (Abrahams 2002; Panagos et al. 2013).

Changes in the quality of the human environment in the city lead to a decrease in the comfort of the population, which is witnessed by medical and demographic indicators, in particular, a high level of morbidity, an increase in genetic diseases and a decrease in life expectancy. Thus, the aim of this study is to investigate the distribution of trace elements in the city of St. Petersburg and to assess the total pollution index of the territory of the city. To achieve this goal, the following objectives were formulated: (1) to evaluate the degree of soil contamination in the city by trace elements; (2) to provide geochemical maps based on the concentration of the pollutants in the soils of Saint Petersburg; (3) to calculate the total soil pollution index (Zc).

## 3.2 Materials and Methods

*Study Area.* The studies were conducted in the city of St. Petersburg in Russia, within the Prinevskaya lowland. The city covers an area of 1 439 $km^2$ with a population of about 4.99 million people. The historical center of the city is located on Litorin marine sediments of various compositions (sands, loams, clays); the northern part of the city is located on limno-glacial deposits of the second Baltic glacial lake. The climate is temperate, transitional from temperate continental to temperate marine. The average annual temperature is +5.8 °C. The annual precipitation is 662 mm (Polyakov et al. 2021).

*Sampling Methodology.* Field studies were conducted in the 2021–2022 years. The study areas belong to different functional zones of the city in areas free from asphalt. The soils were selected from the upper horizons (0–20 cm). Samples were taken in plastic bags and transported to the Applied Ecology Laboratory at SPbSU. The air dry samples were sieved through a 2 mm sieve. 68 soil samples were taken from various functional areas of the city. They include both alluvial territories and soils associated with post-lithogenic soil formation. The sampling points are shown in Fig. 3.1.

*Methods.* To establish the pollution status of the soils and create interactive maps of the city pollution by the risky elements, the following indicators were studied: Zn, Cu, As, Pb, Ni, Hg, Cd. The chemical analysis of the soils was carried out in the Department of Applied Ecology, SPBU. The determination of the content of (Zn, Cu, Pb, Ni, Cd) and metalloids (As) in the soil was carried out in accordance with the inversion voltammetry method on TA analyzers (Soil Quality 1998). The assessment of the risky elements was carried out by comparing the obtained data with the existing hygienic standards and background contents. The geochemical background was used for St. Petersburg according to About the environmental situation in the Leningrad region in 2018 (Region AotL 2018).The total soil pollution index (Zc) is the sum of concentration ratios of toxicants (pollutants) of classes I, II and III of toxicological hazard in relation to background values (3.1).

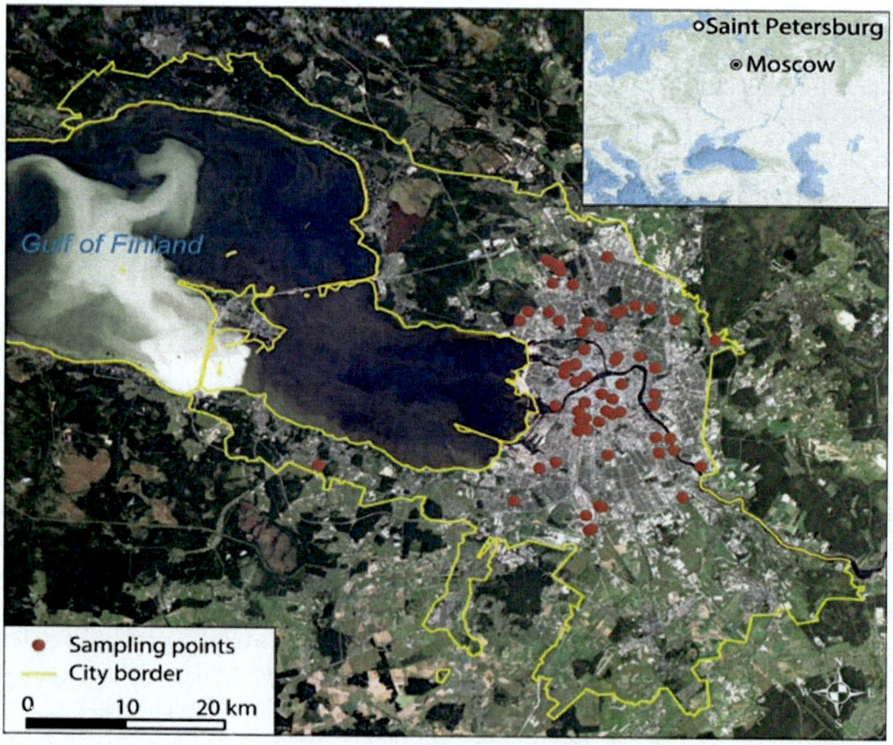


**Fig. 3.1** Study area of St. Petersburg. *Source* Sentinel-2A

$$\mathrm{Zc} = \sum_{i=1}^{n} \mathrm{Kc} - (n - 1) \tag{3.1}$$

where Kc—concentration coefficient of the $i$th chemical element, $n$—the number equal to the number of elements in the geochemical association.

The concentration coefficient (Kc) is calculated by the formula (3.2):

$$\mathrm{Kc} = \frac{\mathrm{Ci}}{\mathrm{Cbc}} \tag{3.2}$$

where Ci—the actual content of the element, Cbc—geochemical background.

*Satellite Data.* The inverse distance-weighting (IDW) method was used to determine the spatial variability and mapping the soil properties on the study plot. The IDW is one of the most widely used tools for interpolating various values, including soil properties (Bhunia et al. 2018). The formula for IDW (3.3) is defined as:

$$Z(x_0) = \frac{\sum_{i=1}^{n} \frac{x_i}{h_{ij}^{\beta}}}{\sum_{i=1}^{n} \frac{x_i}{h_{ij}^{\beta}}} \tag{3.3}$$

where $Z(x_0)$ is the predicted value at the non-sampled location $x_0$; $x_i$ is the $i$th data value; $h_{ij}$ is the separation distance between the interpolated values; and $\beta$ indicates the weighting power (Bhunia et al. 2018).

## 3.3 Results and Discussion

*Soils in the Urban Ecosystem.* To analyze the soil cover of St. Petersburg. The soils of four functional zones were studied: residential, recreational, industrial and traffic. Examples of selected soil profiles of St. Petersburg are shown in Fig. 3.2. Nowadays, in St. Petersburg, soils of fallow lands with a relatively high level of fertility are actively involved to building and city development, as well as, in the west of the city, there is an active alluvial soil formation. In the forest-park zone, the development of soils according to the zonal type of soil formation is noted. The main factors of soil formation in urban environment are humus accumulation, humus formation, overcompaction, overwatering, waterlogging, redeposition of mineral material and salinization (Polyakov et al. 2020). Humus accumulation and humus formation take place mainly in the residential and recreational zones, where soils are less exposed to contamination by priority toxicants. However, due to high anthropogenic load, signs of soil overcompaction can be noted, which negatively affects the hydrothermal regime of soils and may lead to local loss of vegetation (trails and roads). On young soils, formed as a result of alluvial and superficial processes, there are signs of overwatering and waterlogging of the territory (characteristic for park zone) (Maksimova and Abakumov 2015). Soils formed in the industrial and traffic zones, as a result of active anthropogenic impact, are exposed to the greatest transformation. Soils usually have no genetic horizons, or may have a polygenetic structure as a result of redeposition of soil in the course of construction works. These soils have a relatively low organic carbon content, and soil formation processes are poorly observed. The soils forming near linear road objects deserve special attention, and they are subjected to systematic inflow of pollutants together with dust, which, according to some authors (Vlasov et al. 2022; Wei and Yang 2010), is the priority factor of urban soil pollution. Such dust can contain dozens of times more pollutants than in the adjacent soil. Soils of the city of St. Petersburg have a neutral to weakly alkaline pH reaction, and this is due to the alkalinity of the area by introducing reagents, as well as construction dust, which contains lime (Polyakov et al. 2021; Abel et al. 2015). Thus, in the urban environment, we can find both zonal variants of soil formation formed in the recreational zone, weakly transformed soils in the residential zone, alluvial (superficial) soils, as well as completely transformed soils in the industrial and traffic zones.

Natural soils are represented by Histosols (O–$TO_1$–$TO_2$–$TO_3$ profile) with active processes of organic matter accumulation and Hypo-Stagnic Retisol Ruptic (AY–EL–BI–BC–Cg profile) with illuviation of mineral particles. Soils formed in traffic zone—Technosol Entic, Transportic ($C_1$–$C_2$–$C_3$–$C_4$) have poorly developed soil profile in which processes of soil formation and separation of soil horizons are not marked. The sealed soils of the traffic zone—Technosol Transportic Ecranic ($C_1$–$C_2$–BT–$C_3$–$C_4$–G) can retain the structure of the soil profile and serve as buried soils. Relatively young soils of St. Petersburg are alluvial soils formed on the alluvial territories of the city ($C_1$–$C_2$–$C_3$–$C_4$) with signs of the processes of initial soil

**Fig. 3.2** Soil diversity of Saint Petersburg city. **a**—Histosol, recreation zone; **b**—Hypo-Stagnic Retisol Ruptic, recreation zone; **c**—Thapto-Umbric Retisol, recreation/residential zone; **d**—Technosol Entic, Transportic, traffic zone; **e**—Flivosol Antric, alluvial territories, recreation; **f**—Techosol Transportic, recreation/residental zone; **g**—Technosol Transportic Ecranic, traffic zone; **h**—Technosol Entic, industrial zone; **i**—Polygenetic soils: Technosol/Retisol/Gleysol, industrial zone

formation and humus accumulation (AY–$C_1$–$C_2$–Cg). Soils of an industrial zone—Technosol Entic ($C_1$–$C_2$ profile) and Polygenetic soils: Technosol/Retisol/Gleysol (AY–$C_1$–$C_2$–El–BT–$C_3$–$C_4$–$G_1$–$G_2$) as well as the industrial zone are characterized by both complete absence of soil formation processes, and polygenetic profile structure caused by redeposition of material. According to the analysis of the TOC content, it was noted that the highest carbon content is noted in the recreational and residential zones. Soils formed in the industrial and traffic zones are characterized by low content of organic carbon, and this is associated with weak processes of humus accumulation and a low degree of development of soil-forming processes under the influence of anthropogenic pressure. As for cation exchange capacity rate values, they well correspond to clay content and higher in topsoils, also due to increased organic matter percentages. This also good evidence of the soil role in mitigation of urban ecosystems contamination (Lu et al. 2003). In addition to the presented soils, a separate place in urban soils is occupied by park soils and their cultural layers, which reflect those or other processes related to the development of the city (Matinyan et al. 2017). With the gradual increase in the rate of urbanization, more and more lands will be subjected to anthropogenic pressure and transformation, so the study of urban soils is an important task of modern soil science (Blum 2005; Abrahams 2002).

*The soil contamination by heavy metals and metalloid.* We analyzed 68 soil samples at a depth of 0–20 cm (Fig. 3.3). According to the data obtained, we can note that soils have a high degree of contamination by trace elements, i.e., Cu, Zn, Pb, and their average content in the soil exceeds the level of maximum permissible concentrations (MPC) several times. The content of Cd, Ni, Hg, As on average does not exceed the MPC, which indicates a low involvement of these elements in the environment and soil cover. However, all the studied trace elements on average exceed the background concentrations (Table 3.1).

According to the maps of distribution of trace elements, it was found that the pollution of soils with trace elements has a "hot spots" characters, which seems to be associated with the activities of industrial facilities and transport interchanges. Only

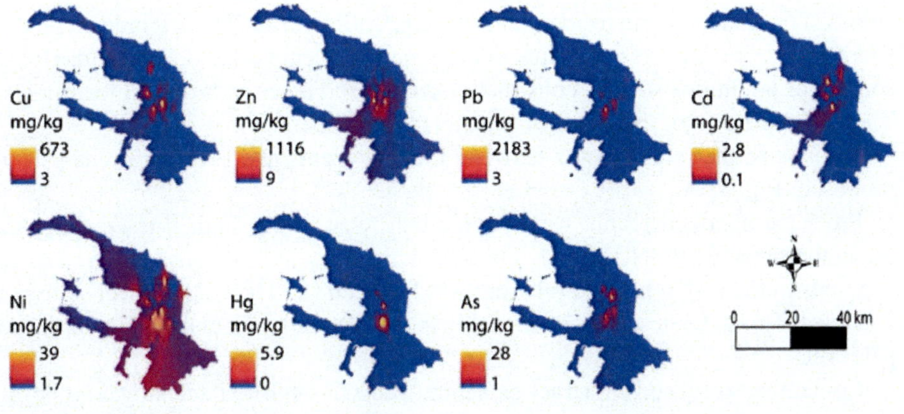


**Fig. 3.3** Geochemical maps of the different heavy metals in St. Petersburg, Russia

**Table 3.1** Descriptive statistics of the risky element concentrations in the soil samples

| Element | n | Mean, mg/kg | Min., mg/kg | Max., mg/kg | SD[a] | CV[b], % | MPC | Background, Mg/kg |
|---|---|---|---|---|---|---|---|---|
| Cu | 68 | 97.18 | 2.80 | 940.20 | 176.40 | 181.51 | 55 | 1.78 |
| Zn | 68 | 192.94 | 9.70 | 1116.50 | 219.48 | 113.75 | 100 | 5.54 |
| Pb | 68 | 124.43 | 2.90 | 2388 | 313.07 | 251.58 | 30 | 5.02 |
| Cd | 68 | 0.42 | 0.10 | 2.77 | 0.60 | 141.63 | 0.5 | 0.07 |
| Ni | 68 | 10.87 | 1.70 | 41 | 8.36 | 76.97 | 85 | 1.48 |
| Hg | 53 | 0.46 | 0.01 | 5.91 | 1.06 | 225.61 | 2.1 | 0.04 |
| As | 68 | 3.60 | 0.99 | 28.05 | 4.77 | 132.42 | 2 | 0.05 |
| TOC[c], % | 38 | 7.18 | 0.12 | 66 | 19.11 | 265.97 | – | – |
| CEC[4], CMP+/kg | 38 | 14.23 | 3 | 78 | 16.84 | 118.35 | – | – |
| Clay, % to fine earth | 34 | 32.14 | 5 | 59 | 2.69 | 48.92 | – | – |

[a] SD—standard deviation
[b] CV—coefficient of variation
[c] total organic carbon
[d] Cation exchange capacity

with regard to Ni, we can note a relatively uniform distribution. The main sources of soil pollution by heavy metals are emissions of pollutants into the atmosphere from transport, energy facilities and industrial enterprises. Metals accumulate relatively quickly in the soil and are very slow to be removed from the soil (Khan et al. 2008; Ufimtseva et al. 2011; Adama et al. 2016).

Analysis of the distribution of trace elements in different functional areas of the city is shown in Figs. 3.4 and 3.5.

From the obtained graphs, we can note that the accumulation of trace elements occurs mainly in the industrial and traffic zones. It is associated with industrial activities, as well as with automobile emissions. However, we can note single extreme values of trace elements in residential zone (e.g., cadmium and lead), it indicates local sources of trace elements in this functional zone. This could be associated with a local sources of pollutions such as construction rubble and other anthropogenic artifacts. The lowest content of trace elements is noted in the recreational zone.

To analyze the relationships between trace elements and pH, factor analysis is proposed (Fig. 3.6).

From the obtained graph, we can note that most of the studied trace elements are in the same block (Ni, Zn, As, Pb, Cu), which may indicate one source of these elements in the soil, while the relationships between Cd, Hg and pH level were not observed. This indicates a low relationship in the accumulation of trace elements and pH level.

Contamination of soil with trace elements occurs through their technogenic dispersion: with emissions of combustion products and dust into the atmosphere during

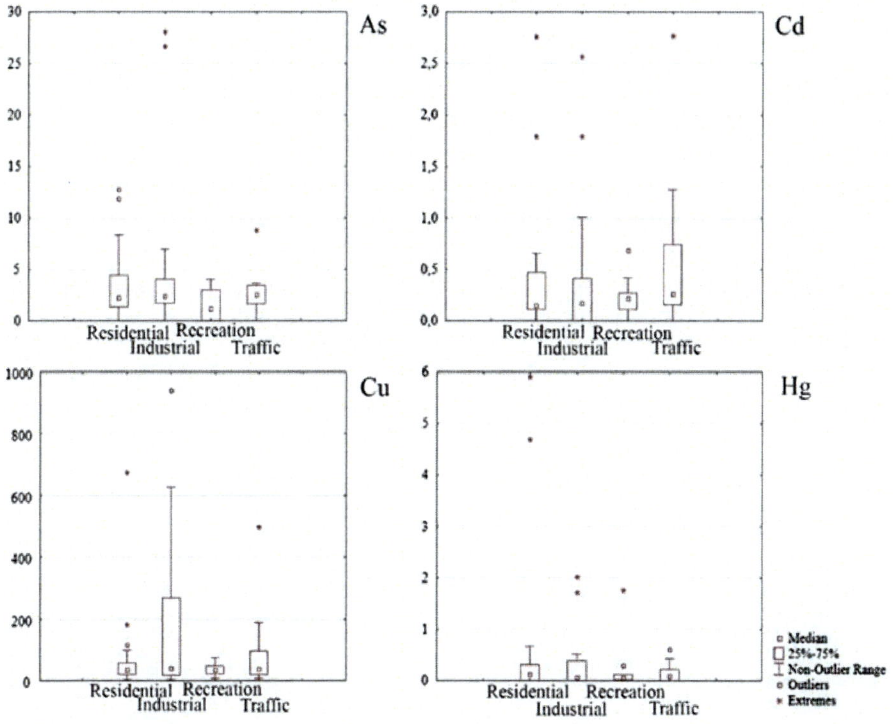


**Fig. 3.4** Distribution of trace elements in different functional zones. The concentration is presented in mg*kg$^{-1}$

high-temperature technological processes (metallurgy, power industry). Heavy metals together with precipitation relatively quickly reach the surface of the soil cover. A significant part of them is included in the soil-forming process, a certain amount is absorbed by plants, as well as carried out with surface runoff of groundwater. As a result, anthropogenic geochemical hotspots of trace elements arise, which are further characterized by a rapid decrease in the concentration of metals from the source of pollution to the periphery (Wei and Yang 2010; Vodyanitskii and Yakovlev 2011; Chen et al. 2005).

*The Total Soil Pollution Index (Zc).* Calculation of Zc pollution index revealed that most of studied samples were extremely highly polluted with trace elements. According to our calculation, 39 samples (57.35% of the total number of samples) showed extremely high pollution level values; 22 samples (32.35%) showed high pollution level values; 4 samples (5.88%) showed moderate pollution level values and 3 samples (4.42%) had low values of pollution. In general, studied soils could be characterized as extremely highly or highly polluted (Table 3.2).

The highest value of Zc index (2741.84, extremely polluted) was observed, it is probably due to the impact of automobile highway, railway and industrial zone, located nearby. The following soil which was sampled close to the previous location,

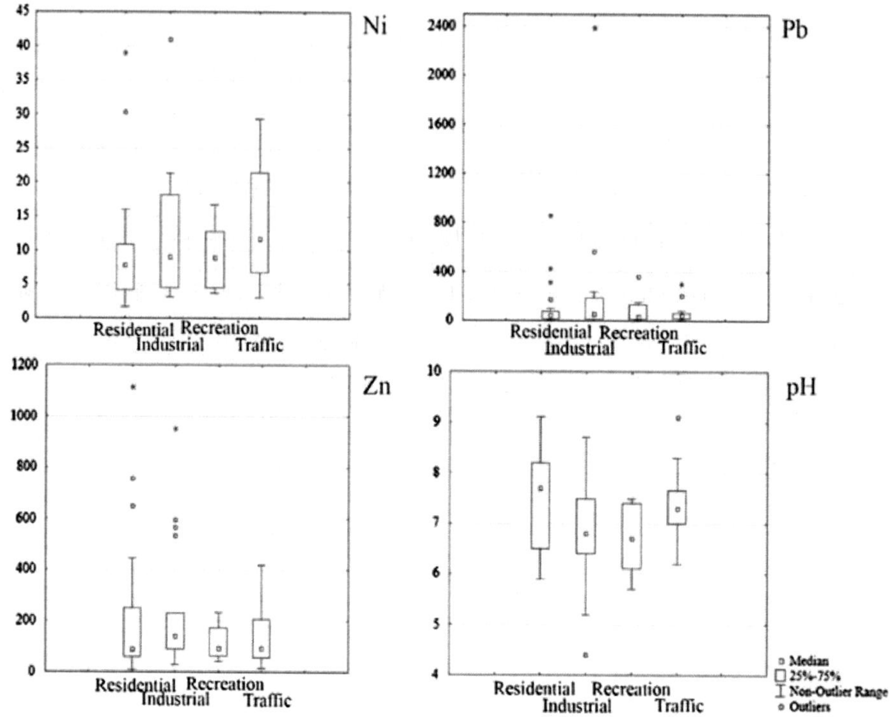


**Fig. 3.5** Distribution of trace elements and pH in different functional zones. The concentration is presented in mg*kg$^{-1}$

also showed extremely high values of pollution load (913.6). The site which is located close to the road with high-density traffic showed high pollution load (125.84). On the other hand, the lowest value of Zc index (3.39) was observed in the sample which is located in the northern part of Saint Petersburg, far away from factories and major roads, on the coast of the lake which is used by locals for recreational purposes. Thus, we can say that industries and traffic considerably increase pollution load of studied soils in the city, mostly up to extremely polluted level, while in remote locations, soils showed low or moderate pollution levels.

Saint Petersburg is a city with many rivers and canals, and it would be interesting to see, how these water objects affect pollution load of adjacent soils and sediments. For instance, sample which is a bottom sediment of the Chyornaya River, in the city center, showed extremely high degree of pollution (365.57). On the other hand, sample of bottom sediment sampled from the Neva River, in the south-eastern part of the city, showed high level of pollution according to Zc index (40.93). This high pollution load level is probably connected with presence of industrial zone nearby, however, this value is much lower than in the Chyornaya River which may be

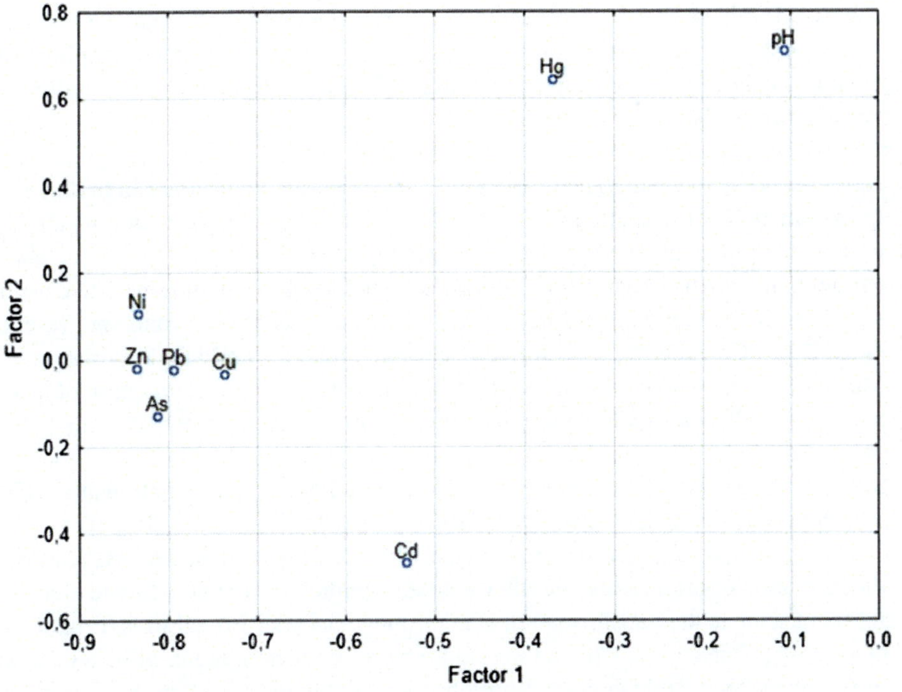


**Fig. 3.6** Factor analysis among trace elements and pH level

**Table 3.2** Levels of soil contamination by total soil pollution index (Revich et al. 1990)

| Contamination level | Total soil pollution index (Zc) | Number of study sites | Impact on human health |
|---|---|---|---|
| Low | 8–16 | 3 | The lowest rates of morbidity in children, the frequency of functional abnormalities is minimal |
| Moderate | 16–32 | 4 | Increase in the general morbidity rate of the population |
| High | 32–128 | 22 | High level of general morbidity, increase in the number of frequently ill children, children with chronic diseases, disorders of the functional state of the cardiovascular system |
| Extremely high | > 128 | 39 | High rate of child morbidity, violation of women's reproductive function (increased toxicity of pregnancy, premature birth, stillbirth, hypotrophy of newborns) |

connected with the fact that Neva River flows from the south-east to north-west direction. Relatively clean water flows to the city and, further, become more polluted in direction to the city center which may cause additional pollution of bottom sediments and adjacent soils in case of flooding.

Soils formed in the urban environment have significant differences from natural soils, this is due to the high heterogeneity of urban development and a high degree of transformation of the landscape (Bullock 1991). Thus, soils in Moscow were characterized by significant diversity and heterogeneity of soil cover, and the coefficient of variation in carbon content varies within 90%, indicating the requirement to develop significantly new approaches for the study of soils in an urban environment (Vasenev et al. 2012). Considering soils formed near large road networks in the west part of Moscow, it can be noted that there are different dynamics of accumulation of heavy metals, in soils near roads accumulated mainly Zn (up to 1364 mg*kg$^{-1}$ in PM 10 of road dust), while other heavy metals do not exceed MPC. Significant increase in concentration of heavy metals is noted in road dust and particulate matter (PM) 10 fraction, which is associated with large specific surface area of PM 10 (Vlasov et al. 2022). Aerial sediments (as a dust PM 10, 2.5) play an important role in the formation of urban soils, and they contain significant amounts of trace elements that contribute to the transformation of anthropogenic-transformed soils (Prokof'eva et al. 2015). In Nanjing, China, the same features of accumulation of heavy metals in soil as in large Russian cities were observed, the highest content corresponds to roadsides (Pb content here reaches up to 308.5 mg*kg$^{-1}$ and the lowest in the park area; Lu et al. (2003) assume that road transport emissions are the main source of Pb in the city. In the eastern district of Moscow, the most polluted soils are in the industrial zone, mean content of Ni is 30.7 mg*kg$^{-1}$, Cu is 329 mg*kg$^{-1}$, Zn is 150 mg*kg$^{-1}$, As is 6.34 mg*kg$^{-1}$, Cd is 0.93 mg*kg$^{-1}$ and Pb is 143 mg*kg$^{-1}$, and the level of pollution here is higher than in the residential, recreational and traffic zones, 3.5 times higher than in the sealed soils in the same functional areas of the city (Nikiforova et al. 2022) and associate with the level of pollution in the studied soils of St. Petersburg. Urban soils have local anthropogenic pollution by heavy metals, this is mainly due to the functioning of industrial centers, as well as the development of the road network (Gorbov et al. 2015; Paltseva et al. 2022). Urban soils are underestimated in terms of ecosystem services, but in an urban environment have an important regulating function, as well as providing and cultural functions (Romzaykina et al. 2021; Deeb et al. 2020). Distribution of trace elements and other pollutants in the urban environment occurs by other mechanisms than in natural soils (Shamilishvili et al. 2016; Vodyanitskii and Yakovlev 2011). This is due to the fact that the soil in the urban environment is anthropogenic-transformed and the natural soil is buried under it, and the cultural layer of such soil can reach several meters (Matinyan et al. 2017). In natural conditions, pollutants mainly accumulate in the upper layers of soils and weakly migrate along the profile, while in the urban environment, soils may have many peaks of heavy metals in the soil profile as a result of the introduction of fresh soil (Polyakov et al. 2020). In Europe, according to data from the European Soil Data Centre, the proportion of potentially polluted sites is

1170*1.000 in urban area, 127*1.000 sites polluted and only 58*1.000 sites remediated. However, according to the predicted calculations, the number of potentially polluted sites will increase up to 2553*1.000, and the number of polluted sites up to 342*1.000 (Panagos et al. 2013). Metallurgy takes the first place among other industries participating in the pollution of urbanized territories (13% of all pollution sources). An important source of urban soil pollution is the input of rubble as a result of construction work, construction rubble is a major source of Pb (219.5 mg*$kg^{-1}$), Cu (119 mg*$kg^{-1}$), Zn (282.2 mg*$kg^{-1}$), Hg (1.56 mg*$kg^{-1}$), which can later enter groundwater (Abel et al. 2015). The same conclusion was reached by scientist from Poland, who noted that anthropogenic artifacts play a significant role in the differentiation of the urban soil (Greinert 2015). Among the priority, pollutants are heavy metals (34.8%), polycyclic aromatic hydrocarbons (10.9%) and aromatic hydrocarbons (10.2%) (Panagos et al. 2013). In Europe, each country spends about 1483.2 million euros per year on remediation measures (80.6%), 15.1% on site investigation and only 4.3% on follow-up care and reclamation measures (Panagos et al. 2013). According to earlier studies conducted in Saint Petersburg, the Summer Garden soils show increased content of heavy metals in the upper soil horizon, high humus content and alkaline reaction, all this indicates an active anthropogenic influence and formation of specific urban soils in the garden park ensembles of the city (Matinyan et al. 2017). As a result of the urbanization, in St. Petersburg were formed artificial alluvial areas, which are a model of soil development in the urban environment, in conditions of active anthropogenic load was revealed that soils are characterized by an increased degree of hydromorphism and the formation of gley conditions, low content of biogenic elements, which negatively affects the vegetation cover area (Maksimova and Abakumov 2015). According to the data on pollution of the territory of the city with heavy metals and polycyclic aromatic hydrocarbon, it was noted that the studied soils of the city are characterized by a dangerous and extremely dangerous level of pollution (Shamilishvili et al. 2016).

## 3.4 Conclusions

Accumulation of trace elements in the soils of urbanized areas is a significant problem in St. Petersburg. It was noted that the content of trace elements in the city exceeds the background content. It was found that Cu, Zn, Pb, which have "hot spots" distribution, indicating local sources of trace elements, makes the largest contribution to pollution status of the city. Based on the data obtained of trace elements distribution, geochemical maps of the pollution of the city were made. It was revealed a high anthropogenic load, which indicates that the active pace of urbanization has a negative impact on the ecotoxicological state of the soil of the city. Pollution of soils is mainly noted in the central part of the city, where there are major transport interchanges. Out of seven trace elements studied, the average values of 4 elements (Cu, Zn, Pb, As) exceed the values of MPC. According to the calculations the total soil pollution index (Zc), it was found that 57.35% of the total number of samples showed extremely high pollution level, which may adversely affect the health of the city population and quality of life in general. In the territory of the city, in areas with high man-made load is necessary to carry out activities to remediate the soil, as high levels of pollution can lead to negative consequences for public health. The data obtained can be used for further development of sustainable soil management systems in the highly urbanized ecosystem of St. Petersburg, located near the Russian–Finnish transboundary zone.

**Acknowledgements** This work was supported by the Russian Scientific Foundation in accordance with agreement from 20.04.2023 № 23-16-20003 and Saint-Petersburg Scientific Foundation in accordance with agreement from 05.05.2023 № 23-16-20003.

## References

11047:1998 I. Soil quality (2020) Determination of cadmium, chromium, cobalt, copper, lead, manganese, nickel and zinc—Flame and electrothermal atomic absorption spectrometric methods. pp 1–18

Abel S, Nehls T, Mekiffer B, Wessolek G (2015) Heavy metals and benzo[a]pyrene in soils from construction and demolition rubble. J Soils Sed 15(8):1771–1780. https://doi.org/10.1007/s11368-014-0959-4

Abrahams PW (2002) Soils: their implications to human health. Sci Total Environ 291(1):1–32. https://doi.org/10.1016/S0048-9697(01)01102-0

Adama M, Esena R, Fosu-Mensah B, Yirenya-Tawiah D (2016) Heavy metal contamination of soils around a hospital waste incinerator bottom ash dumps site. J Environ Public Health 2016:8926453. https://doi.org/10.1155/2016/8926453

Bhunia GS, Shit PK, Maiti R (2018) Comparison of GIS-based interpolation methods for spatial distribution of soil organic carbon (SOC). J Saudi Soc Agric Sci 17(2):114–126. https://doi.org/10.1016/j.jssas.2016.02.001

Blum WEH (2005) Functions of soil for society and the environment. Rev Environ Sci Bio/technol 4(3):75–79. https://doi.org/10.1007/s11157-005-2236-x

Bullock P (1991) Soils in the urban environment. Wiley, Blackwell

Chen T-B, Zheng Y-M, Lei M, Huang Z-C, Wu H-T, Chen H et al (2005) Assessment of heavy metal pollution in surface soils of urban parks in Beijing, China. Chemosphere 60(4):542–551. https://doi.org/10.1016/j.chemosphere.2004.12.072

Deeb M, Groffman PM, Blouin M, Egendorf SP, Vergnes A, Vasenev V et al (2020) Using constructed soils for green infrastructure—challenges and limitations. Soil 6(2):413–434. https://doi.org/10.5194/soil-6-413-2020

ESDAC: The European Soil Data Centre (2022). https://esdac.jrc.ec.europa.eu/. Accessed

Gorbov SN, Bezuglova OS, Varduni TV, Gorovtsov AV, Tagiverdiev SS, Hildebrant YA (2015) Genotoxicity and contamination of natural and anthropogenically transformed soils of the city of Rostov-on-Don with heavy metals. Eurasian Soil Sci 48(12):1383–1392. https://doi.org/10.1134/S106422931512008X

Greinert A (2015) The heterogeneity of urban soils in the light of their properties. J Soils Sed 15(8):1725–1737. https://doi.org/10.1007/s11368-014-1054-6

Ji X, Abakumov E, Polyakov V (2019) Assessments of pollution status and human health risk of heavy metals in permafrost-affected soils and lichens: a case-study in Yamal Peninsula, Russia Arctic. Hum Ecol Risk Assess Int J 25(8):2142–2159. https://doi.org/10.1080/10807039.2018.1490887

Kapelkina LP (2010) Pollutants in soils of megacities. Problems and paradoxes of rationing. Ecol Urban Areas 3:13–19

Khan S, Cao Q, Zheng YM, Huang YZ, Zhu YG (2008) Health risks of heavy metals in contaminated soils and food crops irrigated with wastewater in Beijing, China. Environ Pollut 152(3):686–692. https://doi.org/10.1016/j.envpol.2007.06.056

Korchagina KV, Smagin AV, Reshetina TV (2014) Assessing the technogenic contamination of urban soils from the profile distribution of heavy metals and the soil bulk density. Eurasian Soil Sci 47(8):824–833. https://doi.org/10.1134/S1064229314080080

Lado LR, Hengl T, Reuter HI (2008) Heavy metals in European soils: a geostatistical analysis of the FOREGS Geochemical database. Geoderma 148(2):189–199. https://doi.org/10.1016/j.geoderma.2008.09.020

Lu Y, Gong Z, Zhang G, Burghardt W (2003) Concentrations and chemical speciations of Cu, Zn, Pb and Cr of urban soils in Nanjing, China. Geoderma 115(1):101–111. https://doi.org/10.1016/S0016-7061(03)00079-X

Lu T, Xu CC, Pan GW, Qin Y, Tang H (2014) The summarized study of heavy metal pollution in the city soil. Appl Mech Mater 587–589:842–848. https://doi.org/10.4028/www.scientific.net/AMM.587-589.842

Maksimova E, Abakumov E (2015) Alluviated soils of the Saint-Petersburg city. Bull St Petersburg Univ Ser 3 Biol 4:93–102

Matinyan NN, Bakhmatova KA, Korentsvit VA (2017) Soils of the summer garden (Saint Petersburg). Eurasian Soil Sci 50(6):637–645. https://doi.org/10.1134/S1064229317060060

Nikiforova EM, Kasimov NS, Kosheleva NE, Timofeev IV (2022) Main features and contamination of sealed soils in the east of Moscow city. Environ Geochem Health 44(6):1697–1711. https://doi.org/10.1007/s10653-021-01132-5

Paltseva AA, Cheng Z, McBride M, Deeb M, Egendorf SP, Groffman PM (2022) Legacy lead in urban garden soils: communicating risk and limiting exposure. Front Ecol Evol, vol 10

Panagos P, Van Liedekerke M, Yigini Y, Montanarella L (2013) Contaminated sites in Europe: review of the current situation based on data collected through a European network. J Environ Public Health 2013:158764. https://doi.org/10.1155/2013/158764

Polyakov V, Reznichenko O, Abakumov E, Kostecki J (2020) Ecotoxicological state and pollution status of alluvial soils of St. Petersburg, Russian Federation. Soil Sci Ann 71(3):221–235. https://doi.org/10.37501/soilsa/127089

Polyakov V, Nizamutdinov T, Abakumov E, Morgun E (2021) Soil diversity and key functional characteristics of Yakutsk City: largest urbanized cryogenic world's ecosystem. Energies 14(13). https://doi.org/10.3390/en14133819

Polyakov V, Kozlov A, Suleymanov A, Abakumov E (2021) Soil pollution status of urban soils in St. Petersburg city, North-west of Russia. Soil Water Res 16:164–173

Prokof'eva TV, Shishkov VA, Kiryushin AV, Kalushin IY (2015) Properties of atmospheric solid fallouts in roadside areas of Moscow. Izvestiya Rossiiskoi Akademii Nauk Seriya Geograficheskaya 3:107–120

Region AotL (2019) About the environmental situation in the Leningrad region in 2018. Leningrad region. pp 1–143

Revich BA, Saet YE, Smirnova RS (1990) Methodological recommendations for assessing the degree of pollution of atmospheric air in populated areas by metals in snow and soil. IMGRE, Moscow

Romzaykina ON, Vasenev VI, Paltseva A, Kuzyakov YV, Neaman A, Dovletyarova EA (2021) Assessing and mapping urban soils as geochemical barriers for contamination by heavy metal(loid)s in Moscow megapolis. J Environ Qual 50(1):22–37. https://doi.org/10.1002/jeq2.20142

Shamilishvili GA, Abakumov EV, Gabov DN, Alekseev II (2016) Peculiarities of the fractional composition of polycyclic aromatic hydrocarbons and polyelement pollution of soils of urbanized territories and their hygienic characteristics (on the example of soils of functional zones of St. Petersburg). Hyg Sanitation 95(9):827–837

Ufimtseva MD, Terekhina NV, Abakumov EV (2011) Physico-chemical characteristics of urban soils of the Central district of St Petersburg. Bull St Petersburg Univ 4:85–97

Vasenev VI, Fatiev MM, Lakeev PS, Ivannikov FA, Riccardo V (2012) An approach to access urban soils' spatial variability at the local level. Bull Peoples' Friendship Univ Russ Ser: Agron Anim Husbandry S5:13–24

Vlasov DV, Kukushkina OV, Kosheleva NE, Kasimov NS (2022) Levels and factors of the accumulation of metals and metalloids in roadside soils, road dust, and their PM10 fraction in the Western Okrug of Moscow. Eurasian Soil Sci 55(5):556–572. https://doi.org/10.1134/S1064229322050118

Vodyanitskii YN, Yakovlev AS (2011) Assessment of soil contamination by the content of heavy metals in the soil profile. Eurasian Soil Sci 44(3):297–303. https://doi.org/10.1134/S1064229311010169

Wei B, Yang L (2010) A review of heavy metal contaminations in urban soils, urban road dusts and agricultural soils from China. Microchem J 94(2):99–107. https://doi.org/10.1016/j.microc.2009.09.014

# Chapter 4
# Assessment of Soil Quality in Urban Green Areas of Two Russian Cities by Means of Chemical and Biological Methods

**A. Kamalova, A. Gordeev, P. Galitskaya, and S. Selivanovskaya**

**Abstract** The quality of urban soils in green zones is a result of combination of natural processes, anthropogenic pollution and management. The elevated presence of heavy metals in soils of the urban environment has been recognized as an important source of metal intake into the human body. The aim of this study was to evaluate the concentration of heavy metals in soil of 24 parks and public gardens obtained in Kazan (with 1 Mio citizens) and its satellite city Zelenodolsk. The soil samples and two different herbaceous perennial plant *Plantago major* and *Trifolium repens* were analyzed for 5 metals (Cd, Cu, Pb, Ni, Zn) using ICP-MS. We also determined the physiological profiles of microbial communities using the Biolog Ecoplate multi-substrate testing method and the diversity of plants in urban green zones. The results from the study showed that the heavy metal concentration in soil samples ranged between 0.06 to 0.70 mg/kg, 0.82 to 30.02 mg/kg, 0.06 to 66.42 mg/kg, 9.91 to 71.33 mg/kg and 4.41 to 24.51 mg/kg for Cd, Pb, Cu, Zn, Ni, respectively. In average, the concentrations of the metals in soils did not exceed the levels recommended by WHO. In plants, the concentration of Zn exceeded the recommended levels by 10–25 times. The Shannon–Wiener diversity index for plant communities varied from 2.32 to 2.96, and for microbial community (AWCD)—from 0.143 to 1.695. No significant differences between the diversity indices for large and satellite cities were found. No correlation between the level of metal contents and diversity indices was found.

**Keywords** Heavy metals · Urban Park · Plant diversity · Functional diversity of soil microbial community

A. Kamalova (✉) · A. Gordeev · P. Galitskaya · S. Selivanovskaya
Kazan (Volga region) Federal University (KFU), Kazan, Russia
e-mail: AlinaRAhtyamova@kpfu.ru

 
M. Korneykova et al. (eds.), *Smart and Sustainable Urban Ecosystems: Challenges and Solutions*, Springer Geography, https://doi.org/10.1007/978-3-031-37216-2_4

## 4.1 Introduction

The urban environmental quality is of importance because the majority of people now live in cities (Tresch et al. 2018). Urbanized areas are rapidly expanding at the expense of natural habitats and the corresponding human population shift. It has resulted in landscapes created by humans–urban green spaces, including parks and public gardens, have become a substitute for the lost environmental spaces (Francini et al. 2018; Yan et al. 2019).

Soil of urban green spaces and provides key ecosystem services for a comfortable living in city (Li et al. 2018). Its quality is one of the components of environmental quality, as with the quality of water and air (Andrews et al. 2002). The world's population is increasingly urbanized, with 55% living in city areas, and predictions of up to 66% by 2050 (Różański et al. 2018). Therefore, interest to urban soils due to the potential impact of this soils on population has recently increased (Morel and Heinrich 2008). This is due to the fact that soils in urban and suburban areas are greatly impacted by human activities. This can have significant consequences for the health of people, plants and soil organisms, as well as the ability of water to penetrate the soil (Page-dumroese 2020).

Urban green spaces (UGS) are defined as areas primarily covered by plants, including trees, grasses and shrubs, that have a direct or indirect positive impact on people's lives by purifying toxic substances and improving the microclimate (Haq 2011). This soils quite different from natural soils due to the high level of human activity, leading to increased pollution, physical disruptions and alterations to the surface. This soils quite different from natural soils due to the high level of human activity, leading to increased pollution, physical disruptions and alterations to the surface. Urban soils are frequently associated with degraded or contaminated soils low levels of soil organic carbon and biological activity (Ajmone-Marsan and Biasioli 2010; Scharenbroch et al. 2005) compared to non-urban soils. Compared to them, urban soils are also influenced by the local microenvironment as well as the process of its formation.

In many places around the world, the chemical environment of soils has been modified due to the rapid urbanization and industrial growth (Johnson and Demetriades 2011). The sources of pollution can be transport (motor exhaust, brake pads, tire wear), domestic activities (such as waste disposal and construction, wastewater), commercial and industrial emissions (including those from metallurgical industry, electronics, chemical plants, fuel combustion) and agricultural activities (such as the use of fertilizers, pesticides and wastewater irrigation) (Argyraki and Kelepertzis 2014; Wei 2018).

Along with polyaromatic hydrocarbons, chlororganic compounds and radionuclides, heavy metals are typical urban pollutants that can be transferred into humans and influence their health (Madrid et al. 2002). Toxic metals, such as Pb, Cd and Cr, are still increasingly accumulating in urban environments, causing concern because urban soils as they easily come in contact with humans, e.g., in the form of suspended dust (Ljung et al. 2006; Sörme et al. 2001). These are pollutants that do not undergo chemical or biological degradation and therefore can be accumulated in soils or living organisms. Human exposure to heavy metals can occur through ingestion, inhalation and skin contact and in different places, mainly in urban parks (Penteado et al. 2021). Urban soil in parks acts as a sink for pollution of potential toxic elements (PTE) because to the presence of large open exposed areas of soil (Gu et al. 2016; Zhao et al. 2014).

Human outdoor and activities have increased in parks on a daily basis, for benefits to people's health. Particularly during the COVID-19 pandemic period in 2020–2021, parks and public gardens provided an important place for residents (especially children) for outdoor relaxation and entertainment purposes (Brtnický et al. 2019). Children are at higher risk of exposure to pollution due to their frequent contact with soil and other surfaces (Brtnický et al. 2019; Gu et al. 2016; Zhang 2019),, which can lead to contamination through skin contact, inhalation and hand-to-mouth ingestion (Li et al. 2014b), even at low concentrations. Children are more vulnerable to pollution due to their smaller size, developing nervous systems and higher absorption rates. For instance, while the portion of ingested Pb which absorbed an adult's body less than 5%, children can absorb up to 50% due to their less developed gastrointestinal tracts (Maddaloni et al. 1998).

Urbanization also has a specific effect on the plant development and bacterial soil communities due to conversion of native habitats to various forms of land use and habitat loss (McIntyre 2000). Outstanding properties of urban soils are limited aeration and drainage, which can lead to anaerobic conditions, disrupt nutrient cycling and result in elevated temperatures (Bae and Ryu 2017). Excessive concentration of metals above the maximum permissible limit can negatively impact the quality of soil organisms, such as microfauna and macrofauna, as well as ground and drinking water. This can ultimately impact the quality of food (Nganje et al. 2020; Wang et al. 2020). The contamination of urban soils with heavy metals was discovered to have an impact on soil microorganisms, including microbial biomass and diversity, basal respiration rate, physiological profile and the activities of dehydrogenase, sulfatase, glucosidase and phosphatase (Yang et al. 2006).

Urbanization results in the expansion of cities' boundaries, causing greater patchiness in the landscape. This, combined with loss of habitats, is the primary factor contributing to the decline of biodiversity (Grimm et al. 2008). Being the primary green areas of urban landscapes, they represent important biodiversity hotspots in cities (Nielsen et al. 2014). Parks and public gardens are generally composed of grassy areas, with tree and shrub species commonly used that support limited biodiversity, particularly if non-native and less capacity for critical ecosystem services (Aronson et al. 2017).

Comparing small and big cities can be difficult at times because the studies' methodological approaches, including soil fraction, contaminants studied and chemical extractions. The size of the city does not directly determine the level of contamination, as small urban areas can be in close proximity to industrial areas (Gallego et al. 2002) and experience the same anthropogenic pressures as bigger cities. Although suburban areas don't contribute significantly to industrial pollution, they do contaminate the environment through the disposal of household and yard waste, storm drains, construction activities and materials and traffic (Callender and Rice 2000).

In the present work, quality of soil of urban green areas situated in a Russian big city (over 1 Mio citizens) as well as its satellite small city (about 0.1 Mio citizens)—Kazan and Zelenodolsk, respectively—was studied. Currently, poor information is available about urban soil quality in Russian cities while this country represents about 11.5% of the land on the planet. The study was conducted in summer 2021. The plant and bacterial biodiversity as well as the level of heavy metals pollution were analyzed in the urban green areas investigated.

## 4.2 Materials and Methods

### *4.2.1 Study Site*

This research was carried out in two Russia cities: Kazan (55°47′19.46″ N 49°07′19.70″ E) and Zelenodolsk (55°50′37.54″ N 48°31′4.22″ E). These cities have significant natural, geographical and industrial aspects. Zelenodolsk is a classic satellite city of the monocentric Kazan agglomeration. The both cities are located on the left bank of the Volga River, and the population of Kazan city is 1,259,173 people (as of January 2022), of Zelenodolsk—99 235 people. The city of Zelenodolsk belongs to the category of medium cities. 145 urban green zones are situated in Kazan, covering about 0.84% of city area. Currently, the area of green common areas in Zelenodolsk is 13.1 $m^2$/person, which corresponds to the established requirements of at least 7 $m^2$/person for medium-sized cities (Zakirova and Khusnutdinova 2018).

### *4.2.2 Estimation of Plant Biodiversity*

The species diversity of (sub)urban habitats, including parks and gardens, can be extremelyhigh (Gilbert 1989). The plants in the plot were identified and cover of individual plant and the whole community in the quadrat. From the estimated dominance–abundance of the herbaceous vegetation and for trees and shrubs taxa, the Shannon–Wiener the standard biodiversity index was computed (Hermy and Cornelis 2000). Major herbaceous species in the region include *Achillea millefolium, Dactylis glomerata, Medicago sativa, Plantago major, Trifolium repens* and *Urtica dioica.*

### 4.2.3 *Soil and Plant Sampling*

Sampling on July—August 2021 involved systematic sampling from 15 parks of Kazan and 9 green zones of Zelenodolsk which selected randomly and situated in different districts of these cities. The characteristics of parks and public gardens both cities are listed in Appendix Table 4.1.

Soils and dust (10—20 g) were taken in three replicates from the right corner, middle and left corner of the main/south entrance of each park from 0 to 5 cm depth. Replicates were stored at ca. 4 °C in zip-lock plastic bags prior to transport back to the laboratory. Soil samples were sieved through 1 mm plastic mesh before analysis. The sizes of the inventory area to determine plant diversity were $10 \times 10\ m^2$. Three leaves clover samples and plantain from each of the three population were collected in separate zipper bags from each sites. The distance was approximately 4 m and at least 2-m.

### 4.2.4 *Estimation of Heavy Metals Content in Plants and Soils*

Before the estimation, plants and soils were air dried till the constant weight. The homogenized plant tissues (0.1 g) were placed in a glass beaker and dissolved with 10 ml of $HNO_3$, gradually heated with a rotation speed of $146 \pm 1$ rpm at 95–98 ˚C for 3 h. After the samples were slightly cooled, 1 ml of 33% $H_2O_2$ was added and heating of the samples for 1 h. The samples were cooled and 2.5 ml of bi-distilled water was added, filtered and brought to 15 ml (Ji 2012).

For the determination of heavy metals concentration in the urban soil, the homogenized samples (0.5 g) were accurately weighed into 50 ml «Falcon» poly(propylene) centrifuge tubes (pre-cleaned with 10% nitric acid followed by repeated rinsing with bi-distilled water) and 20 ml of 0.1 M $HNO_3$ solution added to the tubes. The tubes were shaken on a shaker at 200 °C with a rotation speed of $30 \pm 1$ rpm for 24 h. In each batch, a blank sample containing the same reagents was also subjected to the extraction procedure. The extracts were kept in a fridge at 4 °C prior to ICP-MS analyses (Hutton et al. 2004).

### 4.2.5 *Quantification of Soil Heavy Metal Contamination/ Pollution Index (MPI)*

The contamination/pollution index (MPI) was calculated as a ratio of a heavy metals concentration to the permissible standard of set out by EU and WHO top soils standards (Lacatusu 2000; World Health Organization (WHO) 1996).

### 4.2.6 *Estimation of Bacterial Activity and Functional Diversity*

To investigate of activity bacterial soil communities, the so-called community level physiological profiling has been demonstrated to be effective at observing spatial changes in microbial communities (Doan et al. 2013). Garland and Mills were the first who used the BIOLOG system to characterize microbial communities in soil (Garland and Mills 1991).

Community level physiological profiles (CLPPs) were assessed by the Biolog EcoPlate™ system (Oxoid, Hampshire, England). Biolog Inc., CA, USA). Each 96-well plate consists of three replicates, each one comprising 31 sole carbon sources and a water blank. Soil suspensions (soil 5 g, distilled water 45 ml) were shaken for 30 min at 250 rpm. Tenfold dilutions were performed and aliquots of 140 μl from a $10^{-4}$ dilution were inoculated into the microplates. The plates were incubated at 25 °C, and color development in each well was recorded as optical density (OD) at 595 nm with a plate reader at regular 24 h-intervals (Gomez et al. 2006).

With the help of the results received, the following parameters were calculated: average well color development (AWCD) (1) and alpha biodiversity Shannon index (2). The formulae used to calculate the indices are presented below:

$$\text{AWCD} = \sum \frac{A_i - A_0}{31} \tag{4.1}$$

where $A_i$—the optical density within each well; $A_0$—is the absorbance value of the control well

$$-\sum p_i * \ln_{p_i} \tag{4.2}$$

where $p_i$—proportional color development of the well over total color development of all wells of a plate.

### 4.2.7 *Statistical Analysis*

All measurements were conducted in three replicates. The indexes of alpha diversity calculated for each community were the Shannon–Weaver Index (Takada 1954) and Simpson index (Simpson 1949). Error bars in the figures represent the standard error of means from the replicates. Statistical significance of differences was analyzed using nonparametric criteria such as Mann–Whitney *U* test ($p > 0.05$). Statistical analysis was performed in Statistica 10.0 software (StatSoft Inc., Tulsa, OK, USA). Graphs were prepared using Microsoft Excel 2019 MSO (Microsoft, Redmond, WA, USA).

## 4.3 Result and Discussion

### *4.3.1 Heavy Metals Concentrations in Soils*

The soil has the property of absorbing heavy metals as a result of industrial and anthropogenic activities (Ediene and Umoetok 2017). The main environmental standard for metal soil pollution is so-called maximum permissible limits (MPL), i.e., such a content of a pollutant in the soil that does not affect to animals, plants, microorganisms and natural communities as a whole during constant contact or during interaction over a certain period of time (Ediene and Umoetok 2017; World Health Organization (WHO) 1996). According to WHO prescriptions (1996), the MPL for cadmium, lead, copper, nickel and zinc are 0.8 mg/kg, 85 mg/kg, 36 mg/kg, 68 mg/kg, 50 mg/kg of soil and 0.02 mg/kg, 2 mg/kg, 10 mg/kg, 10 mg/kg, 0.6 mg/kg of plant, respectively. The concentration of heavy metals in soils each parks and public gardens is shown on Fig. 4.1a–e.

Anthropogenic activities such as industrial waste disposals, fertilizer application and sewage sludge disposals on land have led to the accumulation of cadmium (Cd) in soil and its leaching under certain soil and environmental conditions (Alloway 2013). The WHO/FAO recommended maximum tolerable intake of Cd is 70 $\mu$g day$^{-1}$. With the estimated half-life of Cd in soil varying between 15 and 1100 years (Kabata-Pendias 2010), its accumulation in the environment and its entry into the food chain are of great concern. In the present study, the determined concentrations of Cd in soils in Kazan parks were in range of 0.17–0.70 mg/kg, while in Zelenodolsk sites, they varied between 0.06 and 0.39 mg/kg (Fig. 4.1a). The highest Cd concentration was recorded for the sample No. 5 in Kazan. The values for Cd across the sampling points do not exceed the 0.8 mg/kg soil range set by WHO and were below the MPL of 3.0, 1.4 and 400 mg/kg set out by EU, UK and USA Standards, respectively, for Cd in soil (Ediene and Umoetok 2017).

We compared our results with the results of other study which estimated the content of heavy metals in a big city and small cities. According to Li. et al., the average concentration of Cd in the urban soils of Copenhagen was 0.84 mg/kg, which is about 5 times higher than the concentration in the reference soil (Li et al. 2014a). The Cd concentrations in urban parks of the third-largest city in China, Guangzhou were ranged 0.028–2.408 mg/kg (Lu et al. 2007). The study paper Kwaterczak et al. presented results of research of Cd distribution in land on playgrounds situated near busy streets in Cracow (Poland). Samples of sand and soil were collected from the most top layer (0–10 cm). Concentration of examined metal was varied from 1.60 to 2.61 mg/kg. The highest concentrations were found in sampling points near the busiest roads (Aleksander-Kwaterczak and Rajca 2015).

Lead (Pb) is accumulating profoundly in the soil through anthropogenic activities (Mulligan et al. 2001). Being a toxic substance and having high transfer rates (from soil to plant), it is therefore studied broadly especially in context to food safety, quality, and biotesting purposes (Uzu et al. 2009). As follows from the Fig. 4.1b, the highest Pb value was revealed in sample No.15 obtained from Kazan—30 mg/kg,

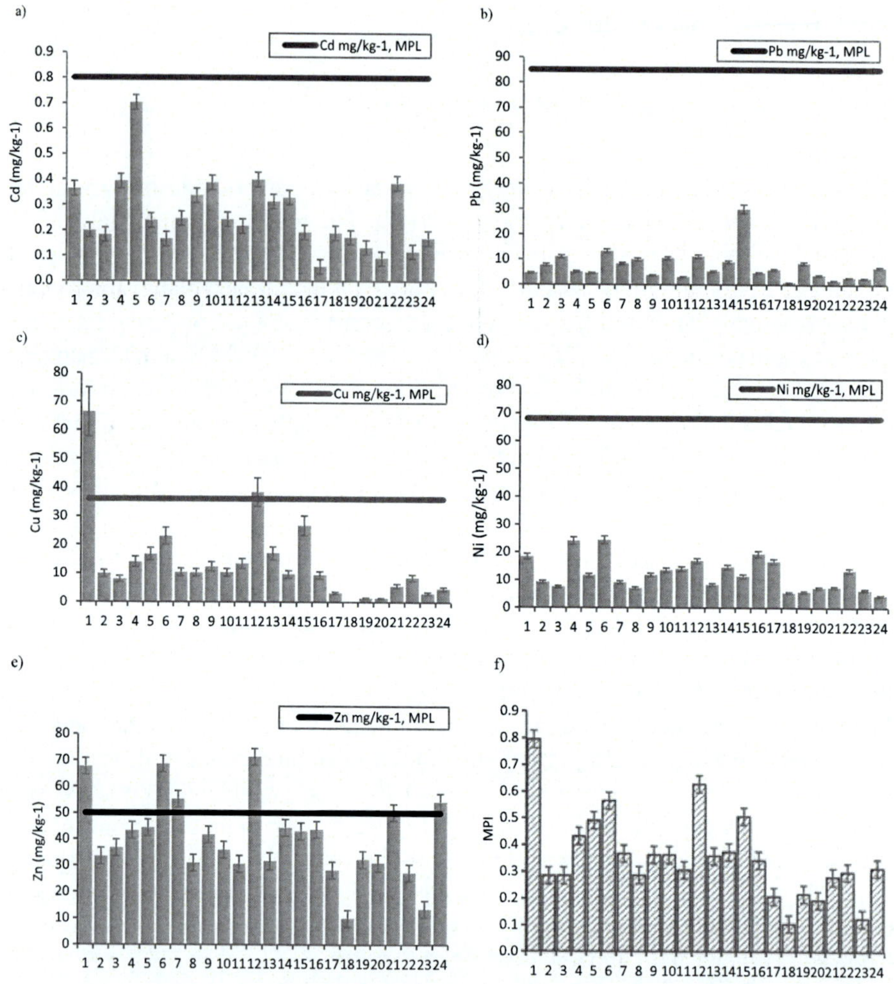


**Fig. 4.1** Concentrations of Cd (**a**), Pb (**b**), Cu (**c**), Ni (**d**), Zn (**e**) and contamination/pollution Index (MPI) (**f**) found in urban soils sampled in Kazan (1–15) and Zelenodolsk (9–24) cities (Russia). The black lines indicate the maximum permissible limits according to the WHO standards

the lowest sample No.18 from Zelenodolsk—0.81 mg/kg. The highest value of Pb concentration recorded in the urban park was not higher than the WHO (85 mg/kg) and the 6.4 mg/kg UK standards (70 mg/kg) (Ediene and Umoetok 2017).

Galušková et al. (2014) conducted a study that compared Pb concentrations in Prague and Ostrava parks. The values open areas ranged between 22–213 mg/kg in and 27–125 mg/kg, respectively. The soil pollution may have been contributed by the former brown coal burning in local heating plants (Galušková et al. 2014). According to literature (American Society of Agronomy 1982), highest Pb concentrations were detected in areas with higher population density and more industrial activity. Additionally, Pb pollution may result from the past use of leaded gasoline and emissions from vehicles (Vig et al. 2003). The soil in Durham city's park (North

Carolina, USA) had a lead concentration of 42.1 ± 25.0 mg/kg, which is consistent with other U.S. cities such as Indianapolis and Greensboro, where soils near homes had the highest lead concentration due to exterior house paint. However, in cities like Detroit and New Orleans, soils near roadways pose the highest lead exposure risk due to the contamination of soil being resuspended. Despite this, the soils from city parks and suburban areas showed lower concentrations of metals(loids) that were comparable to the natural levels found in the environment (Wang et al. 2022).

Copper (Cu) is considered as a micronutrient for plants however, it may be toxic in excess quantities (Ediene and Umoetok 2017). Cu particulates are released into the atmosphere by windblown dust and anthropogenic sources, including transport and industrial emissions. Evaluation of Cu in soils of parks and public gardens in Kazan (Fig. 4.1c) showed that values were ranged from 7.97 to 66.33 mg/kg, while the Zelenodolsk soils recorded a mean of 4.12 mg/kg. In soil samples No.1 and No.12, concentrations of Cu were recorded above the maximum permissible limit set by WHO, but were below the MPL of 140, 1.4 and 3.0 mg/kg set out by EU, UK and USA, respectively.

According to Paltseva et al., the concentration of Cu in urban soils (0–20 cm) in cities with over 1 Mio citizens such as Vienna, Moscow, Mumbai (Bombay) was equaled to: 18 mg/kg, 30 mg/kg and 147 mg/kg, respectively. At the same time, in cities with less than 1 Mio people, this value was recorded to 24 mg/kg in Oslo (Tijhuis et al. 2002) and 16 mg/kg in New Orleans (Mielke et al. 2019). In general, in European cities, the concentration of Cu was lower than recommended by WHO, with the exception of the soils from Mumbai (Bombay) (Wang et al. 2022). High concentrations of Cu in the parks of Kazan may be due to the presence in the region of a developed oil production and oil refining industry.

Nickel (Ni) has been considered to be an essential trace element for human and animal health (Nazir et al. 2015). The concentration values obtained for Ni at the urban green areas (Fig. 4.1d) revealed a range of 7.09–24.33 mg/kg, while a mean of 9.56 mg/kg was recorded in Zelenodolsk parks. The overall results indicate that in all the collected soil samples, there was low concentration of Ni found.

According to the literature (Zhong and Jiang 2017), contamination of Ni in urban soils of small cities is higher than in big ones. The value of Ni in Berlin and Melbourne was recorded 8 mg/kg and 15 mg/kg, respectively. While in soils in Baltimore and Novi Sad and Izmit (Turkey), it was 2.8, 28.7 and 39 mg/kg. This suggests that although heavy metal contamination of soil and water has been limited by some government policies, historically produced heavy metals are retained in soil due to their perdurability and degradation durability. In some places, the concentration of nickel has surpassed even the Chinese soil criteria (grade II), and its concentrations also exceed the national quality soil standards (Chen et al. 2020).

Zinc (Zn) is one of the important trace elements that play a vital role in the physiological and metabolic process of many organisms. Nevertheless, higher concentrations of Zn can be toxic to the organism (Nazir et al. 2015).

Concentration of Zn in soils samples from Kazan ranged between 30.67 mg/kg to 71.33 mg/kg and 9.97 mg/kg to 54.33 mg/kg obtained from Zelenodolsk, respectively

(Fig. 4.1e). The highest value of 71.33 mg/kg was recorded at No.12, closely followed by values of 68.67 mg/kg of No. 6 and 67.67 mg/kg from the ground No.1.

The quantity of Zn observed at Kazan parks and public gardens greatly exceeded the value of 50 mg/kg recommended by WHO. The level of Zn in sample No. 24 from Zelenodolsk was as well as higher than WHO's recommended limit—54.33 mg/kg.

According to Lee with co-authors, in Hong Kong (about 7.4 Mio citizens), concentrations of Zn ranged from 23 to 930 mg/kg, while in Karachi (about 14 Mio people), it was 99.5 mg/kg (Karim and Qureshi 2014). In small cities such as Zagreb (Croatia), soil contamination with Zn averaged 70 mg/kg (Romic and Romic 2003). The probable sources of Zn contamination primarily include wastewater discharges, followed by automobile transportation and unauthorized dumping of household waste.

To determine the pollution loading of the soil at the various sampling points in parks and public gardens, the MPI was used. According to Lacatusu, the MPI index is the ratio of analyzed concentration of the heavy metal to the maximum allowable limit in the soil (Lacatusu 2000). Despite on the fact that the content of Zn and Cu was exceeded the level recommended by WHO in 5 and 2 parks of 24, respectively, the MPI index in all samples did not exceed 1.0. Therefore, that the level of metal pollution is acceptable. It should be noted though, that the average value of MPI for all metals was higher in Kazan than that in Zelenodolsk (Fig. 4.1f).

### *4.3.2 Heavy Metals Concentrations in Plants*

According to the literature, there is a certain relationship between the chemical composition of plants and the elemental composition of the environment. Different plant parts contain different heavy metals quantities; the highest quantities are in roots and leaves, and the lowest are in fruits and seeds. Some plants species were found to have lower tolerance to toxic metals uptake in parks of soil as they were found to accumulate high concentrations of Cd, Cu, Ni, Pb and Zn (Guala et al. 2010).

In the present study, metals were determined in the two plant species present in all the parks—plantain (*Plantago major* L.) and white clover (*Trifolium repens* L.). The results are presented on Fig. 4.2 and compared with the corresponding WHO standards for plants (Bani et al. 2015). It was found that the concentration of Ni was below the limits in all the plant samples. The concentrations of Cd, Pb and Cu exceeded the limits in 5, 1 and 1 samples, respectively. And the concentrations of Pb were by 10–25 times above the limits in all the plant samples analyzed.

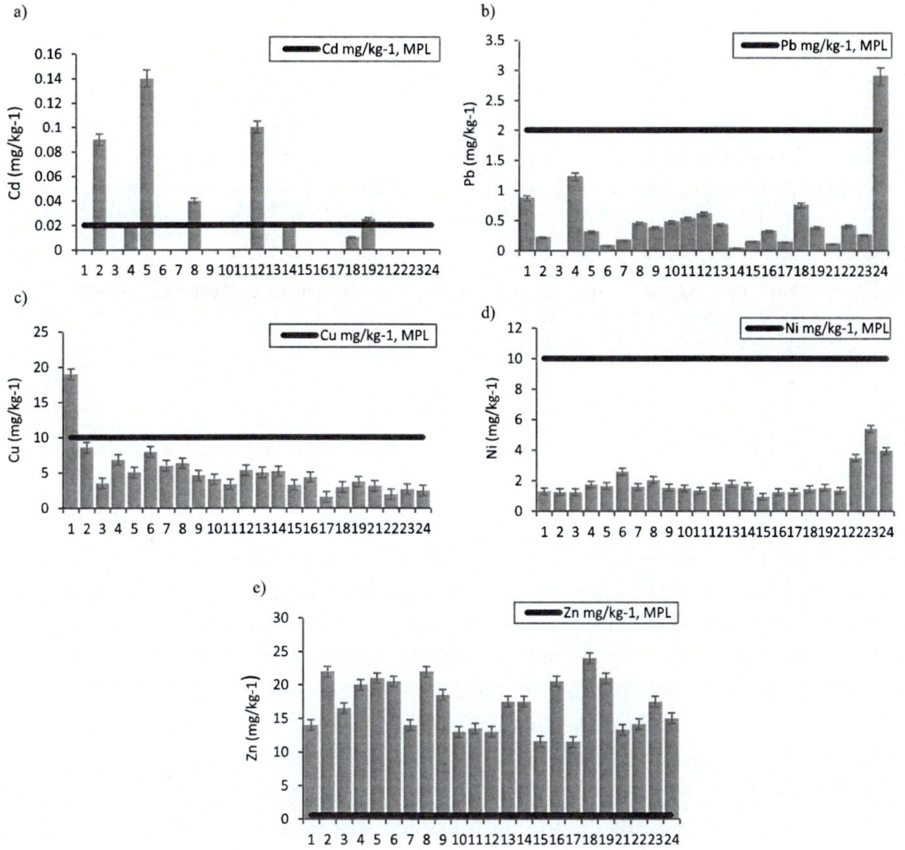


**Fig. 4.2** Concentrations of Cu (**a**), Pb (**b**), Cu (**c**), Ni (**d**), Zn (**e**) and Transfer Factor (TF) (Appendix Fig. 4.5) found in plants samples in Kazan (1–15) and Zelenodolsk (9–24) cities (Russia). The black lines indicate the maximum permissible value according to the WHO standards

### *4.3.3 Plant Diversity and Cover at Various Study Sites*

Species diversity is a crucial index mostly employed to evaluate the sustainability at various ecosystem scales and is a multifaceted factor used to estimate a variety of plant indicators (Eshaghi et al. 2009; Ricotta 2005).

In our study, we assessed the biodiversity of plants in parks and squares in Kazan and Zelenodolsk. The amount of grass coverage on the site was determined by the relative occurrence of each species and the number of species in each park.

The Shannon–Wiener index across the parks ranged from 1.51 to 2.43 of Kazan, and from 0 to 2.53 of Zelenodolsk (see Fig. 4.3). According to the literature, urban parks included annual and perennial plants species such as *Zoysia species* (with no petals), *Trifolium repens* (with white petals) and with subcomponent species such as *Sonchus oleraceus, Oxalis corniculate, Cirsium japonicum, Trifolium pretense and Ajuga nipponensis* in Yokohama (Japan) (Tomitaka et al. 2021). In the study,

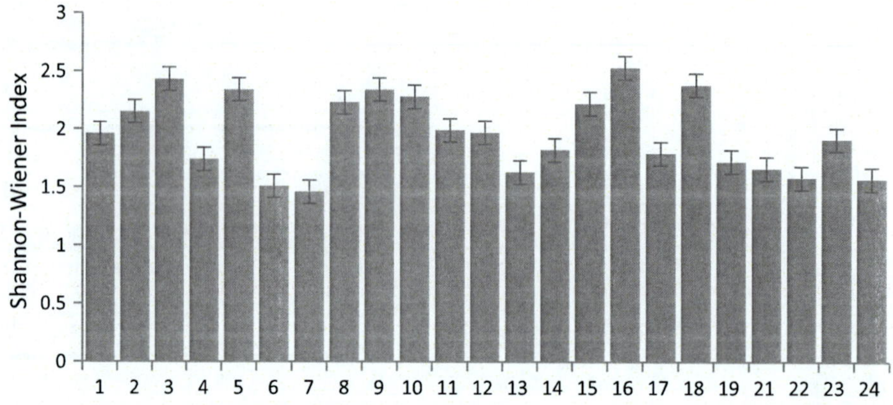


**Fig. 4.3** Shannon–Wiener index indicating plant biodiversity in urban green zones in Kazan and Zelenodolsk

Muratet et al. noted that in the park «Parc des Buttes Chaumont» (France), that covers 25 ha. among the native species recorded in the park, the majority were spontaneous species generally considered to be weeds, including *Trifolium repens, Hordeum murinum* and *Veronica chamaedrys* (Scharenbroch et al. 2005). Dominant species in Kazan were *Achillea millefolium, Arctium lappa, Chenopodium album, Convolvulus arvensis, Plantago major and Poa trivialis.* Dominant species in satellite city were *Achillea millefolium, Trifolium repens, Artemisia vulgaris, Cichorium intybus, Dactylis glomerata and Plantago major.* Dominant species in Zelenodolsk were the same. No significant differences in the species list as well as in the level of Shannon–Wiener indexes between the two cities were found.

We made in attempt to compare the level of metals' pollution of soils with the plant diversity indexes in urban green zones. However, no correlation between the MPI and Shannon–Wiener index was found.

### *4.3.4 Bacterial Activity and Functional Diversity in the Soils from Urban Green Zones*

Currently, many ways to estimate the state of the microbiomes in the soils are used. These might be wide range of methods, from estimations of soil microbial biomass and respiration to analysis of soil microbial diversity using next generation sequencing methods. All these methods have their own advantages and disadvantages and should be chosen according to the investigation's main goal. In our study, quick, reproducible method semi-independent on climate conditions were required, and the Biolog Ecoplate® profiling met the requirements (Use of Biolog for the community level physiological profiling (CLPP) of). Using this method, two characteristics of the soil microbial communities were determined—(i) the AWCD indicating the

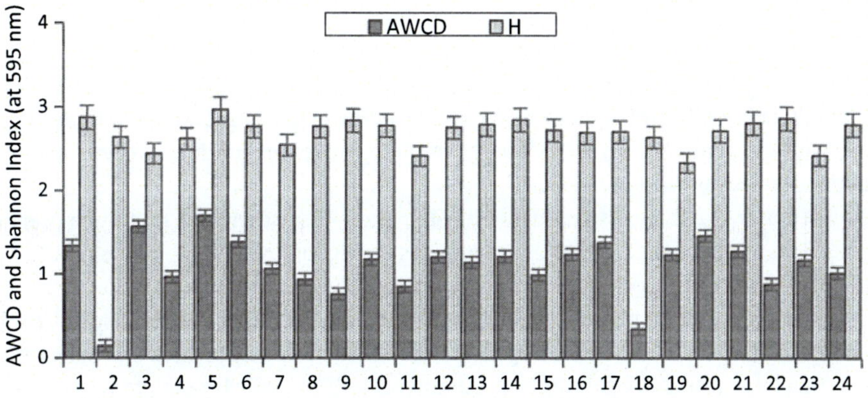


**Fig. 4.4** AWCD and the biodiversity Shannon index of microbial communities in the soils of urban green zones as revealed by Biolog Ecoplate® method

intensity of microbial decomposition of different organic substrates means overall microbial activity in the soil samples; and (ii) the Shannon H-index indicating the functional diversity of the microbial community (Fig. 4.4). (Li et al. 2007; Monard et al. 2016).

The highest AWCD index was found in sample No. 5 (1.695 ± 0.340), the lowest one—in sample No. 2 (0.143 ± 0.250), while no correlation was revealed between the level of metal pollution or vegetation diversity and AWCD. The values received correspond with the data obtained by other authors which analyzed soil samples from native vegetation site (roadside verge) and topsoil (0–15 cm deep) (Liu et al. 2012; Xue et al. 2008). It can be suggested that the differences in AWCD are due to differences in organic matter content in the samples, indeed, according to high microbial counts and diversity are usually attributed to rich soils. Thus, in the investigations carried out by Dobrovolskaya et al. and Kovda et al., tight direct correlation between the biological activity of microorganisms and organic matter content in soil was shown. Shannon index showing the level of functional alpha diversity in the community ranged between 2.32 and 2.96. No correlation between the Shannon indices and the AWCD or the level of metal pollution of soil was found (Kumar et al. 2017; Li et al. 2013).

## 4.4 Conclusion

It can be concluded that in overall, soils from urban green zones in a large (Kazan) and satellite (Zelenodolsk) cities contain Pb, Zn, Cu, Cd and Ni in the concentrations below the WHO standards. Unique cases of exceeding were determined for Cd, Cu and Zn, but the pollution was not complex.

No differences between the green zones and in large and satellite city (biodiversity of plants and microorganisms) as well as no correlation between the metal content and diversity indices were found. It was found that the migration of Pb, Ni, Cu for plantain and white clover don't as intense as in the case of Zn and Cd. Perhaps these urban plants, which grow everywhere, are able to accumulate these metals and also get on the leaves with urban dust.

The absence of a direct stable relationship allows us to conclude that a comprehensive assessment of the state of green areas is applicable, where the content of heavy metals is lower and does not have a significant impact on biodiversity. Accordingly, these measures relate not only to human health, but also more indirect aspects of the quality of life: the structural integrity of the surrounding ecosystems. This work is part of a large research cycle and measurements will be made for other environmental components.

**Acknowledgements** The work is carried out in accordance with th Strategic Academic Leadership Program «Priority 2030» of the Kazan Federal University of the Government of the Russian Federation. We are grateful to Kone Foundation (Finland) for funding the «LesRus» project that supported soil sampling and experimental planning. We thank the project coordinator Ph. D. Aki Sinkkonen.

## Appendix

See Table 4.1 and Fig. 4.5.

**Table 4.1** Characteristics of urban green space at Kazan and Zelenodolsk (Russia)

| No | Park name | Location coordinates | Area, % Natural, manicured, unkempt | Park age | Size, $m^2$ | Park description |
|---|---|---|---|---|---|---|
| 1 | M. Karima public garden | 55.763388° 49.14939° | 40 | 1 | 5500 | Public garden near a car highway, people walk through every day |
| | | | 50 | | | |
| | | | 10 | | | |
| 2 | Public garden(Ippodromnaya street) | 55.773302° 49.147741° | 60 | < 10 | 2680 | Small public garden, people walk through every day, next to the kindergarten |
| | | | 10 | | | |
| | | | 30 | | | |
| 3 | Public garden of Labour Glory | 55.755792° 49.122546° | 50 | 25–50 | 6500 | Local garden next to the Kazan Synthetic Rubber Plant (Founded in 1936), realized for the workers. The park is located in a remote area of the city |
| | | | 10 | | | |
| | | | 40 | | | |
| 4 | Park Molodozhenov | 55.774184° 49.09985° | 55 | 25–50 | 25,000 | It rather a big park, but not popular. Next to the river port, there are no playgrounds in the park. This is a remote area of the city |
| | | | 30 | | | |
| | | | 15 | | | |
| 5 | Gorkiy Central Park of Culture and Recreation | 55.799874° 49.147063° | 60 | 85 | 244,794 | Quite big, will require 2–3 h at least to get around. It seems like forest than urban park. Clean and cozy, there are many playgrounds for kids, many places to eat, great for picnic |
| | | | 40 | | | |
| | | | 0 | | | |

(continued)

**Table 4.1** (continued)

| No | Park name | Location coordinates | Area, % Natural, manicured, unkempt | Park age | Size, $m^2$ | Park description |
|---|---|---|---|---|---|---|
| 6 | Kirov Garden | 55.785683° 49.112507° | 30<br>40<br>30 | 61 | 10,286 | Central public garden with the fountain, a lot of people walk through every day. Clean park, located next to an elite residential complex |
| 7 | Shkolnikov public garden | 55.822214° 49.116643° | 40<br>50<br>10 | 5 | 13,637 | Near to the school, there are small lake too. Many types of ornamental shrubs are grown in the park. The lawn is mowed regularly |
| 8 | «Sosnovaya Roshcha» Park | 55.824225° 49.089368° | 70<br>30<br>0 | ~ 50 | 86,011 | Clean and cozy, there are many playgrounds for kids, comfortable for walking and doing yoga |
| 9 | «Aktash» Park | 55.794808° 49.230825° | 40<br>15<br>45 | 4 | 9144 | Small park next to multi-story apartments and the private sector. it‘s not popular park. There are dog‘s training ground too |
| 10 | «Muzykalnyi» public garden | 55.790177° 49.222738° | 15<br>70<br>15 | ~ 10 | 10,117 | The public garden is located next to the sports center; many young people train here. Football, basketball playgrounds. Near garden the multi-story buildings |

(continued)

**Table 4.1** (continued)

| No | Park name | Location coordinates | Area, % Natural, manicured, unkempt | Park age | Size, $m^2$ | Park description |
|---|---|---|---|---|---|---|
| 11 | «Dubovyi kardon» public garden | 55.750428° 49.236008° | 30 | ~ 15 | 5333 | Small public garden, only oak trees grow in this park, there are no playgrounds |
| | | | 10 | | | |
| | | | 60 | | | |
| 12 | Tat'yanin public garden | 55.783665° 49.110683° | 30 | < 10 | 1475 | Public garden near a car highway, there are no playgrounds, near to church |
| | | | 10 | | | |
| | | | 60 | | | |
| 13 | «Millenium» Park | 55.78359° 49.123183° | 15 | 16 | 52,172 | Park is located at the center of Kazan. The big park with fountain. The park has lots of benches to relax on and admire the view. No playgrounds |
| | | | 65 | | | |
| | | | 20 | | | |
| 14 | Public garden (Peterburgskaya street) | 55.782852° 49.130877° | 30 | 10 | 1957 | Small public garden next to five-story office building |
| | | | 40 | | | |
| | | | 30 | | | |
| 15 | Leninskiy Garden | 55.794019° 49.123481° | 30 | 131 | 25,608 | Park is located at the center of Kazan. Leninsky park is situated next to university. There is a beautiful fountain in the middle of the park. No playgrounds |
| | | | 50 | | | |
| | | | 20 | | | |
| 16 | Memorial of Memory and Glory (Aisha village, Zelenodolsk region) | 55.869668° 48.629383° | 35 | – | 4172 | Small public garden in the center of Aisha village |
| | | | 15 | | | |
| | | | 50 | | | |

(continued)

**Table 4.1** (continued)

| No | Park name | Location coordinates | Area, % Natural, manicured, unkempt | Park age | Size, m$^2$ | Park description |
|---|---|---|---|---|---|---|
| 17 | Public garden next to multistorey building (Komarova street) | 55.85801° 48.564893° | 30 | – | 3142 | Public garden in the center of Zelenodolsk, there is a little fountain |
| | | | 10 | | | |
| | | | 60 | | | |
| 18 | N.Altynov public garden | 55.854286° 48.553947° | 40 | 6 | 6533 | This small park famous for a tank weighing 40 tons, which stand there. Next to multi-story buildings |
| | | | 20 | | | |
| | | | 40 | | | |
| 19 | Green area opposite «Temple of the First Apostles Peter and Paul» | 55.871510° 48.548155° | 15 | 130 | 3137 | Small green area next to the temple |
| | | | 85 | | | |
| | | | 0 | | | |
| 20 | Temple of the First Apostles Peter and Paul | 55.871269° 48.547893° | 10 | 130 | 4489 | Area of the temple is small but well-organized |
| | | | 90 | | | |
| | | | 0 | | | |
| 21 | Public park (Kosmonavtov street) | 55.848502° 48.505101° | 15 | > 10 | ~ 7000 | Most-visited park in Zelenodolsk |
| | | | 85 | | | |
| | | | 0 | | | |
| 22 | Summer cafe (Karl Marks street) | 55.849247° 48.510100° | 40 | > 10 | 300 | It is a part of big park. This summer cafe has been closed for a long time. So no one walks here, as the grass is tall |
| | | | 0 | | | |
| | | | 60 | | | |

(continued)

**Table 4.1** (continued)

| No | Park name | Location coordinates | Area, % Natural, manicured, unkempt | Park age | Size, m$^2$ | Park description |
|---|---|---|---|---|---|---|
| 23 | Public park (near lake) | 55.851756° 48.508549° | 40 | 5–10 | ~ 2500 | This green area can hardly be called a city park. This place is not completed, has a bench and gazebo, but it is difficult to get here. The park is very unkempt |
| | | | 0 | | | |
| | | | 60 | | | |
| 24 | Park «Rodina» | 55.845695° 48.490492° | 30 | – | 22,040 | Clean and cozy, there are many playgrounds for kids, comfortable for walking. Next to school |
| | | | 70 | | | |
| | | | 0 | | | |

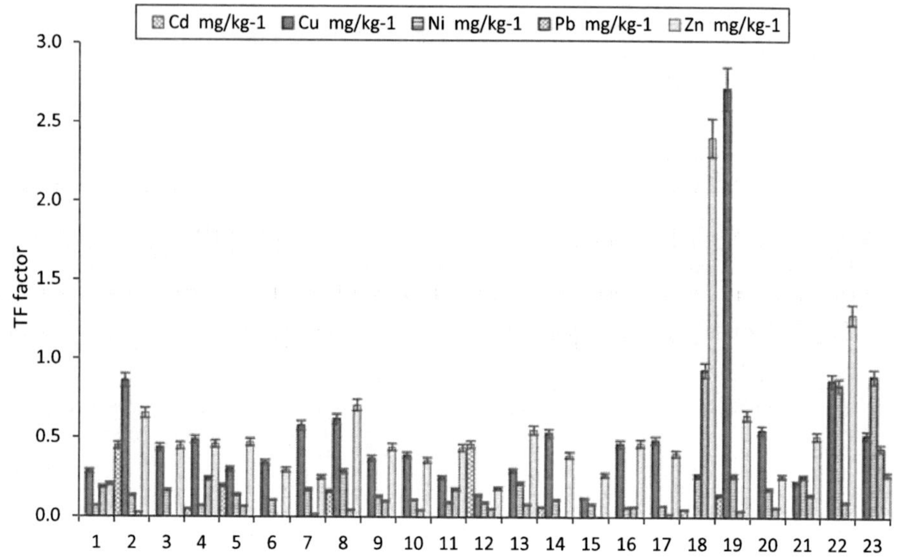


**Fig. 4.5** Transfer factor (TF) found in plants samples in Kazan (1–15) and Zelenodolsk (9–24) cities (Russia)

# References


Ajmone-Marsan F, Biasioli M (2010) Trace elements in soils of urban areas. Water Air Soil Pollut 1–4(213):121–143

Aleksander-Kwaterczak U, Rajca A (2015) Urban soil contamination with lead and cadmium in the playgrounds located near busy streets in Cracow (South Poland). Geol Geophys Environ 1(41):7

Alloway BJ (2013) Sources of heavy metals and metalloids in soils, pp 11–50

American Society of Agronomy (1982) Methods of soil analysis part 2. Chemical and microbiological properties. American Society of Agronomy, pp 785–787

Andrews SS, Karlen DL, Mitchell JP (2002) A comparison of soil quality indexing methods for vegetable production systems in Northern California. Agr Ecosyst Environ 1(90):25–45

Argyraki A, Kelepertzis E (2014) Urban soil geochemistry in Athens, Greece: the importance of local geology in controlling the distribution of potentially harmful trace elements. Sci Total Environ 1(482–483):366–377

Aronson MFJ et al (2017) Biodiversity in the city: key challenges for urban green space management. Front Ecol Environ 15(4):189–196

Bae J, Ryu Y (2017) Spatial and temporal variations in soil respiration among different land cover types under wet and dry years in an urban park. Landsc Urban Plan 167:378–385

Bani A, Osmani M, Hoxha B (2015) Heavy metals and Ni phytoextractionin in the metallurgical area soils in Elbasan. Albanian J Agric Sci 4(14):414–419

Brtnický M et al (2019) Assessment of phytotoxicity, environmental and health risks of historical urban park soils. Chemosphere 220:678–686

Callender E, Rice KC (2000) The urban environmental gradient: anthropogenic influences on the spatial and temporal distributions of lead and zinc in sediments. Environ Sci Technol 2(34):232–238

Chen X et al (2020) A review on remediation technologies for nickel-contaminated soil. Hum Ecol Risk Assess 3(26):571–585

Doan HN et al (2013) Analysis community level physiological profile of bacteria in Nha Trang Bay in dry season applying Biolog Ecoplates. Proceedings Of Vast - Ird Symposium On Marine Science: 403–410

Ediene V, Umoetok S (2017) Concentration of heavy metals in soils at the municipal dumpsite in Calabar metropolis. Asian J Environ Ecol 2(3):1–11

Eshaghi RJ, Manthey M, Mataji A (2009) Comparison of plant species diversity with different plant communities in deciduous forests. Int J Environ Sci Technol 3(6):389–394

Francini G et al (2018) Soil biota in boreal urban greenspace: Responses to plant type and age. Soil Biol Biochem 118:145–155

Gallego JLR, Ordóñez A, Loredo J (2002) Investigation of trace element sources from an industrialized area (Avilés, northern Spain) using multivariate statistical methods. Environ Int 7(27):589–596

Galušková I et al (2014) Lead isotope composition and risk elements distribution in urban soils of historically different cities Ostrava and Prague, the Czech Republic. J Geochem Explor PB 147:215–221

Garland JL, Mills AL (1991) Classification and characterization of heterotrophic microbial communities on the basis of patterns of community-level sole-carbon-source utilization. Appl Environ Microbiol 8(57):2351–2359

Gilbert OL (1989) The ecology of urban habitats. O. L. Gilbert, London, p 450

Gomez E, Ferreras L, Toresani S (2006) Soil bacterial functional diversity as influenced by organic amendment application. Biores Technol 13(97):1484–1489

Grimm NB et al (2008) Global change and the ecology of cities. Science 319(5864):756–760

Gu YG, Gao YP, Lin Q (2016) Contamination, bioaccessibility and human health risk of heavy metals in exposed-lawn soils from 28 urban parks in southern China's largest city, Guangzhou. Appl Geochem 67:52–58

Guala SD, Vega FA, Covelo EF (2010) The dynamics of heavy metals in plant-soil interactions. Ecol Model 8(221):1148–1152

Haq SMA (2011) Urban green spaces and an integrative approach to sustainable environment. J Environ Prot 5(2):601–608

Hermy M, Cornelis J (2000) Towards a monitoring method and a number of multifaceted and hierarchical biodiversity indicators for urban and suburban parks. Landsc Urban Plan 3–4(49):149–162

Hutton EA et al (2004) Validation of bismuth film electrode for determination of cobalt and cadmium in soil extracts using ICP-MS. Talanta 4(63):849–855

Ji W et al (2012) Identifying the criteria of cadmium pollution in paddy soils based on a field survey. Energy Proc 16:27–31

Johnson CC, Demetriades A (2011) Urban geochemical mapping: a review of case studies in this volume, pp 7–27

Kabata-Pendias A (2010) Trace elements in soils and plants, 4th edn. A. Kabata-Pendias, pp 1–520

Karim Z, Qureshi BA (2014) Health risk assessment of heavy metals in urban soil of Karachi, Pakistan. Human Ecol Risk Assess 3(20):658–667

Kumar U et al (2017) Variation of functional diversity of soil microbial community in sub-humid tropical rice-rice cropping system under long-term organic and inorganic fertilization. Ecol Ind 73:536–543

Lacatusu R (2000) Appraising levels of soil contamination and pollution with heavy metals

Li F et al (2013) Changes in soil microbial biomass and functional diversity with a nitrogen gradient in soil columns. Appl Soil Ecol 64:1–6

Li L et al (2014a) Release of cadmium, copper and lead from urban soils of Copenhagen. Environ Pollut 187:90–97

Li P et al (2014b) Pollution evaluation and health risk assessment of heavy metals from atmospheric deposition in Lanzhou. Huanjing Kexue/environ Sci 3(35):1021–1028

Li Z, Wu X, Chen B (2007) Changes in transformation of soil organic c and functional diversity of soil microbial community under different land uses. Agric Sci China 10(6):1235–1245

Li G et al (2018) Urban soil and human health: a review. Eur J Soil Sci 69(1):196–215
Liu W et al (2012) Effect of veterinary oxytetracycline on functional diversity of soil microbial community. Plant Soil Environ 7(58):295–301
Ljung K, Selinus O, Otabbong E (2006) Metals in soils of children's urban environments in the small northern European city of Uppsala. Sci Total Environ 2–3(366):749–759
Lu Y et al (2007) Chemical fractionation of heavy metals in urban soils of Guangzhou, China. Environ Monit Assess 1–3(134):429–439
Maddaloni M et al (1998) Bioavailability of soilborne lead in adults, by stable isotope dilution. Environ Health Perspect 6:1589–1594
Madrid L, Díaz-Barrientos E, Madrid F (2002) Distribution of heavy metal contents of urban soils in parks of Seville. Chemosphere 10(49):1301–1308
McIntyre NE (2000) Ecology of urban arthropods: a review and a call to action. Ann Entomol Soc Am 93(4):825–835
Mielke HW et al (2019) The concurrent decline of soil lead and children's blood lead in New Orleans. Proc Natl Acad Sci USA 44(116):22058–22064
Monard C et al (2016) Habitat generalists and specialists in microbial communities across a terrestrial-freshwater gradient. Sci Rep 6:1–10
Morel J-L, Heinrich AB (2008) SUITMA—soils in urban, industrial, traffic, mining and military areas. J Soils Sediments 4(8):206–207
Mulligan CN, Yong RN, Gibbs BF (2001) Remediation technologies for metal-contaminated soils and groundwater: an evaluation. Eng Geol 1–4(60):193–207
Nazir R et al (2015) Accumulation of heavy metals (Ni, Cu, Cd, Cr, Pb, Zn, Fe) in the soil, water and plants and analysis of physico-chemical parameters of soil and water collected from Tanda Dam Kohat. J Pharm Sci Res 3(7):89–97
Nganje TN et al (2020) The concentration, distribution and health risk from potentially toxic elements in the soil–plant–water system developed on black shales in SE Nigeria. J Afr Earth Sci 165:103806
Nielsen AB et al (2014) Species richness in urban parks and its drivers: a review of empirical evidence. Urban Ecosyst 1(17):305–327
Page-dumroese DS (2020) Forest and rangeland soils of the United States under changing conditions. D.S. Page-dumroese
Penteado JO et al (2021) Health risk assessment in urban parks soils contaminated by metals, Rio Grande city (Brazil) case study. Ecotoxicol Environ Safety 208:111737
Ricotta C (2005) Through the jungle of biological diversity. Acta Biotheor 1(53):29–38
Romic M, Romic D (2003) Heavy metals distribution in agricultural topsoils in urban area. Environ Geol 7(43):795–805
Różański SŁ et al (2018) Heavy metal content and mobility in urban soils of public playgrounds and sport facility areas, Poland. Chemosphere 212:456–466
Scharenbroch BC, Lloyd JE, Johnson-Maynard JL (2005) Distinguishing urban soils with physical, chemical, and biological properties. Pedobiologia 4(49):283–296
Simpson EH (1949) Measurement of diversity. Nature 163(4148):688
Sörme L, Bergbäck B, Lohm U (2001) Century perspective of metal use in urban areas—a case study of Stockholm. Water Air Soil Pollut Focus 1:197–211
Takada Y (1954) On the mathematical theory of communication. Jpn J Psychol 2(25):110–117
Tijhuis L, Brattli B, Sæther OM (2002) A geochemical survey of topsoil in the City of Oslo, Norway. Environ Geochem Health 1(24):67–94
Tomitaka M et al (2021) Species richness and flower color diversity determine aesthetic preferences of natural-park and urban-park visitors for plant communities. Environ Sustain Indic 11(2021):100130
Tresch S et al (2018) Urban soil quality assessment-a comprehensive case study dataset of urban garden soils. Front Environ Sci 6(136)
Uzu G et al (2009) Study of lead phytoavailability for atmospheric industrial micronic and submicronic particles in relation with lead speciation. Environ Pollut 4(157):1178–1185

Vig K et al (2003) Bioavailability and toxicity of cadmium to microorganisms and their activities in soil: a review. Adv Environ Res 1(8):121–135
Wang F et al (2020) Contamination characteristics, source apportionment, and health risk assessment of heavy metals in agricultural soil in the Hexi Corridor. CATENA 191:104573
Wang Z et al (2022) Legacy of anthropogenic lead in urban soils: Co-occurrence with metal(loids) and fallout radionuclides, isotopic fingerprinting, and in vitro bioaccessibility. Sci Total Environ 806:151276
Wei X et al (2018) Expansion of rice enzymatic rhizosphere: temporal dynamics in response to phosphorus and cellulose application. Plant Soil 445:169–181
World Health Organization (WHO) (1996) Permissible limits of heavy metals in soil and plants. Geneva, Switzerland
Xue D et al (2008) Soil microbial community structure in diverse land use systems: a comparative study using biolog, DGGE, and PLFA Analyses. Pedosphere 5(18):653–663
Yan Z et al (2019) Impervious surface area is a key predictor for urban plant diversity in a city undergone rapid urbanization. Sci Total Environ 650:335–342
Yang Y et al (2006) Microbial indicators of heavy metal contamination in urban and rural soils. Chemosphere 11(63):1942–1952
Zakirova J, Khusnutdinova S (2018) The problems of formation and conservation of the green frame (green carcass) of the satellite city (on the example of Zelenodolsk). IOP Conf Ser Earth Environ Sci 107:012139.
Zhang Q et al (2019) Variation of soil anaerobic microorganisms connected with anammox processes by 13C-phospholipid fatty acid analysis among long-term fertilization regimes in a crop rotation system. Appl Soil Ecol 133:34–43
Zhao L et al (2014) Source identification and health risk assessment of metals in urban soils around the Tanggu chemical industrial district, Tianjin, China. Sci Total Environ 468–469:654–662
Zhong M, Jiang L (2017) Refining health risk assessment by incorporating site-specific background concentration and bioaccessibility data of Nickel in soil. Sci Total Environ 581–582:866–873

# Chapter 5
# Assessment of Urban Soil Pollution by Heavy Metals (Russian Federation, Republic of Bashkortostan)

**R. F. Khasanova, I. N. Semenova, Ya. T. Suyundukov, and G. R. Ilbulova**

**Abstract** Introduction. This chapter presents the results of a study of the quality of soil cover in cities located on the territory of the mining districts of the Trans-Urals of the Republic of Bashkortostan. The object of the study was the soils of the lands of various purposes of the cities of Uchaly, Sibay and Baymak. During the study of the soils of industrial zones, samples were selected adjacent to the dumps and tailings dumps of mining production, at a distance of up to 500 m. The recreational zone included the territories of parks, all residential neighborhoods where the city's population lives were included in the residential zone. The territories of collective gardens located near pollution sources were also studied. Material and methods. The total content of heavy metals was determined by atomic absorption method. To assess the degree of soil pollution, the values of the maximum permissible concentrations and the regional geochemical background generally accepted in ecology were used. To assess the degree of soil contamination, the total Zc index was calculated according to the formula of Yu. E. Saet. Results. As a result of studies, in all studied soil samples, an excess of maximum permissible concentration (MPC) in total forms was revealed in terms of Cu, Zn, in some cases Mn, Cd, Co. The soils of industrial territories are the most polluted. According to the general indicator Zc, calculated by the total content of heavy metals, urban soils of all types of land have an acceptable category of pollution. Conclusion. The level of polyelement pollution of the soil cover of urbanized areas in the conditions of the mining region depends on the intended purpose of urban lands. Soils of industrial zones, which are located close to industrial enterprises, are subject to the greatest pollution. In all the studied cities, the lack of sanitary protection zones of enterprises led to the development of residential buildings. In these residential microdistricts, there are household plots where the population grows vegetables and fruit and berry crops.

R. F. Khasanova
Experimental station "Ufimskaya" of the Ufa Federal Research Center of the Russian Academy of Sciences, Ufa, Russia

R. F. Khasanova (✉) · I. N. Semenova · Ya. T. Suyundukov · G. R. Ilbulova
Sibay Institute (branch) of Ufa University of Science and Technology, Sibay, Russia
e-mail: lncs@springer.com



M. Korneykova et al. (eds.), *Smart and Sustainable Urban Ecosystems: Challenges and Solutions*, Springer Geography, https://doi.org/10.1007/978-3-031-37216-2_5

**Keywords** Heavy metals · Mining industry · Trans-Urals of the Republic of Bashkortostan

## 5.1 Introduction

The Republic of Bashkortostan is a subject of the Russian Federation, which is part of the Volga Federal District. The Republic is located in the southern part of the Ural Mountains, on the border of Europe and Asia. It occupies most of the Southern Urals and the adjacent plains of the Bashkir Cis-Urals and the high-plain strip of the Bashkir Trans-Urals.

The southeastern territory of the republic or the Trans-Ural zone is the center of the mining and manufacturing industries. On its territory, there are a number of administrative districts and cities: Baymaksky (with the cities of Sibay and Baymak), Uchalinsky (with the city of Uchaly), Khaibullinsky, Abzelilovsky, Burzyansky, Zilairsky, Zianchurinsky (Suyundukov et al. 2014).

In terms of relief and geology, the region represents the fourth belt of the Southern Urals, enriched with ore-bearing igneous rocks. The climate of the territory is sharply continental. Winters are cold and snowless, summers are warm and dry.

Industrial activity is a source of negative impact on the environment, including the soil cover (Kovalsky 1991; Ilyin 1990). The ore is mined by open and underground methods. As a result of the functioning of these enterprises, technogenic provinces were formed, with a high content of toxic elements in the soils. An anomalous content of chemical elements was found in the soils of the environs of more than 40 depleted deposits (Belan 1998). As a result of mining and enrichment of ore material, huge masses of rocks are extracted and processed. Some of them are used, and the remaining processed raw materials are stored in dumps and tailings. Huge dumps are sources of mineral dust and effluents, which are characterized by high concentrations of toxic elements. Every year, about 1000 tons of dust is blown off the dumps, which includes about 20 different toxic elements (Opekunova et al. 2002; Fatkullin 2002). Roshydromet data, presented in the State Report "On the state and protection of the environment of the Russian Federation in 2016", emphasize that Sibay and Baymak are included in the list of points in the Russian Federation that fall into the moderately hazardous category of soil pollution. The total indicator of pollution (Zc) of the soil cover in a kilometer zone from the source of pollution is assessed as hazardous. Priority pollutants include copper, cadmium, zinc and lead (State Report "On the State and environmental protection of the Russian Federation in 2016 2017). In the studies of (Opekunova et al. 2002), it is emphasized that in the area of the tailing dump of the Bashkir copper-sulfur plant, which is located in the city of Sibay, an excess of approximately permissible concentrations for copper by 1.2–2.4 times, zinc—1.6–2.6 times, arsenic—1.1–1.8 times. The high values of copper, zinc, iron and sometimes cadmium in soils adjacent to mining enterprises are confirmed by the studies of (Shagieva and Suyundukov 2001; Suyundukov et al. 2013; Baimova et al. 2007).

The soils around the dumps of the Sibai quarry have a high content of total forms of copper, zinc, iron, manganese, cadmium and mobile forms of cadmium, copper and zinc. Evaluation of soils by total damage index Zc by detection total forms are rated as "moderately dangerous", for mobile—"dangerous" (Abakumov, et al. 2016). Polyelement pollution of the urban environment in man-made areas, the association of lead, cadmium, arsenic, zinc, copper and other toxic elements. There is a high risk of metal poisoning in the human and animal body, entering the body with food and water, often deposited in organs and tissues, which will be a potential hazard to public health (Rafikova et al. 2017; Teregulova et al. 2009).

The concentration levels of heavy metals in the dry matter of plants growing in the soils of the Trans-Ural zone of the Republic of Bashkortostan in some cases exceeded the permissible limits, in particular lead (more than 10 times), cadmium (more than 6 times) and zinc (2 times) (Semenova et al. 2016).

The soil cover of household and collective gardens in the town of Sibay is characterized by a high content of acid-soluble and mobile forms of copper, zinc, lead and cadmium, within a radius of 5 km from the industrial zone. Suyundukov et al. (2013) noted that for the majority of cultivated plants, an excess content of zinc, a deficiency of iron and manganese was revealed. Belan (1998) states that for the town of Uchaly, out of the total amount of toxic elements entering the human body, local vegetables receive 9% of copper and zinc, cadmium—40%, lead—up to 73%. This is many times higher than the recommended standards.

To date, quite a few works have been published showing the features environmentally dependent diseases of the population of the mining region (Zakharenkov et al. 2015; Savilov et al. 2016; Vekovshinina et al. 2018). So, in the work of Starovoy et al. (1998), there is a high level of overall morbidity of the population, the spread of diseases of the circulatory and nervous systems, respiratory and digestive organs, skin inflammation, an increase is observed by 1.3–2.5 times. In children, there is an increase of 1.5–3.0 times in diseases of the respiratory system, digestion, endocrine system and metabolism, nervous system and sensory organs. In the works of L.N. Belan in the Uchaly district, there is a high level, compared with the average republican indicators, of primary morbidity in the adult population, diseases of the endocrine system, blood and hematopoietic organs, circulatory system, diseases of the nervous system and oncological diseases (Belan 1998). In the study of Shaykhlislamova et al. (2017), the features of diseases of workers of the Uchalinsky mining and processing plant showed violations of the musculoskeletal system of workers. In the blood of residents of the town Uchaly, there is an excess of the norm for cadmium and nickel and a deficit for copper, zinc and manganese. High levels of chromium, zinc, lead, copper, arsenic, nickel, and manganese were found in hair (Suleymanov et al. 2008). An excess of iron, manganese and zinc is also noted in the hair of children in the Baimaksky district.

Excess Fe, Mn and Zn were found in hair samples of the children's population of the Baymak district: in 100, 72.5 and 54.5% of cases, respectively, as well as excess Ni, Cd and Pb in a small number of children (in 18.2, 13.6 and 9.1% of the studied population, respectively). Deficiency of Cu and Co, essential trace elements for human health, was found in 27.3% of the surveyed population (Semenova et al. 2018).

The specificity of soil pollution with the development of the mining industry requires research on hygienic control and the development of preventive measures.

The aim of the study is to study the ecological and geochemical state of urbanized soils in the mining region of the Bashkir Trans-Urals.

## 5.2 Materials and Methods

The research was carried out on the territory of towns Sibay, Baymak and Uchaly. In the period from 2016 to 2022, soil sampling was carried out on the territory of industrial, recreational and residential areas of the city, as well as in collective gardens located in suburban areas. The soil cover of industrial zones is mainly bulk (technogenic surface formations), although chemically transformed natural soils of varying degrees violations persisted in the peripheral parts of the industrial zone.

Soil samples were taken according to the methods generally accepted in geochemistry and soil science. At each polygon, soil and soil samples were taken from the upper layer of the soil cover (from 0 to 20 cm) at three equidistant points. These samples were thoroughly mixed and formed the average sample. Soil samples were ground in a mortar and passed through a sieve with a hole diameter of 1 mm. The total content of heavy metals was determined by atomic absorption method. To assess the degree of soil contamination with heavy metals for copper, zinc, nickel, manganese, lead, the values of MPC or ODC generally accepted in ecology for soils close to neutral, neutral (loamy and clay), pH KCl > 5.5 (MPC and ODC 2006), cobalt and cadmium content regional background values were used (Usmanov et al. 2014; Taipova and Semenova 2012). To assess the level of soil contamination with heavy metals, the total $Z_c$ index was calculated for formula proposed by Saet et al. (1990):

$$Z_c = \sum (K_i / K_f) - (n - 1);$$

where $Z_c$ is the total indicator of heavy metal pollution; $K_i$–specific element concentration in soil (mg/kg); $K_f$—is the background concentration of a particular element (mg/kg); $n$—is the number of summed elements.

The following critical values are distinguished, which characterize the level of soil contamination with toxic metals: permissible contamination ($Z_c < 16$); moderately dangerous ($16 < Z_c < 32$); especially dangerous ($32 < Z_c < 128$); extremely dangerous at ($Z_c > 128$).

## 5.3 Results

In the conditions of the city, the conditions of soil formation regimes, which determine urban pedogenesis, change. The soil cover of cities is represented by urbanozems. In the towns of Sibay and Uchaly, they make up to 60–70% of the area, in Baimak up to 40%. The recreational zone accounts for about 15% of the territory, they are characterized by minimal anthropogenic impact.

Figure 5.1 (A, B) shows the content of total forms of heavy metals in the soil cover of towns

Surveys of the soil cover of the functional zones of towns showed the excess of the maximum permissible concentrations for copper, zinc, in some cases for manganese

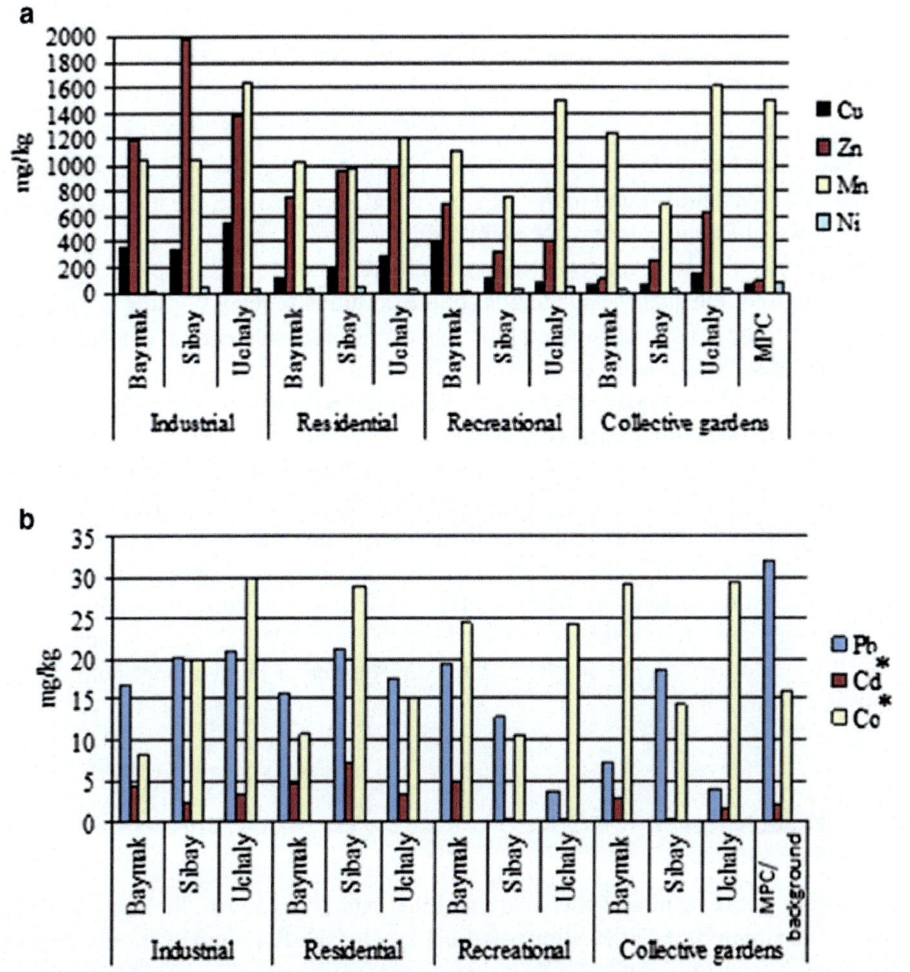


**Fig. 5.1** Concentrations of total forms of heavy metals in the soil cover of towns in the Trans-Urals. Note. For Cu, Zn, Mn, Ni, Pb—MPC, for Co, Cd—background values

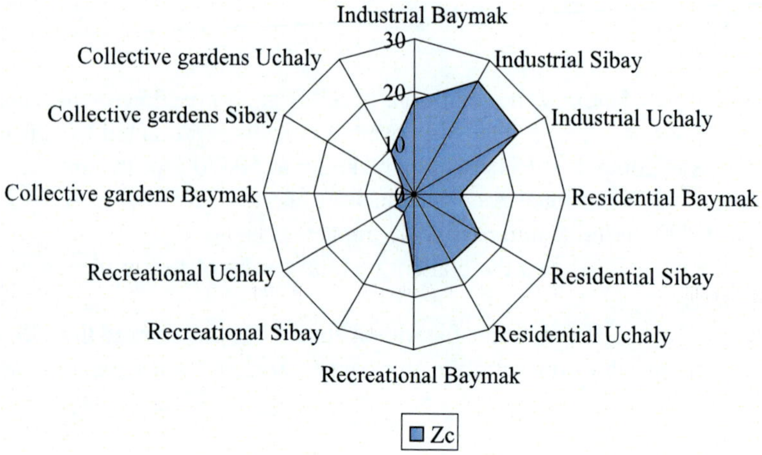


**Fig. 5.2** The total indicator of contamination with heavy metals $Z_c$ of urban soils in the mining area in terms of total forms

and cadmium. Exceeding the permissible concentrations by 9.7 times, manganese by 1.9, cobalt by 1.1 was found in the soils of the industrial zone of the town of Uchaly. In the town of Sibay, an excess of permissible norms for zinc by 19.8 times and for cadmium by 3.6 times was revealed in a residential area located in close proximity to the dumps of the quarry. Calculations of the pollution index according to Saet ($Z_c$) showed that the soils of all studied functional zones in terms of total content belong to the permissible category of pollution. The level of contamination of the surface layer of the soil with heavy metals in all the studied towns is higher in industrial areas, then in residential areas (Fig. 5.2).

In the residential areas of Uchaly and Sibay, a high content of toxic metals was found in suburban areas (Uchaly-2, Buransy in Uchaly, Zoloto and Gorny in Sibay), which are located along quarry dumps. In the town of Baymak, high concentrations of toxic metals have been found in a recreational area. There is a large city park located along the city, on its territory there is a Baimak foundry and mechanical plant with a huge warehouse (more than 1 million tons) of metallurgical slag, the sources of which were former copper smelters. They are sources of polyelement pollution of soils.

## 5.4 Discussion

According to the results of this study and the data presented in the reports of the Territorial Committee of the Ministry of Ecology of the Republic of Bashkortostan, the main sources of pollution of the urban environment in the Trans-Urals are motor vehicles and mining enterprises that determine the flow of heavy metals into various environmental objects. The priority polluting metals in the soils of Sibay, Baymak and Uchaly are Zn, Cu and Cd, in some cases Mn, the level of accumulation and contrast of the areal distribution of which is determined by the functional use of the territory. A comparative analysis of soil pollution inside urban areas showed that the most polluted soils are located in the center and other old areas of the city located directly near industrial enterprises. Relatively clean are the soils of recreational areas of Uchay.

The study of the specificity of polyelement pollution with heavy metals of urban soils in the mining region of the Trans-Urals of Bashkortostan confirms the opinion of other researchers about the presence of high mosaicity (Usmanov et al. 2014; Taipova and Semenova 2012). Several reasons contribute to this. The presence of different geochemical provinces, which differ in the elemental composition of rocks. Dispersion of heavy metals by aerogenic means during the extraction, transportation of raw materials, storage of waste in the course of mining enterprises. Movement of substrates with a high content of toxic metals over landscape elements as a result of water and wind erosion. For cities, additional sources of toxic metals are the presence of vehicles, highways, thermal power plants and other facilities.

## 5.5 Conclusion

The level of polyelement pollution of the soil cover of urbanized areas in the conditions of the mining region depends on the intended purpose of urban lands. Soils of industrial zones, which are located close to industrial enterprises, are subject to the greatest pollution. In all the studied cities, the lack of sanitary protection zones of enterprises led to the development of residential buildings. In these residential microdistricts, there are household plots where the population grows vegetables and fruit and berry crops.

When solving the problems associated with polyelement pollution of urban soils with heavy metals and improving the ecological situation of the urban environment, it is necessary to conduct periodic studies to control the quality of soils in functional zones for various purposes. The obtained data should be taken into account when planning land use, placing children's educational, sports and medical institutions, as well as when developing strategies for carrying out work on the reclamation of disturbed and polluted lands.

**Acknowledgements** The chapter was prepared at the expense of financial support for the implementation of the state task of the State Institution of Strategic Studies of the Republic of Bashkortostan (topic leader—YaT. Suyundukov).

## References

Abakumov EV, Suyundukov YT, Pigareva TA et al (2016) Biological and sanitary assessment of the dumps of the Sibaysky quarry of the Republic of Bashkortostan. Hyg Sanitation 95(10):929–934. https://doi.org/10.18821/0016-9900-2016-95-10-929-934

Baimova SR, Redkina NN, Lykasova IA (2007) The content of heavy metals in the organs of animals in the Bashkir TRANS-Urals. Bull Bashkir State Univ 2:27–28

Belan LN (1998) Heavy metals in technogenic-contaminated soils. Ufa

Fatkullin IR (2002) Assessment of technogenic resources of mining enterprises of the Republic of Bashkortostan: report on the topic. Ufa 93–100

Ilyin VV (1990) Content of heavy metals in garden soils and crops, polluted mining enterprises. Izvestie SO AN SSSR. Biol Sci Ser 58–63

Kovalsky VV (1991) Geochemical environment, trace elements, reactions of organisms. In: Nauka M (ed) Proceedings of the biochemical laboratory, pp 5–23

Maximum permissible concentrations (MPC) and approximate permissible concentrations (ODC) of chemicals in the soil. Hygienic standards. M.: Federal Center for Hygiene and Epidemiology of Rospotrebnadzor

Opekunova MG, Alekseeva-Popova NV, Arestova IJ (2002) i dr. Heavy metals in soils and plants of the southern Urals. The ecological status of anthropogenically disturbed areas. Bull St. Petersburg State Univ 7, 1(7):63–71

Rafikova YS, Semenova IN, Suyundukov YT, Rafikov SS, Biktimerova GY (2017) Regional ecological and hygienic features of the environment and the health status of the population of the Bashkir Zauralye. Sibay

Saet JE, Revich BA, Yanin EP and others (1990) Geochemistry of the environment. M.: Nedra

Savilov ED, Anganova EV, Ilina SV, Stepanenko LA (2016) Technogenic pollution of the environment and public health: situation analysis and forecast. Hyg Sanitation 95(6):507–512

Semenova IN, Rafikova YS, Suyundukov YT, Biktimerova GY (2016) Regional peculiarities of micro-element accumulation in objects in the Trans-Ural region of the Republic of Bashkortostan. Lect Notes Earth Syst Sci 9783319249858:179–187. https://doi.org/10.1007/978-3-319-24987-2_15

Semenova IN, Rafikova YS, Khasanova RF, Suyundukov YT (2018) Analysis of metal content in soils near abandoned mines of Bashkir Trans-Urals and in the hair of children living in this territory. J Trace Elem Med Biol 50:664–6702018. https://doi.org/10.1016/j.jtemb.2018.06.017

Shagieva YA, Suyundukov YT (2001) Technogenesis and the problem of environmental safety in the Bashkir TRANS-Urals. Sbornik dokladov nauchno-prakticheskoj konferencii, posvjashhennoj 95-letiju so dnja rozhdenija professora Bashkir State Agrarian University, S.N. Tajchinova. Ufa

Shaykhlislamova ER, Valeyeva ET, Karimova LK, Galimov RR, Galiullina DM (2017) Multi-year analysis of occupational morbidity in workers of the mining industry of the Republic of Bashkortostan. 1:37–43

Starova NB, Teregulova ZS, Borisova NA (1998) The complex decision of environmental problems of the Republic of Bashkortostan in the space-time unity. In: Proceedings: international. M, 115–139

State Report "On the State and environmental protection of the Russian Federation in 2016" (2017) Ministry of Natural Resources of Russia. NIA-Priroda, Moscow

Suleymanov RA, Allayarova GR, Karimova LK, Valeev TK, Daukaev RA (2008) Organization of the system of social and hygienic monitoring in the territories with developed mining industry of the Republic of Bashkortostan. Hyg Sanitation 1:84–87

Suyundukov YT, Yanturin SI, Singizova GS (2013) Accumulation and migration of heavy metals in the main components of anthropogenic ecosystems of the Bashkir Zauralye in the zone of influence of objects of the mining complex Gilem, Ufa

Suyundukov IT, Semenova IN, Zulkarnaev AB, Khabirov IK (2014) Anthropogenic transformation of soils in the city of Sibay in the zone of influence of the mining industry Ufa, Gilem

Taipova OA, Semenova IN (2012) Ecological and toxicological assessment of the quality of potatoes grown in areas adjacent to the quarry dumps. Modern problems of science and education. 1. http://www.science-education.ru/101-5399

Teregulova ZS, Belan LN, Askarov RA, Teregulova ZF, Altynbayeva AI (2009) Features of environmental pollution and morbidity in the mining region of the Republic of Bashkortostan. Med Bull Bashkortostan 4(6):20–25

Usmanov IY, Semenova IN, Shcherbakov AV, Suyundukov YT (2014) Endemic the ecological niches of the South (Bashkirian) of the TRANS-Urals: a multidimensional and fluctuating modes. Bull Bashkir State Agrarian State Agrarian Univ 1:16–22

Vekovshinina SA, Klein SV, Zhdanov-Tableswitch IG, Chetverkin KV (2018) The quality of the environment and the risks to the health of the population living under the influence of emission of ferrous metallurgy and woodworking industry. Hyg Sanitation 97(1):16–20

Zakharenkov VV, Strelnikova TN, Oleshchenko AM, Surzhikov DV, Kislitsyn VV, Korsakova TG (2015) Prevention of occupational morbidity in workers of the mining industry. Vestnik Rossijskoj akademii estestvennyh nauk. Zapadno-Sibirskoe Otdelenie. 17:151–153

# Chapter 6
# Distribution of Microelements in the Mineral Part of Chernozems Under Different Types of Plant Communities in the Rostov Agglomeration

Check for updates

**N. V. Salnik, S. N. Gorbov, E. N. Minaeva, A. K. Sherstnev, I. P. Melnikova, and S. S. Tagiverdiev**

**Abstract** The features of the main trace elements distribution in the natural soils profile of the Rostov agglomeration under various types of woody plantings have been studied. Artificial forest communities in the steppe zone create conditions for the accumulation of zinc and lead in the upper humus-accumulative horizons and copper at the level of the carbonate barrier in the illuvial-carbonate horizons of chernozem.

**Keywords** Microelements · Calcic Chernozems · Park and recreational areas of the city

## 6.1 Introduction

Urban soils are same unique natural buffer and at the same time represent an area for the accumulation of major pollutants (Gorbov et al. 2015). Currently, the process of urbanization and related environmental problems cause the need for indication and assessment of the urban environment state (Korel'skaya and Popova 2012). Human influence on the soil cover of urban areas leads to the emergence of technogenic geochemical anomalies, which necessitates the inventory and systematics of such soils, as well as the study of the features of their ecological functions (Sereda et al. 2018).

N. V. Salnik (✉) · S. N. Gorbov · E. N. Minaeva · A. K. Sherstnev · I. P. Melnikova · S. S. Tagiverdiev
Academy of Biology and Biotechnology of D.I. Ivanovsky, Southern Federal University, Rostov-on-Don, Russia
e-mail: salnik@sfedu.ru

S. S. Tagiverdiev
e-mail: stagiverdiev@sfedu.ru



M. Korneykova et al. (eds.), *Smart and Sustainable Urban Ecosystems: Challenges and Solutions*, Springer Geography, https://doi.org/10.1007/978-3-031-37216-2_6

The influence of artificial woods, so-called "forest belts" on the characteristic of chernozem was studied in the "Stone Steppe" by Dokuchaev and noted in subsequent scientific work by others scientists (Gromovik et al. 2013; Trofimov et al. 2013; Ukenov and Verkhoshentseva 2017). However, the widely discussed problem of assessing changes in the basic properties of chernozems under the influence of atypical woody vegetation for them remains unresolved (Cheverdin et al. 2014).

The Rostov region is located in the steppe zone, which is the most transformed type of zonal landscapes of Northern Eurasia, but even with a significant anthropogenic load on natural complexes, it is distinguished by the preservation of a relatively high level of biodiversity (Fedyaeva et al. 2021). To create a microclimate in the steppe zone around cities, tree arrays are planted, which can also be considered as an anthropogenic impact on the soils of the urban environment. The planting of trees atypical for the steppe is accompanied by a series of changes. As tree crops grow and develop, forest litter begins to form in the surface horizons of these soils, which is a medium for the accumulation of atmospheric dust and heavy metals contained in it (Salnik et al. 2021).

With the growth and development of tree crops, the ecological conditions of functioning of steppe soils change, biological diversity increases, the water balance of the territory undergoes a significant change and, ultimately, the direction of the flow of natural soil-forming processes (Cheverdin et al. 2014). As a consequence, descending moisture currents begin to dominate in such soils, the distribution of humus in the soil profile and the depth of calcium carbonate occurrence change (Gorbov 2018).

The purpose of our research is to assess changes in the microelements composition of individual horizons of Calcic Chernozem under the influence of woody plant associations.

The aim of the study was to study the accumulation and profile distribution of microelements in soils under the influence of tree associations in the agglomeration conditions of the steppe zone.

## 6.2 Objects and Methods

The objects of the study were the natural soils of the park and recreational zones of the Rostov agglomeration, applying to one subtype—Calcic Chernozems, experiencing minimal influence of the anthropogenic factor (the soil cover has been preserved since the period of agrogenic use, the absence of modern human economic activity)—Temernitsky grove (soil profile cut 2103p), Shchepkin forest (soil profile cut 2101p), the Botanical Garden of the Southern Federal University—nursery of spruce planting (soil profile cut 2005p) and fallow (soil profile cut 2002p). Geobotanical descriptions were carried out for each forest park to assess species biodiversity. The projective coverage of a single geobotanical description site was determined visually (Fig. 6.1).

The total content of microelements in soils, namely Ni, Cu, Zn, and Pb was determined using an X-ray spectrometer "Spectroscan MAKS-GVM" (Spectron,

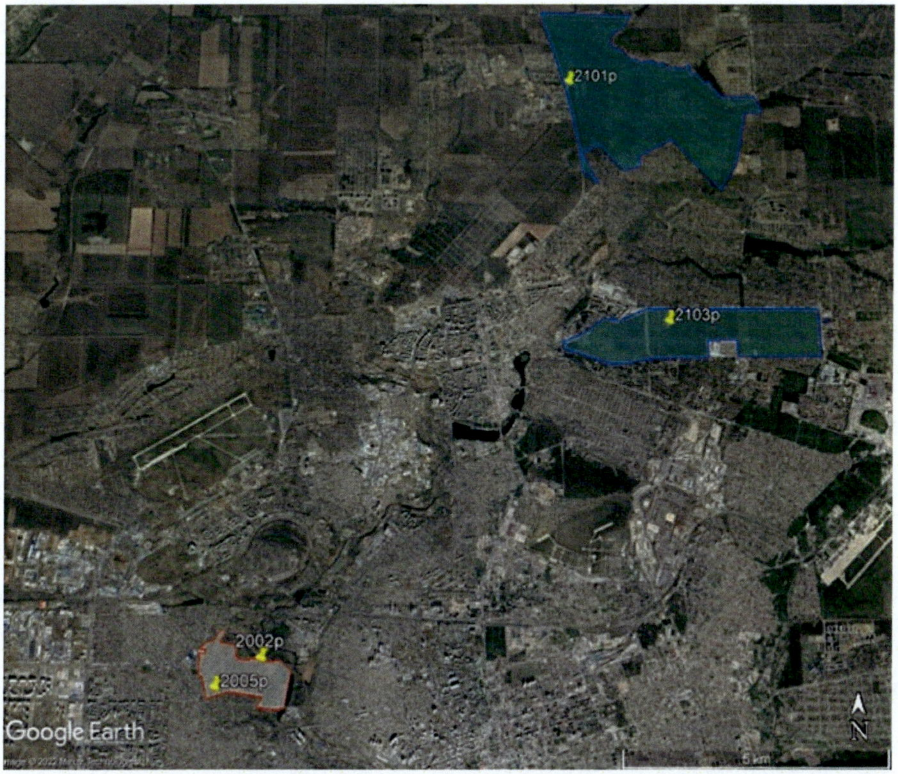


**Fig. 6.1** Map of the location of the studied soil profiles: under steppe vegetation (fallow) (2002p) and under forest massive (2103p—Temernitskaya grove; 2101p—Shchepkinsky forest; 2005p—Spruce tree nursery)

Russia) in accordance with the certified methodology. The dried soil samples are crushed in an agate mortar to particle size of $\leq 250\ \mu$m. Boric acid is poured into a mold and a cup of at least 3 mm depth is formed using a shaped punch. The shaped punch is removed, the sample material is poured and pressed with the smooth punch. The tablet is placed into the sample holder and then into the spectrometer. The samples are analyzed automatically according to the preset program (16.1.42-04 2010). Statistical data processing was performed using the software Statistica for Windows 10.0, MS Exel.

The maximum species diversity of vegetation under geobotanical description was found on the territory of the Temernitskaya grove. It is dominated by representatives of families *Fabaceae* and *Aceraceae* (*Robinia pseudoacacia* L., *Gleditsia triacanthos* L., *Acer negundo* L., *A. tataricum* L., *Amorpha fruticosa* L.). In addition, representatives of weed-shrub, weed-meadow and forest-meadow vegetation with a predominance of 3–5 species take an important part in the herbaceous cover of the Temernitsky grove (*Glechoma hederacea* L., *Galium aparine* L., *Elytrigia repens* L., *Artemisia vulgaris* L., *Sonchus arvensis* L.).

On the territory of the protected area "Shchepkinsky forest", the main area is occupied by plantings *Quercus robur* L., *Acer platanoides* L., *Fraxinus excelsior* L. Other species are also mixed, mainly from families *Aceraceae* и *Rosaceae* (*A. tataricum* L., *A. negundo* L., *Prunus mahaleb* L., *Rosa canina* L., *Pyrus communis* L.). In the nursery of the Botanical Garden of the Southern Federal University planted exclusively *Picea abies* (L.) H. *Karst)* (Nalivaychenko et al. 2021).

## 6.3 Results and Discussions

The quantitative content of microelements in soils is determined by significant factors, which include the direction and intensity of soil formation processes, as well as their content in the parent rock. Recently, the amount of mineral elements in soils has been increasingly influenced by anthropogenic pressure. In this regard, the work on the study of the trace element composition of soils, carried out before the stages of large-scale industrial growth and development of industrial enterprises in a single geochemical province, becomes especially important. Therefore, when studying the soils of the Rostov agglomeration, we take the data given in the work as background values (Akimtsev et al. 1962).

The results of the studies are shown in the Table 6.1. They indicate that the gross content of microelements in fallow Calcic Chernozem (soil profile cut 2002p) can be arranged in the following row: Cr>Zn>V>Ni>Pb>Cu>Co.

The highest content of microelements was observed on virgin soil in the humus-accumulative horizon, and the lowest—in the soil-forming rock. The increased values of individual microelements are associated with the concentration function of both individual plant species of the steppe zone and the entire plant association as a whole, since at least 17–19 chemical elements are known to concentrate in living matter, primarily in plant organisms (Vernadskiy 1983). However, in an urban environment, the high content of microelements in natural soils can also be explained by the cumulative effect of heavy metals caused by the processes of industrialization (the presence of a large paint and varnish enterprise CJSC Empils in Rostov-on-Don) and the likely migration of heavy metals along the soil profile. (Dubinina et al. 2016). It is also impossible to exclude the fact of pedo-geochemical specificity of loess-like loams themselves as a soil-forming rock for most of the region. (Akimtsev et al. 1962).

Thus, for the whole range of elements considered, nickel, copper, lead, and zinc are the most significant. This is caused both by the nature of the elements themselves and by the presence or absence of technogenic emissions characteristic of the Rostov agglomeration.

The profile distribution of nickel in the horizons of fallow chernozem and woodlands shows a gradual decrease in the content of this element with a depth in a narrow range of values, which indicates the stability of the compounds of these elements and their weak migration down the profile.

The distribution of copper along the profile of the natural soils of the city differs from the distribution of nickel. If there is a tendency for nickel to consolidate and accumulate in the humus-accumulative horizon, then copper compounds in the soil profile are more mobile. This is reflected, on the one hand, in a certain depletion of the surface horizons, on the other—in the occurrence of peaks in the concentration of this element in the middle part of the profile (Fig. 6.2). Thus, the amount of copper in the chernozem of the Shchepkin forest in the upper horizons lies in the range of 23–29 ± 3.18 mg/kg of soil, in the middle part of the profile—32–37 ± 3.33. A similar distribution is observed in the chernozem of the Temernitsky grove (upper horizons—24–27 ± 3.18 mg/kg, the middle part of the profile—37–40 ± 3.31). Analysis of the profile distribution of copper in agglomeration soils showed that the copper content in the middle part of the profile is the highest. Perhaps this indicates the accumulation of copper on the carbonate barrier during eluviation and confirms the possibility of movement of compounds of this element with descending moisture currents. However, it is necessary to take into account the fact that the parent rocks, represented by loess-like clays and loams, are initially rich in copper.

In the conditions of the Rostov agglomeration, the main source of zinc intake into the surface horizons of urban soils were large paint and varnish enterprises operating from the mid-twentieth to the beginning of the XXI century. Zinc shows a pronounced tendency to increase concentrations in the sod and humus-accumulative horizons, while in the underlying thickness there is an equalization of its content (Fig. 6.3). Humus substances of chernozems bind microelements with the formation of complex heteropolar salts (Bodeeva 2012), and zinc refers specifically to the

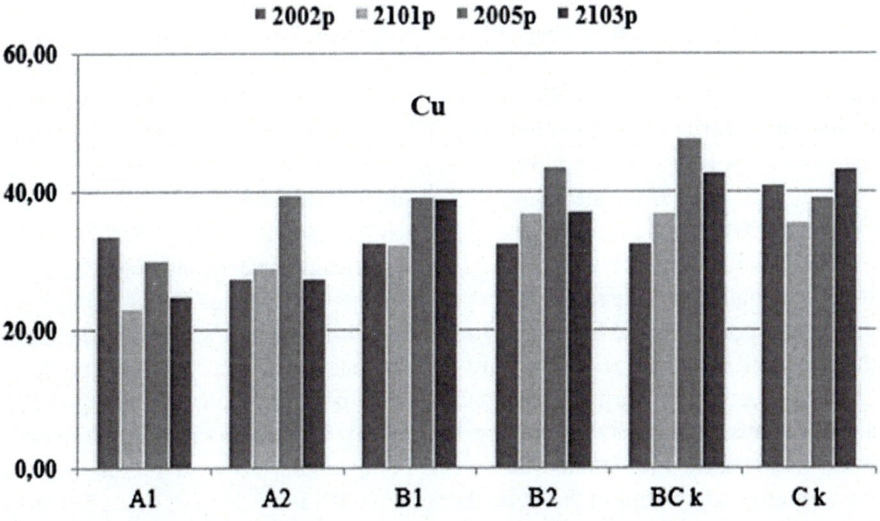


**Fig. 6.2** Profile distribution of copper in Calcic Chernozem of Rostov agglomeration: under steppe vegetation (fallow) (2002p) and under forest massive (2103p—Temernitskaya grove; 2101p—Shchepkinsky forest; 2005p—Spruce tree nursery)

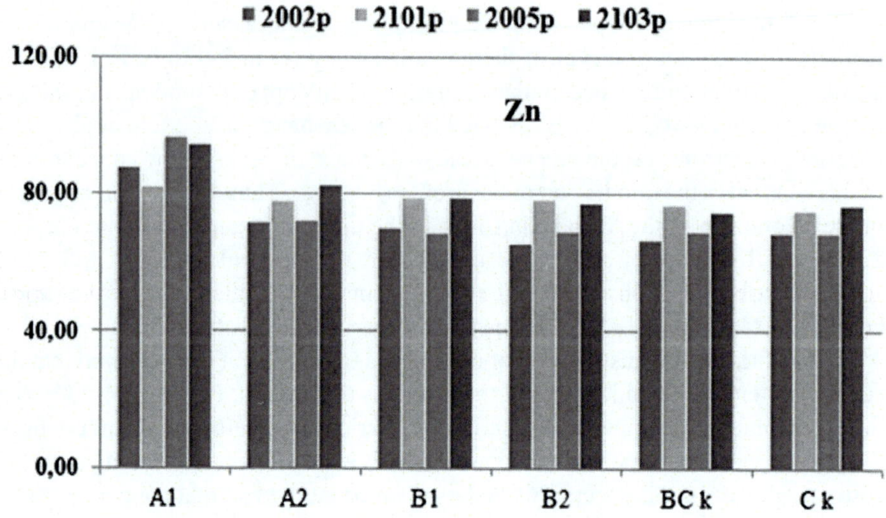


**Fig. 6.3** Profile distribution of zinc in Calcic Chernozem of Rostov agglomeration: steppe vegetation (fallow) (2002p) and under forest massive (2103p—Temernitskaya grove; 2101p—Shchepkinsky forest; 2005p—Spruce tree nursery)

elements of biogenic accumulation, and there is a relationship between the content of organic matter and its concentration in soils (Yakhiyaev et al. 2011).

So, if on average for urban soils of the Rostov agglomeration, the gross zinc content is 65 mg/kg, then in the surface layer of forest soils, zinc concentrations significantly exceed the values given (Table 6.1). With an average zinc content of 85.72 mg/kg, the oscillation range is in the range of 73.95–96.73 mg/kg. The maximum amount of zinc is observed in the humus-accumulative horizon of Calcic Chernozem under the plantings of common spruce (Picea abies (L.) H. Karst)). This is due to the fact that among the coniferous plants introduced in the region, Picea abies (L.) H. Karst) is the most sensitive to air pollution, characterized by a high gas absorption capacity. Even at a distance from the sources of pollution, zinc accumulates in its litterfall (Borodina 2012).

The content and distribution of lead differ significantly in concentrations (Fig. 6.4). If the accumulation of lead in the surface horizons is explained by its proximity to major transport hubs and highways, then the wave-like migration of this element in the soil profile is due to its high affinity with humus substances. If the accumulation of lead in the surface horizons is explained by its proximity to major transport hubs and highways, then the wave-like migration of this element in the soil profile is due to its high affinity with humus substances. For example, in the Temernitsky grove, the average lead content in the upper horizon (0–15) is 25.64 ± 3.34 mg/kg, in the Shchepkin forest—18.56 ± 3.35, in the spruce nursery—36.22 ± 3.25 mg/kg. In the middle part of the profile, a low lead content is observed in the soil under forest vegetation, which may be due to the peculiarities of the granulometric composition and the specifics of the fractional group composition of humus in these horizons.

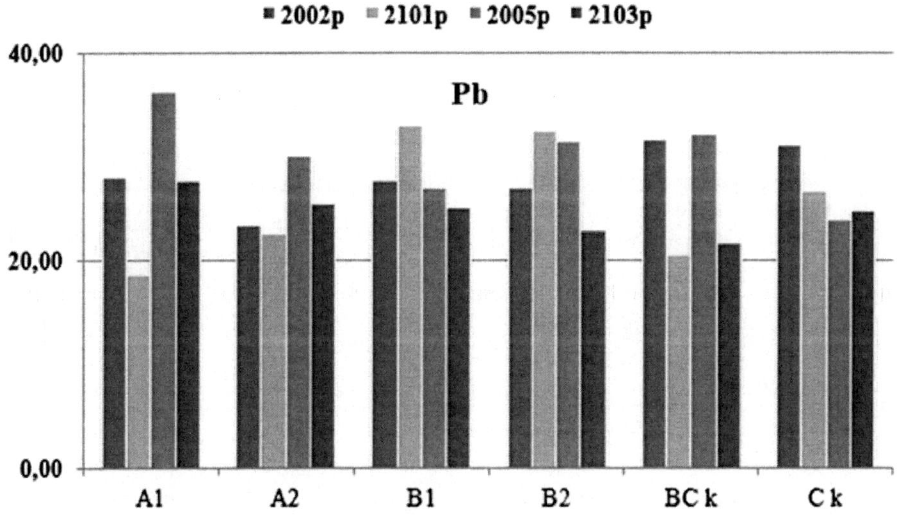


**Fig. 6.4** Profile distribution of lead in Calcic Chernozem of Rostov agglomeration: steppe vegetation (fallow) (2002p) and under forest massive (2103p—Temernitskaya grove; 2101p—Shchepkinsky forest; 2005p—Spruce tree nursery)

## 6.4 Conclusion

The distribution of microelements in the profile of natural soils (chernozems of fallow areas and forest parks) shows stable patterns. In the Calcic Chernozem—in the fallow area of the Botanical Garden of Southern Federal University—the highest content of microelements is observed in the humus-accumulative horizon with a gradual decrease to the soil-forming rock.

Despite the fact that the soils of the recreation zones experience the impact of woody vegetation, which is unusual for them in genesis, the distribution of microelements remains within the limits characteristic of virgin chernozems forming under grass-grain associations. Regardless of the plant community type, copper compounds in the soil profile are characterized by greater mobility. It leads to peaks in the concentration of this element in the middle part of the profile at the level of the carbonate barrier. Nickel is characterized by a gradual decrease down the soil profile in a narrow range of values. The zinc content tends to increase concentrations in the humus-accumulative horizons, lead is characterized by a dispersion of concentrations in the soil profile with minimal values in its central part.

**Acknowledgements** The study was supported by the Strategic Academic Leadership Program of Southern Federal University ("Priority 2030"). The research was financially supported by the Ministry of Science and Higher Education of the Russian Federation (no. FENW-2023-0008). The study was financially supported by the President's Grant for Young Scientists-Candidates of Sciences MK-3257.2022.1.4.

## Appendix

See Table 6.1.

**Table 6.1** Microelements composition of Calcic Chernozem under steppe vegetation (fallow) and forests of Rostov agglomeration, mg/kg (ppm)

| Horizon | Sampling depth | Microelement composition, mg/kg (ppm) (average values) | | | |
|---|---|---|---|---|---|
| | | Ni | Cu | Zn | Pb |
| Background | | 45 | 30 | 65 | 20 |
| Maximum allowable concentration | | 80 | 132 | 220 | 130 |
| *Calcic Chernozem, steppe vegetation (fallow), Botanical Garden. Soil profile cut 2002p* | | | | | |
| A1 | 0–10 | 48.45 ± 2.65 | 33.49 ± 3.31 | 87.70 ± 2.71 | 28.03 ± 3.34 |
| A2 | 10–35 | 47.04 ± 2.63 | 27.51 ± 3.28 | 71.92 ± 2.69 | 23.38 ± 3.39 |
| B1 | 35–60 | 45.94 ± 2.63 | 32.59 ± 3.26 | 70.10 ± 2.60 | 27.70 ± 3.41 |
| B2 | 60–75 | 41.35 ± 2.62 | 32.44 ± 3.31 | 65.54 ± 2.57 | 27.05 ± 3.41 |
| BC k | 75–95 | 42.26 ± 2.61 | 32.64 ± 3.34 | 66.67 ± 2.57 | 31.60 ± 3.40 |
| C k | 95–135 | 43.98 ± 2.65 | 40.91 ± 3.44 | 68.70 ± 2.59 | 31.10 ± 3.41 |
| *Calcic Chernozem, forest park, Temernitskaya grove. Soil profile cut 2103p* | | | | | |
| A1 | 0–10 | 52.51 ± 2.70 | 24.77 ± 3.14 | 94.29 ± 2.70 | 25.64 ± 3.34 |
| A2’ | 10–30 | 55.15 ± 2.62 | 27.48 ± 3.19 | 82.60 ± 2.62 | 27.39 ± 3.40 |
| A2” | 30–50 | 55.68 ± 2.61 | 29.66 ± 3.21 | 81.76 ± 2.61 | 32.45 ± 3.39 |
| B1 | 50–70 | 60.62 ± 2.67 | 38.81 ± 3.38 | 79.08 ± 2.67 | 25.02 ± 3.46 |
| B2 | 70–90 | 57.31 ± 2.78 | 37.08 ± 3.38 | 77.50 ± 2.66 | 22.93 ± 3.43 |
| BC k’ | 90–110 | 54.75 ± 2.81 | 40.73 ± 3.45 | 76.15 ± 2.67 | 28.73 ± 3.47 |
| BC k” | 110–130 | 51.01 ± 2.76 | 42.75 ± 3.47 | 74.59 ± 2.66 | 21.61 ± 3.42 |
| C k | 130–150 | 53.83 ± 2.76 | 43.76 ± 3.46 | 76.98 ± 2.68 | 24.68 ± 3.42 |
| *Calcic Chernozem, forest park, Shchepkinsky forest. Soil profile cut 2101p* | | | | | |
| A1 | 0–15 | 51.43 ± 2.54 | 23.12 ± 3.11 | 81.84 ± 2.56 | 18.56 ± 3.35 |
| A2’ | 15–70 (15–35) | 55.88 ± 2.61 | 28.59 ± 3.19 | 78.31 ± 2.56 | 10.51 ± 3.42 |
| A2” | 15–70 (35–70) | 60.84 ± 2.76 | 29.43 ± 3.24 | 77.94 ± 2.62 | 34.64 ± 3.46 |
| B1 | 70–95 | 62.73 ± 2.76 | 32.18 ± 3.27 | 79.20 ± 2.64 | 32.90 ± 3.44 |
| B2 | 95–120 | 58.24 ± 2.76 | 36.87 ± 3.36 | 78.33 ± 2.66 | 32.44 ± 3.48 |
| BC k | 120–145 | 53.10 ± 2.72 | 36.81 ± 3.36 | 77.02 ± 2.64 | 20.41 ± 3.44 |

(continued)

**Table 6.1** (continued)

| Horizon | Sampling depth | Microelement composition, mg/kg (ppm) (average values) | | | |
|---|---|---|---|---|---|
| | | Ni | Cu | Zn | Pb |
| Background | | 45 | 30 | 65 | 20 |
| Maximum allowable concentration | | 80 | 132 | 220 | 130 |
| C k | 145–160 | 43.61 ± 2.57 | 35.54 ± 3.34 | 75.47 ± 2.59 | 26.60 ± 3.38 |
| *Calcic Chernozem, Spruce tree nursery. Soil profile cut 2005p* | | | | | |
| A1 | 0–10 | 48.31 ± 2.60 | 29.84 ± 3.24 | 96.73 ± 2.75 | 36.22 ± 3.25 |
| A2' | 10–50 (10–30) | 53.86 ± 2.72 | 38.63 ± 3.38 | 75.64 ± 2.64 | 44.19 ± 3.41 |
| A2" | 10–50 (30–50) | 54.70 ± 2.74 | 39.08 ± 3.40 | 68.35 ± 2.61 | 34.86 ± 3.41 |
| B1 | 50–70 | 56.03 ± 2.75 | 39.21 ± 3.40 | 68.41 ± 2.61 | 26.98 ± 3.39 |
| B2 | 70–95 | 51.12 ± 2.73 | 43.51 ± 3.48 | 69.21 ± 2.63 | 31.49 ± 3.39 |
| BC k | 95–110 | 47.90 ± 2.72 | 47.68 ± 3.53 | 69.08 ± 2.62 | 32.17 ± 3.38 |
| C k | 110–140 | 47.01 ± 2.60 | 39.12 ± 3.38 | 68.50 ± 4.10 | 23.91 ± 3.33 |

## References


PND F 16.1.42-04 (2010) Methods for measuring the mass fraction of metals and metal oxides in powder soil samples by X-ray fluorescent analysis (approved by the Federal Scientific-Methodical Center for Analysis and Monitoring of the Environment Ministry of Natural Resources of Russia 23.03.2004). Moscow

Akimtsev VV, Boldyreva AV, Golubev, SN, Kudryavtsev MN, Rudenskaya KV, Sadimenko PA, Sobdornikova IG (1962) Content of trace elements in the soils of the Rostov region. In: Trace elements and natural radioactivity. Rostov State University Press, Rostov-on-Don, pp 37–42

Bodeeva EA (2012) Microelements (Cu, Zn, Ni, Pb) in humus substances of chernozems and chestnut soils of Western Transbaikalia. Dissertation. Novosibirsk, 127 pp

Borodina NA (2012) Accumulation of heavy metals by pine needles in urban ecosystem of Blagoveshchensk city. Izv Samarskogo NTs RAN 14(1–8):1958–1962

Cheverdin YuI, Vavin VS, Akhtyamov AG, Voronin DA (2014) Role of wood plantings in change of properties of chernozems. Achiev Sci Technol AIC 2:11–14

Dokuchaev VV (1949) The works of the expedition equipped by the Forest Department, under the guidance of Prof. Dokuchaeva. Selected works, vol. 2: works on geology and agriculture. Gosudarstvennoe izd-vo s/kh literatury, Moskow, pp 231–247

Dubinina MN, Gorbov SN, Bezuglova OS, Sherstnev AK (2016) Content and profile distribution of heavy metals total forms in soils of urban territories of Russia south. Izvestia Samara Sci Center Russian Acad Sci 18(2–2):349–354

Fedyaeva VV, Shmaraeva AN, Shishlova ZN, Kuz'menko IP (2021) Rare plant species in the protected landscapes of Rostov region. Living Biocos Syst 35. https://doi.org/10.18522/2308-9709-2021-35-2

Gorbov SN, Bezuglova OS, Aleksikova AS, Tagiverdiev SS, Dubinina MN, Sherstnev AK (2015) Content and distribution of heavy metals and arsenic in the soils of Rostov-on-Don. Modern Probl Sci Educ 4:543

Gorbov, SN (2018) Genesis, classification, ecological role of urban soils in the south of the European part of Russia (on the example of the Rostov aglomeration). Dissertation. Moscow, 448 pp

Gromovik AI, Korolev VA, Yonko OA (2013) The influence of field-protecting forest belts on the basic parameters of soils fertility in Stone steppe. Proc Voronezh State Univ Ser Chem Biol Pharm 1:105–114

IUSS Working Group WRB (2022) World reference base for soil resources. In: International soil classification system for naming soils and creating legends for soil maps, 4th edn. International Union of Soil Sciences (IUSS), Vienna, Austria

Korel'skaya TA, Popova LF (2012) Heavy metals in the soil and vegetation cover of the residential landscape of Arkhangelsk. Arctic North 7:136–152

Nalivaychenko AA, Skripnikov PN, Gorbov SN, Matetskaya AY (2021) Assessment of the state of the vegetation cover of artificial forestations in Rostov-on-Don. In: Biodiversity of natural and anthropogenic landscapes: study and protection. Astrakhan State University Press, Astrakhan, pp 56–60

Salnik NV, Gorbov SN, Bezuglova OS (2021) Accumulation and distribution of Ni, Cu, Zn, and Pb in chernozems of the recreational park zone of Rostov-on-Don. In: Training of personnel in the context of the transition to an innovative path of development of forestry: materials of scientific and practical conference. Voronezh, pp 368–371

Sereda LO, Kurolap SA, Yablonskikh LA (2018) Ecological and geochemical assessment of anthropogenic pollution of the soil cover of industrial cities. Nauchnaya kniga, Voronezh

Trofimov IT, Bekhovykh YuV, Bolotov AG, Sizov EG (2013) Physical properties of chernozem soils under coniferous windbreaks. Proc Altai State Agric Univ 9(107):23–27

Ukenov BS, Verkhoshentseva YP (2017) Influence of artificial forest belts on the physical properties of chernozems of the Urals. In: Ecological problems of the Southern Urals and ways of their solution: proceedings of the All-Russian scientific and practical conference, pp 173–179

Vernadskiy VI (1983) Essays on geochemistry. Nauka Publ, Moscow, 422 pp

Yakhiyaev MA, Ramazanov AS, Adzhieva AI (2011) Gross and mobile forms of zinc in the soils of the southern exposition of the foothills of Dagestan. Herald of Dagestan State University. Nat Sci 6:162–165

# Chapter 7
# Seasonal Dynamics of Mobile Phosphorus and Potassium in Podzol Soils Within Slope Catena of Different Anthropogenic Load at the RSAU-MTAA Forest Experimental Station

**Solomon Melaku Melese, Ivan Ivanovich Vasenev, and Dmitry Vladimirovich Morev**

**Abstract** Monitoring studies on the spatiotemporal variability of mobile phosphorus and potassium Podzol soils were carried out at the Forest experimental Station, RSAU-MTAA, which was established 160 years ago. The spatial and temporal variability of mobile phosphorus and potassium studied soils have considered the following: (a) their position (top of hillslope → middle slope → lower part (sole); (b) severity of the path network (c) soil horizons; (d) stock of mobile phosphorus and potassium. Three levels of anthropogenic load are considered. Soil samples were taken during three seasons of 2021: in spring, summer and autumn from a depth of 0–5, 5–10 and 10–15 cm. Mobile phosphorus and potassium were determined using a photo-electro-colorimeter and a flame photometer, respectively. Maximum mobile $P_2O_5$ (241 mg/kg) and $K_2O$ (170 mg/kg) were found at the lower part of the slope and top of the hill slope, respectively, in the minimum anthropogenic load (0–5 cm). In contrast to phosphorus, the distribution of potassium along the slope is reversed. Mobile phosphorus and potassium are higher in spring than in summer and early autumn. During the growing season, there is a noticeable trend of a decline in the content of mobile phosphorus (on average 20%). As the anthropogenic load increases, the amount of mobile potassium fractions drops by about 30%. In all plots, mobile phosphorus and potassium stocks in the upper soil depth significantly decrease with an increase in anthropogenic load.

**Keywords** Mobile phosphorus · Mobile potassium · Soil stock · Podzol soils · Season · Anthropogenic load

---

S. Melaku Melese (✉) · I. Ivanovich Vasenev · D. Vladimirovich Morev
Department of Ecology, Russian State Agrarian University - Moscow Timiryazev Agricultural Academy, Moscow, Russian Federation
e-mail: solyeme@gmail.com

S. Melaku Melese
Department of Forestry, College of Agriculture, Wollo University, Dessie, Ethiopia



M. Korneykova et al. (eds.), *Smart and Sustainable Urban Ecosystems: Challenges and Solutions*, Springer Geography, https://doi.org/10.1007/978-3-031-37216-2_7

## Abbreviations

| | |
|---|---|
| GLM | General Linear Model |
| ANOVA | Analysis of Variance |
| SPSS | Statistical Package for Social Sciences |
| THS | Top Hillslope |
| MSNE | Middle part of slope with a northeastern exposure |
| LPNES | Lower part of slope base with northeastern exposure |
| RSAU-MTAA | Russian State Agrarian University—Moscow Agricultural Academy named after K.A. Timiryazev |

## 7.1 Introduction

Forest ecosystems accumulate huge amounts of soil nutrients in various basins. Soil is the main component of the urban ecosystem, which interacts with the air, water, and bioinert components of the environment (Nizamutdinov and Abakumov 2021; Burdukovskii et al. 2020). The soil cover is the basic element of any landscape, which most fully reflects the current state of all ecosystems and plays an important role in ensuring biological diversity and resistance to the anthropogenic load of forest and urban biogeocenoses (Gracheva 1992; Yimer et al. 2015; Kizilkaya and Dengiz 2010). The main tasks of modern ecology become the maintenance of favorable environmental conditions and the stability of soil functioning under conditions of increased anthropogenic load in urban areas by taking into attention the regional and local regularities of nutrient and seasonal dynamics (Wang 1999; Yesilonis et al. 2016; Li et al. 2013).

The rate of accumulation of mobile fractions of phosphorus (P) and potassium (K) in different soils varied significantly (Emiru and Gebrekidan 2013). These differences were noticeable even within the same soil type and were influenced by many factors (Lemanowicz et al. 2014). Mobile phosphorus and potassium of the upper soil horizons are key variables in regulating the exchange of nutrients between soil, plants, and atmosphere. It is largely determined by the seasonal dynamics of soil respiration, microbiological activity, nutrient reserves in the soil, and microclimate of forest ecosystems in urban conditions (Wright 2009; Demin and Barabanshchikova 2021; Borzenko et al. 2007; Xu et al. 2012; Afanas'ev and Merzlaya 2013).

Despite the fact that there is a soil problem for forest ecosystems in urban areas, forest soils provide a variety of urban ecosystem services and ecological functions. The main environmental problem of large cities is a sharp increase in anthropogenic pressure on their natural environment, including soil cover of the natural forest ecosystem (Sorokina 2010). As a result, serious changes occur in the ecological functions and ecosystem services of urban soils. Thus, a systematic monitoring observation is required that takes into account both the initial spatiotemporal variability of soils and their changes in seasons (Krzyzaniak-Sitarz 2008; Song et al. 2006; Polupan and Velichko 2014; Land Monitoring Forum 2009).

The city's forest ecosystem is used as functions of regulating the composition of nutrients, immobilizing pollutants and regulating the water regime. Moscow is the largest metropolis in Europe, characterized by a high diversity of soil cover and anthropogenic load cities (Voronina et al. 2019). Among them is Forest Experimental Station, which is well-stocked podzolic soils (Baliuk et al. 2015; Sorokina 2010; Naumov et al. 2020). The purpose of this work is to analyze the seasonal dynamics of mobile phosphorus and potassium and its stock in the upper horizons of Podzol soils in different level of anthropogenic load with a representative slope of the Forest experimental Station of RSAU-MTAA.

## 7.2 Objects and Methods

The research was conducted at the Forest Experimental Station of RSAU-MTAA, named for K.A. Timiryazev. In the study area, there is a spatial heterogeneity, with a greater prevalence of windfall and windbreak. The average value of early spring snow cover thickness has increased on the lower part of slope (Eremina and Kayugina 2022). Investigation has been done monthly since April to September 2021. Topsoil samples from the depths 0–5, 5–10, 10–15 cm of Podzol soils were collected at the representative plots within the slope catena (Top hillslope (THS) → Middle part of northeastern slope (MNES) → Lower part of northeastern slope (LPNES)), with three levels of anthropogenic loads: minimum (Background), medium, and high (Table 7.1). The soil analysis was carried out in the soil testing laboratory of RSAU-MTAA, Department of Ecology. Quantitative mobile compounds of phosphorus and potassium were determined using a photo-electro-colorimeter and a flame photometer, respectively. Data were statistically processed using ANOVA according to the GLM procedure (IBM SPSS) version 25 and Microsoft Excel 2010.

## 7.3 Results and Discussions

In the studied area, mobile phosphorus distribution was affected by growing season, profile, anthropogenic load, and relief (Fig. 7.1). Mobile phosphorus is more abundant in the upper horizons of the profile, reaching up to 241 mg/kg in the 0–5 cm. Mobile phosphorus content was found lower in the podzolic soil at the depth of 5–10 cm and 10–15 cm than at the depth of 0–5 cm. These are due to the fact that the higher content of mobile phosphorus in the illuvial layer is apparently connected with the result of mineralogical composition and decomposition of tree leaves and others (herb and mosses etc.) (Zarishnyak et al. 2012; Kovaleva et al. 2021; Melese and Vasenev 2020).

**Table 7.1** Brief description of the key areas of the studied area (Vasenev et al. 2020)

| Slope | Relief | Dominant tree species | Soil: profile | Snow cover* |
|---|---|---|---|---|
| 1 (TMH) | Top hillslope | The dominant tree species on the slope are *Tilia cordata and pine.* Crown density—60% | Podzol soils: O + 2 – AY8 – AEL16 – EL1(f)42 – EL2(g)53 – BEL(g)64 – BTg80 – BC120↓ | 29 cm |
| 2 (MSNE) | Middle part of slope with a northeastern exposure | The dominant tree species on the slope are *Betula pendula, Tilia cordata, Acer Platanoides* and *Rough elm.* Crown density is 50–70% | Podzol soils: O + 1 – AY11 – AEL22 – EL30 BEL(g)65 – BTg80 – BC120↓ | 32 cm |
| 3 (LSNE) | Lower part of slope base with northeastern exposure | The dominant tree species on the slope are *Quercus robur, Acer Platanoides* and *Pinus sylvestris* Crown density is 40–45% | Podzol soils gleyic: O + 2 – AY6 – AEL20 – EL(g)25 – BEL(g)55 – BTg80 – BCg120↓ | 40 cm |

[a]Average value of early spring snow cover from 2009 to 2015 years, cm (Vasenev et al. 2020)

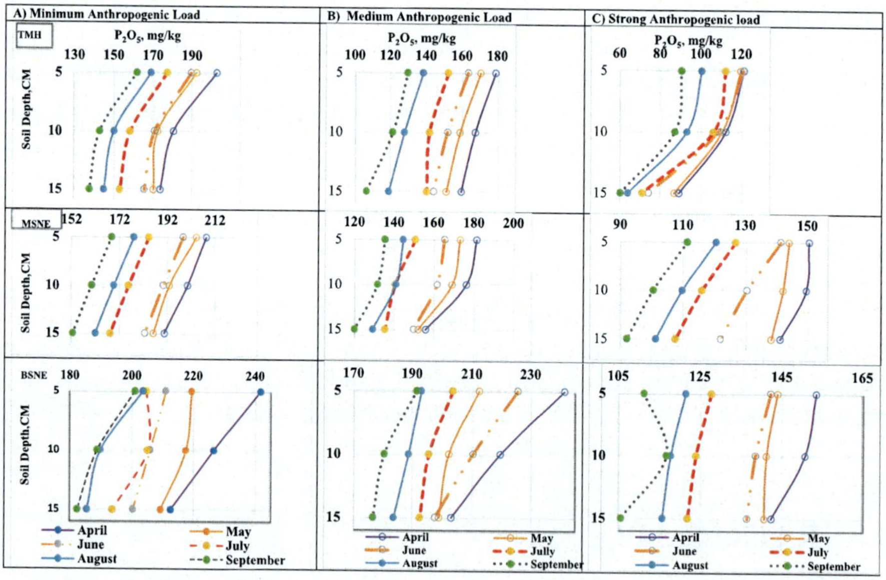


**Fig. 7.1** Seasonal dynamics of $P_2O_5$ stock (mg/kg day$^{-1}$) in the slope catena of Podzol soils at the Forest Experimental Station RSAU-MTAA in a condition of different anthropogenic load

Seasonal dynamics of mobile phosphorus in soils for the months of April to September under different anthropogenic load conditions revealed significant variation. Thus, there is a pronounced trend of a decrease in the content of available fractions of this element during the growing season (about 20% on average) in all studied areas. There are obvious seasonal dynamics of the mobile $P_2O_5$ with maximum values in April and gradual decrease in September (Fig. 7.1). It is clearly expressed that the average mobile phosphorus in the summer months is approximately 1–1.5 times lower than in the spring (Fig. 7.1). High phosphorus concentrations are brought about in the spring by weathering, mineralization, desorption (immobilization), precipitation, runoff from lawns and urban areas, leaking septic systems, and sewage treatment of plant discharges (Ye et al. 2014; Condron et al. 2005; Melese and Vasenev 2022).

According to the data obtained, the areas with a high anthropogenic load are characterized by relatively low concentrations of mobile phosphorus and higher rates of its loss during the growing season (up to 35%). Background areas with a minimum anthropogenic load have the maximum content of mobile phosphorus (241 mg/kg, 0–5 cm depth) in April compared to the area with a strong anthropogenic load (150 mg/kg, 10–15 cm depth) in the lower part of northeastern slope (Fig. 7.1). There is a high content mobile phosphorus in a place with low anthropogenic loads due to low development of organophosphorus compounds and high phosphorus linkages with mineral particles (Vasenev and Raskatova 2009; Shen et al. 2011).

The lower part of slope had a higher content of mobile phosphorus (241 mg/kg) than middle slope (210 mg/kg) and the top hill slope (200 mg/kg). That is, in the case of relief, the availability of phosphorus rises down the slope (Vasenev et al. 2019). The soil is characterized by a higher content of mobile phosphorus in the upper horizon (0–5 cm) on the lower part slope and a low content of mobile phosphorus on the top of slope (10–15 cm). The higher content of mobile phosphorus in the lower slope is apparently connected with its leaching effect on the upper slope (Rupp et al. 2018). On average, mobile phosphorus of the humus-accumulative horizon (0–5 cm) in the soil with a minimum level of anthropogenic load was consistently 1–1.6 times higher than in the same soils of 5–10 and 10–15 cm depth. This may be due to the surface transfer of mobile fractions of nutrients and their accumulation is slow in profile (Lemanowicz 2018).

Findings have revealed that there are significant differences in the rate of accumulation of mobile fractions of potassium on different slopes, growing season, anthropogenic loads, and soil depth (Fig. 7.2). The spatial differentiation of the mobile potassium composition of the studied soils was lowest in the summer season compared to spring (Fig. 7.2). As seen in Fig. 7.2, during the summer period (June–August) and early autumn (September), there is a regular decrease in the value of mobile potassium in the soil. According to the growing season, there is a decrease in mobile potassium under all plots and was recorded in September. This may be due to the increase in soil moisture causing higher mobile potassium levels (April), which in turn improves availability and boosts potassium transport to plant roots (Borzenko et al. 2007; Afanas'ev and Merzlaya 2013; Vasenev et al. 2020). The

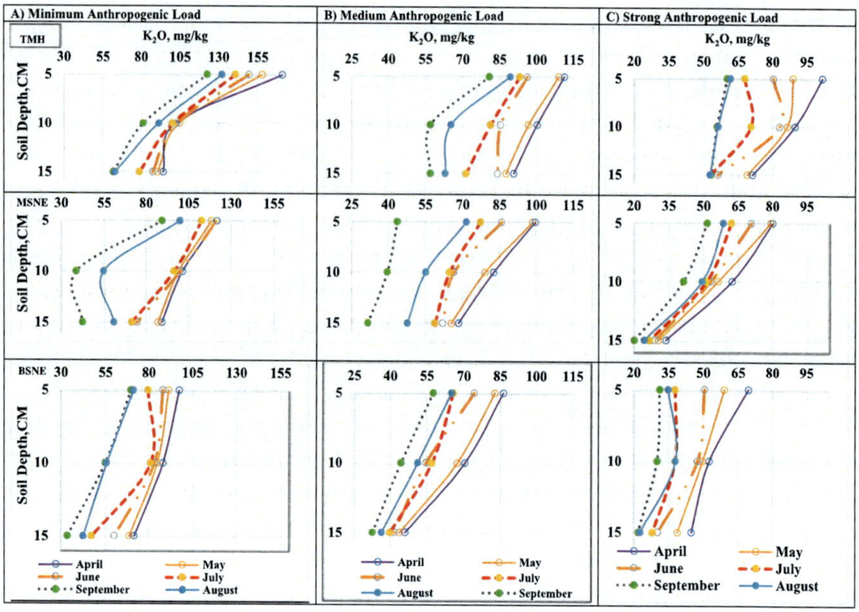


**Fig. 7.2** Seasonal dynamics of $K_2O$ stock (g $m^{-2}$ $day^{-1}$) in the slope catena of Podzol soils at the Forest Experimental Station of RSAU-MTAA in a condition of different anthropogenic load

properties of the forest floor are also in close genetic relationship with the composition of plant residues from which forest litter is formed, and the conditions under which humification occurs (Vasenev et al. 2020).

The content of mobile potassium decreases as the anthropogenic load increases; approximately 30% is lost during the observation period. Particularly high rates of loss (up to 55%) have been observed in areas with the highest degree of anthropogenic load. An area with minimal anthropogenic load has the maximum content of available potassium (165 mg/kg, horizon A1) in April compared to the plot with medium load (117 mg/kg, horizon A1) and strong load (95 mg/kg, horizon A1) on the top of the moraine hill slope (Fig. 7.2). In contrast to phosphorus, the distribution of potassium along the slope has an inverse character: a higher content is noted in the areas at the top of the moraine hill slope. It is connected with the variations in physiological functions of potassium in plants. The increased concentration of $K_2O$ in the background (areas with the lowest anthropogenic load) may be related to young trees and their root systems, which act as a nutrient pump, collecting nutrients from deep soil layers and transporting them to the surface layer via leaf fall (Vasenev et al. 2012). Generally, in all minimum anthropogenic loads, higher values of mobile potassium were recorded in the top hillslopes relative to middle slope and lower part slope. Composition and structure of the litter varies depending on the composition of the forest stand, development of undergrowth, age, density and sanitary condition of the forest (Afanas'ev and Merzlaya 2013; Vasenev et al. 2007, 2012; Eremina and

Kayugina 2022). According to Fig. 7.2, the $K_2O$ content of the soil at the "RSAU-MTAA Forest Experimental Station" ranged from 28.2 to 121.6 mg/kg. Topsoil (illuvial) potassium concentration (0–5 cm) is higher than that at the lower level (5–10, 10–15 cm). It is suggested that the soil profile is greatly affected by the nature of the forest litter in the upper layers (Vasenev et al. 2007).

The soils on the lower part of the slope have higher mobile phosphorus stock in the spring (Table 7.2). According to Condron et al. (2005), mobile phosphorus stocks decline across slopes during the least wet and most hot seasons. On average, the lower part slope in the mobile phosphorus stock is 1–1.8 times higher than the top hillslope (Table 7.2). The lower part slope base with northeastern exposure (April) of minimum anthropogenic load (44.93 g $m^{-2}$) had a higher mobile phosphorus stock than the medium (42.2 g $m^{-2}$) and strong anthropogenic loads (32.39 g $m^{-2}$). Due to the poor solubility of phosphate compounds and the high P-binding capacity of soil components, the stock of mobile phosphorus is reduced at higher anthropogenic loads (Ye et al. 2014; Condron et al. 2005). The mobility of phosphorus stocks is higher with a lower anthropogenic load. It could be due to the availability of plants with sufficient phosphates, the ratio of the mobile fractions of phosphorus, the activity of microbiological processes in the soil, etc. (Ye et al. 2014). Minimum anthropogenic load, on the other hand, may have longer-lasting residual effects due to composition and preceding phosphorus application (Zavišić et al. 2018). It has been noted that as soil depth increases, soil phosphorus stock decreases as a result of hydrolysis in the soil profile. This finding is consistent with (Condron et al. 2005; Melese and Vasenev 2022; Zavišić et al. 2018).

The analysis of seasonal changes in the upper horizons of the studied Podzol gleyic soils showed highly significant differences in mobile potassium stocks (Table 7.3). The increase in anthropogenic load is associated mainly with a decrease in the mobile form of potassium stocks in the 0–15 cm layer of the studied soils. Changes in potassium stock with different levels of anthropogenic load are most pronounced in Podzol gleyic soils (Naumov et al. 2020; Melese and Vasenev 2020). The lowest potassium stock was found on the lower part of the slope (10–15 cm, 7.18 g $m^{-2}$) in September, while the maximum mobile potassium was found at the top of the hill slope (0–5 cm, 20.77 g $m^{-2}$) in April. The low available $K_2O$ concentration in the lower part slope could be due to the inherently low soil K and/or the presence of K in unavailable form (Afanas'ev and Merzlaya 2013; Eremina and Kayugina 2022; Kour et al. 2020).

## 7.4 Conclusions

The seasonal dynamics of mobile phosphorus and potassium in Podzol soils of the representative slope catena with different levels of anthropogenic load were observed during the monitoring campaigns in the spring, summer, and autumn in the conditions of Forest Experimental Station, Northern District of Moscow. It was found that Timiryzev Forest Experimental Station is characterized by high variability in the

**Table 7.2** Seasonal dynamics of mobile form of phosphorus stocks (g $m^{-2}$) in the upper part (0–15 cm) of Podzol soils slope catena at the Forest Experimental Station of RSAU-MTAA in conditions of different anthropogenic load

| Key plot | Level loads | Depth (cm) | April | | May | | June | | July | | August | | September | |
|---|---|---|---|---|---|---|---|---|---|---|---|---|---|---|
| | | | Stocks of $P_2O_5$ (g $m^{-2}$) in horizons in the 0–15 cm layer | | | | | | | | | | | |
| Top hillslope | Minimum | 0–5 | 9.98 | 31.95 | 9.63 | 30.94 | 9.63 | 30.68 | 9.17 | 29.23 | 8.86 | 28.25 | 8.43 | 26.71 |
| | | 5–10 | 10.60 | | 10.17 | | 10.13 | | 9.73 | | 9.46 | | 8.98 | |
| | | 10–15 | 11.38 | | 11.13 | | 10.91 | | 10.33 | | 9.93 | | 9.29 | |
| | Medium | 0–5 | 9.92 | 30.51 | 9.47 | 30.23 | 9.29 | 29.29 | 8.82 | 28.45 | 8.02 | 25.40 | 7.48 | 23.28 |
| | | 5–10 | 9.87 | | 9.55 | | 9.24 | | 8.85 | | 8.18 | | 7.62 | |
| | | 10–15 | 10.73 | | 11.21 | | 10.76 | | 10.78 | | 9.19 | | 8.18 | |
| | Strong | 0–5 | 6.87 | 20.49 | 6.88 | 21.31 | 7.02 | 20.37 | 6.59 | 19.63 | 5.94 | 17.81 | 5.22 | 16.38 |
| | | 5–10 | 6.91 | | 7.46 | | 7.39 | | 7.29 | | 6.54 | | 6.17 | |
| | | 10–15 | 6.71 | | 6.97 | | 5.97 | | 5.76 | | 5.34 | | 5.00 | |
| Middle part of slope with a northeastern exposure | Minimum | 0–5 | 11.05 | 35.06 | 10.94 | 34.44 | 10.74 | 34.06 | 10.24 | 32.27 | 10.09 | 31.67 | 10.80 | 34.24 |
| | | 5–10 | 11.62 | | 11.29 | | 11.25 | | 10.66 | | 10.46 | | 11.15 | |
| | | 10–15 | 12.40 | | 12.21 | | 12.07 | | 11.37 | | 11.12 | | 12.29 | |
| | Medium | 0–5 | 10.08 | 31.34 | 10.58 | 33.07 | 10.19 | 32.17 | 9.53 | 29.46 | 9.30 | 29.05 | 8.48 | 26.78 |
| | | 5–10 | 10.77 | | 11.02 | | 10.62 | | 9.45 | | 9.63 | | 8.78 | |
| | | 10–15 | 10.48 | | 11.46 | | 11.36 | | 10.47 | | 10.13 | | 9.51 | |
| | Strong | 0–5 | 9.11 | 29.09 | 10.00 | 31.71 | 9.89 | 29.57 | 9.07 | 26.85 | 8.76 | 25.77 | 7.25 | 22.01 |
| | | 5–10 | 9.81 | | 10.58 | | 9.79 | | 8.89 | | 8.53 | | 7.46 | |
| | | 10–15 | 10.17 | | 11.13 | | 9.89 | | 8.89 | | 8.48 | | 7.30 | |
| Lower part slope base with northeastern exposure | Minimum | 0–5 | 14.97 | 44.93 | 14.10 | 44.86 | 13.41 | 43.39 | 13.15 | 42.85 | 12.58 | 41.45 | 11.56 | 36.99 |
| | | 5–10 | 14.40 | | 15.03 | | 14.57 | | 14.38 | | 14.07 | | 12.34 | |

(continued)

**Table 7.2** (continued)

| Key plot | Level loads | Depth (cm) | April | | May | | June | | July | | August | | September | |
|---|---|---|---|---|---|---|---|---|---|---|---|---|---|---|
| | | | Stocks of $P_2O_5$ (g $m^{-2}$) in horizons in the 0–15 cm layer | | | | | | | | | | | |
| | | 10–15 | 15.57 | | 15.73 | | 15.42 | | 15.32 | | 14.80 | | 13.08 | |
| | Medium | 0–5 | 12.74 | 42.2 | 12.14 | 40.48 | 11.79 | 39.01 | 11.7 | 39.02 | 11.84 | 37.65 | 10.70 | 34.01 |
| | | 5–10 | 14.14 | | 13.6 | | 13 | | 13.28 | | 12.39 | | 11.38 | |
| | | 10–15 | 15.32 | | 14.74 | | 14.22 | | 14.04 | | 13.42 | | 11.93 | |
| | Strong | 0–5 | 10.06 | 32.39 | 8.64 | 27.22 | 9.15 | 28.86 | 10.40 | 32.84 | 10.69 | 34.00 | 6.52 | 21.80 |
| | | 5–10 | 10.86 | | 8.90 | | 9.43 | | 10.74 | | 11.11 | | 7.72 | |
| | | 10–15 | 11.47 | | 9.69 | | 10.27 | | 11.69 | | 12.21 | | 7.56 | |

**Table 7.3** Seasonal dynamics of mobile form of potassium stocks (g $m^{-2}$) in the upper part (0–15 cm) of Podzol soil slope catena of the Forest Experimental Station RSAU-MTAA in conditions of different anthropogenic load

| Key plot | Level loads | Depth (cm) | April | | May | | June | | July | | August | | September | |
|---|---|---|---|---|---|---|---|---|---|---|---|---|---|---|
| | | | Stocks of $K_2O$ (g $m^{-2}$) in horizons of the 0–15 cm layer | | | | | | | | | | | |
| Top hillslope | Minimum | 0–5 | 8.93 | 20.77 | 8.36 | 20.15 | 7.98 | 19.54 | 7.74 | 18.87 | 7.40 | 17.10 | 6.80 | 15.76 |
| | | 5–10 | 5.66 | | 5.85 | | 5.79 | | 5.81 | | 5.40 | | 4.80 | |
| | | 10–15 | 6.18 | | 5.94 | | 5.77 | | 5.32 | | 4.30 | | 4.16 | |
| | Medium | 0–5 | 6.15 | 18.12 | 6.09 | 18.31 | 5.44 | 16.90 | 5.40 | 15.90 | 5.27 | 14.19 | 4.73 | 12.60 |
| | | 5–10 | 5.89 | | 5.75 | | 5.21 | | 5.04 | | 4.08 | | 3.51 | |
| | | 10–15 | 6.08 | | 6.47 | | 6.26 | | 5.46 | | 4.84 | | 4.36 | |
| | Strong | 0–5 | 5.78 | 16.60 | 5.11 | 16.43 | 4.66 | 14.81 | 3.97 | 13.06 | 3.64 | 11.96 | 3.46 | 11.78 |
| | | 5–10 | 5.48 | | 5.82 | | 5.65 | | 4.82 | | 3.94 | | 3.93 | |
| | | 10–15 | 5.33 | | 5.50 | | 4.50 | | 4.27 | | 4.38 | | 4.39 | |
| Middle part of slope with a northeastern exposure | Minimum | 0–5 | 5.52 | 15.14 | 6.02 | 16.07 | 5.29 | 14.33 | 4.87 | 13.76 | 4.59 | 12.08 | 4.46 | 11.00 |
| | | 5–10 | 5.04 | | 5.15 | | 4.36 | | 4.34 | | 3.76 | | 3.82 | |
| | | 10–15 | 4.58 | | 4.91 | | 4.69 | | 4.54 | | 3.74 | | 2.72 | |
| | Medium | 0–5 | 5.17 | 14.95 | 4.91 | 14.41 | 4.78 | 13.68 | 4.43 | 12.56 | 4.04 | 10.42 | 2.73 | 7.88 |
| | | 5–10 | 5.12 | | 4.99 | | 4.93 | | 4.92 | | 3.44 | | 2.64 | |
| | | 10–15 | 4.65 | | 4.51 | | 3.98 | | 3.21 | | 2.93 | | 2.51 | |
| | Strong | 0–5 | 4.20 | 10.89 | 4.11 | 10.94 | 3.54 | 9.56 | 2.71 | 7.93 | 2.52 | 7.38 | 2.03 | 5.99 |
| | | 5–10 | 3.45 | | 3.70 | | 3.57 | | 2.90 | | 2.95 | | 2.23 | |
| | | 10–15 | 3.23 | | 3.13 | | 2.46 | | 2.32 | | 1.91 | | 1.74 | |
| Lower part slope base with northeastern exposure | Minimum | 0–5 | 6.66 | 19.20 | 6.55 | 18.91 | 6.77 | 18.37 | 6.47 | 17.89 | 5.87 | 13.93 | 5.15 | 10.68 |
| | | 5–10 | 6.29 | | 6.21 | | 6.31 | | 6.24 | | 3.59 | | 2.53 | |

(continued)

**Table 7.3** (continued)

| Key plot | Level loads | Depth (cm) | April | | May | | June | | July | | August | | September | |
|---|---|---|---|---|---|---|---|---|---|---|---|---|---|---|
| | | | Stocks of $K_2O$ (g $m^{-2}$) in horizons of the 0–15 cm layer | | | | | | | | | | | |
| | | 10–15 | 6.25 | | 6.15 | | 5.29 | | 5.18 | | 4.47 | | 3.00 | |
| | Medium | 0–5 | 5.34 | 13.42 | 5.18 | 13.38 | 4.68 | 11.86 | 4.24 | 11.60 | 4.26 | 11.01 | 3.23 | 8.24 |
| | | 5–10 | 4.62 | | 4.81 | | 3.93 | | 4.19 | | 3.81 | | 2.80 | |
| | | 10–15 | 3.46 | | 3.40 | | 3.25 | | 3.17 | | 2.93 | | 2.21 | |
| | Strong | 0–5 | 5.25 | 12.47 | 5.63 | 12.46 | 5.06 | 11.52 | 4.54 | 10.80 | 4.35 | 10.36 | 3.02 | 7.18 |
| | | 5–10 | 4.49 | | 4.26 | | 4.03 | | 3.98 | | 3.89 | | 2.71 | |
| | | 10–15 | 2.72 | | 2.56 | | 2.43 | | 2.29 | | 2.12 | | 1.44 | |

content of mobile phosphorus and mobile potassium. As a result, mobile phosphorus and potassium were higher in the illuvial (0–5 cm) layer and lower in the eluvial layer (5–10, 10–15 cm). In the spring months (April–May), the mobile phosphorus and potassium and their stock were higher in comparison with summer and the beginning of Autumn. Season has a direct impact on the mobile phosphorus, potassium and their stock in the soil. In all cases of anthropogenic load, the maximum mobile phosphorus was recorded in the lower part of the slope, in contrast to mobile potassium. At the same time, mobile phosphorus stock was higher in the lower part of the slope with the minimum anthropogenic load on it, whereas mobile potassium stock was higher on top of the hill slope. We concluded that the established seasonal dynamics of phosphorus, potassium, and their stocks depended on the soil horizons in different parts of the urban slope landscape, even with a slight slope (up to 3–4°) and the presence of vegetation. It was observed that management practices and certain types of vegetation exert a profound influence on soil nutrients. Due to the changeability of mobile soil nutrients under the slope catena of diverse anthropogenic loads and the cyclical nature of the seasons in urban environments, it is crucial to promptly prepare and reviewing soil monitoring data as well as environmental impact assessments in a timely manner.

**Acknowledgments** The authors are grateful for the cooperation of Timiryazev Ecology Laboratory workers. The research was financed by the Ministry of Science and Higher Education of the Russian Federation under Agreement No. 075-15-2021-1030.

## References

Afanas'ev RA, Merzlaya GY (2013) Dynamics of the mobile fractions of phosphorus and potassium in the soils of prolonged trials. Russ Agric Sci 39(4):333–337

Baliuk SA, Nosco BS, Zarishnyak AS, Lisovoj MV (2015) Agro-chemical service in the system of factors of preservation and increase of soil fertility (to the 50th anniversary of the foundation of the Agrochemical Service of Ukraine) scientific report. In common. Ed. Acad. NAAN S.A. Baliuk, Moscow, 64 p.

Borzenko SG, Sokolova TA, Prokoshev VV (2007) Effect of potassium fertilizer composition on the potassium status of Podzol soils under model experimental conditions. Moscow Univ Soil Sci Bull 62:208–214

Burdukovskii ML, Perepelkina PA, Kiseleva IV (2020) Dynamics of vegetation and soil properties of fallow ecosystems. Theor Appl Ecol 3:79–83

Condron LM, Turner BL, Cade-Menun BJ (2005) Chemistry and dynamics of soil organic phosphorus. In: Sims JT, Sharpley AN (eds) Phosphorus: agriculture and the environment

Demin E, Barabanshchikova L (2021) Mineral fertilizers influence on the dynamics of nitrogen, phosphorus and potassium in corn area grown in the forest-steppe zone of Trans-Urals. IOP Conf Ser: Earth Environ Sci 839:022080. https://doi.org/10.1088/1755-1315/839/2/022080

Emiru N, Gebrekidan H (2013) Effect of land use changes and soil depth on soil organic matter, total nitrogen and available phosphorus contents of soils in Senbat Watershed, Western Ethiopia. ARPN J Agric Biol Sci 8(3):1990–6145

Eremina DV, Kayugina SM (2022) Phosphorus-potassium state as an element of natural fertility of gray forest soils of Western Siberia. In: IOP conference series: earth and environmental science, vol 1043, no 1. IOP Publishing, p 012017

Gracheva NM (1992) The influence of anthropogenic pollution on the forest-growing properties of Podzol. Moscow Order of Lenin and The Order of Labor of The Red Banner Timiryazev Agricultural Academy, Moscow

Kizilkaya R, Dengiz O (2010) Variation of land use and land cover effects on soil some physicochemical characteristics and soil enzyme activity. Zemdirbyste-Agriculture 97:15–24

Kour D, Rana KL, Kaur T, Yadav N, Halder SK, Yadav AN, Sachan SG, Saxena AK (2020) Potassium solubilizing and mobilizing microbes: biodiversity, mechanisms of solubilization, and biotechnological implication for alleviations of abiotic stress. In: New and future developments in microbial biotechnology and bioengineering. Elsevier, pp 177–202

Kovaleva NO, Reshetnikova RA, Kovalev IV (2021) Phosphorus in cultural layers and soils of urban ecosystems. Moscow Univ Soil Sci Bull 76(4):217–226

Krzyzaniak-Sitarz M (2008) Influence of anthropopressure on physicochemical properties of soils in the Spa Park in Inowroclaw. Ecol Technol 16:181–189

Land Monitoring Forum (2009) Land and Soil Monitoring: a guide for SOE and regional council reporting. Land Monitoring Forum/Envirolink, Hamilton, New Zealand

Lemanowicz J (2018) Dynamics of phosphorus content and the activity of phosphatase in forest soil in the sustained nitrogen compounds emissions zone. Environ Sci Pollut Res 25(33):33773–33782

Lemanowicz J, Siwik-Ziomek A, Koper J (2014) Effects of farmyard manure and nitrogen fertilizers on mobility of phosphorus and sulphur in wheat and activity of selected hydrolases in soil. Int Agrophys 28:49–55

Li X, Sun Y, Mander Ü, He Y (2013) Effects of land use intensity on soil nutrient distribution after reclamation in an estuary landscape. Landsc Ecol 28(4):699–707

Melese SM, Vasenev II (2020) Impact of overgrown plant deposit on physicochemical properties: SodPodzolic soils during the last 60 years in the Central State Biosphere Forest Stock, Western European Part of Russia. In: Smart and sustainable cities conference. Springer, Cham, pp 132–149

Melese SM, Vasenev II (2022) Seasonal dynamics of humidity and density of soddy-podzolic soils on the slope catena of the Leningrad State Agrarian University-Moscow Agricultural Academy named after V.I. K.A. Timiryazev at different levels of anthropogenic load [Electron. resource]. AgroEcoInfo: Electron Sci Prod J 2. http://agroecoinfo.ru/STATYI/2022/2/st_243.pdf. https://doi.org/10.51419/202122243 (in Russian)

Naumov VD, Kamennyh NL, Lebedev AV, Gemonov AV, Gemonova PS (2020) Heavy metals in Podzol soils under forest stands of Moscow. In: IOP conference series: earth and environmental science, vol 421, no 6. IOP Publishing, p 062036

Nizamutdinov T, Abakumov E (2021) Features of the morphological structure of fallow soils of the Yamal region (in Russian)

Polupan MI, Velichko VA (2014) Nomenclature and diagnostics of ecological and genetic status of Ukraine soils for their large-scale research. Agrarian Science, Kiev, 494 p.

Rupp H, Meissner R, Leinweber P (2018) Plant available phosphorus in soil as predictor for the leaching potential: insights from long-term lysimeter studies. Ambio 47(1):103–113

Shen J, Yuan L, Zhang J, Li H, Bai Z, Chen X, Zhang W, Zhang F (2011) Phosphorus dynamics: from soil to plant. Plant Physiol 156:997–1005

Song T, Wang K, Peng W et al (2006) Ecological effects of intercropping white clover on tea plantation in a subtropical hilly region. Acta Ecol Sin 26(11):3647–3655 (in Chinese)

Sorokina OA (2010) Diagnostic parameters of soil formation in gray forest soils of abandoned fields overgrowing with pine forests in the middle reaches of the Angara River Eurasian. Soil Sci 43(8):867–875. https://doi.org/10.1134/S1064229310080041

Vasenev VI, Yaroslavtsev AM, Vasenev II, Demina SA, Dovltetyarova EA (2019) Land-use change in new Moscow: first outcomes after five years of urbanization. Geogr Environ Sustain 4:24–34

Vasenev II, Raskatova TV (2009) Spatio-temporal variability of the main parameters of background environmental monitoring of soddy-podzolic soils of the Forest Experimental Dacha of the RGAU-MSHA. Bull Mari State Tech Univ Ser: Forest Ecol Nat Manag 2:83–91 (in Russian)

Vasenev II, Naumov VD, Raskatova TV (2007) Structural-functional organization of soil-ecological monitoring of the forest experimental dacha of RGAU – MCHA. Izvestiya TSHA 4, 29–44

Vasenev II, Vizirskaya MM, Vasenev VI (2012) Comparative analysis of principal factors of spatial-temporal variability of $CO_2$ emission from Moscow urban soils with various levels of anthropogenic impact. TSHA. – Spec. Issue, pp 102–112. (in Russian)

Vasenev II, Avilova AA, Tikhonova MV, Ermakov SYu (2020) Assessment of within forest variability in Albeluvisol quality in an urban forest ecosystem for the northern part of the Moscow megalopolis. Springer geography, pp 133–144

Voronina LP, Morachevskaya EV, Akishina MM, Kozlova ON (2019) Evaluation of environmental health of the Kolomenskoye Park under anthropogenic pressure from Moscow City. J Soils Sediments 19(8):3226–3234

Wang Z (1999) Geostatistics and its application in ecology. Science Press, Beijing (in Chinese)

Wright AL (2009) Soil phosphorus stocks and distribution in chemical fractions for long-term sugarcane, pasture, turfgrass, and forest systems in Florida. Nutr Cycl Agroecosyst 83:223–231

Xu G, Shao HB, Sun JN, Chang SX (2012) Phosphorus fractions and profile distribution in newly formed wetland soils along a salinity gradient in the Yellow River Delta in China. J Plant Nutr Soil Sci 175:721–728

Ye X, Bai J, Lu Q, Zhao Q, Wang J (2014) Spatial and seasonal distributions of soil phosphorus in a typical seasonal flooding wetland of the Yellow River Delta, China. Environ Earth Sci 71:4811–4820. https://doi.org/10.1007/s12665-013-2872-3

Yesilonis I et al (2016) Historical land use and stand age effects on forest soil properties in the Mid-Atlantic US. For Ecol Manag 370:83–92

Yimer F, Alemu G, Abdelkadir A (2015) Soil property variations in relation to exclosure and open grazing land use types in the Central Rift Valley area of Ethiopia. Environ Syst Res 4:17. https://doi.org/10.1186/s40068-015-0041-2

Zarishnyak AS, Baliuk SA, Lisovoj MV, Komaristja AV (2012) Balance of humus and nutrients in soils of Ukraine. Bull Agrar Sci 1:28–32

Zavišić A, Yang N, Marhan S, Kandeler E, Polle A (2018) Forest soil phosphorus resources and fertilization affect ectomycorrhizal community composition, beech P uptake efficiency, and photosynthesis. Front Plant Sci 9:463

# Chapter 8
# Grazing Effect on Carbon Stocks and Fluxes in Soils of the Mountainous Pastures

**R. Bekmurzaeva, Z. Iriskhanova, K. Ivashchenko, M. Korneykova, D. Sarzhanov, K. Makhinya, D. Gosse, and M. Nakhaev**

**Abstract** Grazing is known to have a negative impact on soil C stocks; however, the effects depend on the grazing intensity and management practices. This research aimed to study the effect of grazing on carbon stocks and $CO_2$ emissions in the mountainous pastures in the Northwest Caucasus, Chechenia Republic of Russia. WAYCARBON polygon established in Chechenia Republic aims to support investigation searching for the grazing practices which will not disturb soil C stocks. Soil survey and measurements of soil $CO_2$ emission were organized in summer 2022 at the three research sites representing different intensity of grazing: intensive (grazed zone, GZ), light (transit zone, TZ) and reference site, where sheep are not grazed. Topsoil C stocks at the reference site were 10.4 kg $m^{-2}$ which was 10 and 40% higher than at the grazing and transit zones correspondingly. Almost 30% of the total topsoil C stocks at the reference site were distinguished by inorganic C. Soil $CO_2$ emissions were $2.8 \pm 0.9$, $1.7 \pm 0.3$ and $2.4 \pm 0.9$ g C $m^{-2}$ $day^{-1}$ at the grazed, transit and reference site correspondingly. The sustainability of soil C stocks expressed by T0.5 and T0.95 were similar for the grazed and reference sites, whereas soil C stocks at the transit area were almost three times more vulnerable to biodegradation. The absence of the negative effects of grazing on soil C stocks is a positive message for the land-use planners, which tested and verified by the further investigations.

**Keywords** Natural grasslands · Grazing intensity · Microbial respiration · Carbon balance · Carbon neutrality · Chechenia Republic

---

R. Bekmurzaeva · Z. Iriskhanova · M. Nakhaev
Kadyrov Chechen State University, Grozny, Russia

K. Ivashchenko · M. Korneykova (✉) · D. Sarzhanov · K. Makhinya
Smart Urban Nature Research Center, RUDN University, Moscow, Russia
e-mail: Korneykova.maria@mail.ru

K. Ivashchenko
Institute of Physicochemical and Biological Problems in Soil Science, Russian Academy of Sciences, Pushchino, Russia

D. Gosse
Faculty of Soil Science, Lomonosov Moscow State University, Moscow, Russia



M. Korneykova et al. (eds.), *Smart and Sustainable Urban Ecosystems: Challenges and Solutions*, Springer Geography, https://doi.org/10.1007/978-3-031-37216-2_8

## 8.1 Introduction

Together with deforestation and urbanization, the grazing effect on carbon stocks can be considered a classical topic in soil carbon assessments (Elmore and Asner 2006). This topic attracts additional attention for mountainous pastures, where the adjustment of the traditional farm households to a more intensive production has led to changes in grazing practices, and the environmental consequences of these changes remain poorly known (Garcia-Pausas et al. 2017; MacDonald et al. 2000). Soil carbon stocks can be considered a function of the balance between inputs from primary production and outflows from heterotrophic respiration—decomposition of organic matter by soil microbes (Chapin et al. 2006). As a result of overgrazing, the quantity of inputs to the soil may be reduced due to the aboveground biomass removal and belowground biomass disturbance by animals. This reduction in biomass causes decrease in primary productivity in comparison to nongrazed grasslands (Ferraro and Oesterheld 2002). At the same time, light or moderate grazing intensity can stimulate aboveground plant productivity by so-called compensatory growth, when plants respond to defoliation by allocating carbon between above- and belowground biomass (Chen et al. 2006; Zhao et al. 2008). The magnitude of this effect depends on the intensity of defoliation, nutrient availability and water availability (Klumpp et al. 2009; Zhao et al. 2008). In addition, an excess of trampling and continuous overgrazing increases the area of bare soil and correspondingly the risk of erosion. For example, the degradation of grasslands due to overgrazing in the Tibetan plateau caused relevant losses of soil organic carbon stocks during the last 30 years (Xie et al. 2007). In the Alps, degradation of the vegetation cover by overgrazing resulted in annual erosion rates up to 20 Mg ha$^{-1}$ (Meusburger and Alewell 2014).

Although the negative effect of over-grazing on soil carbon stocks is widely accepted, soil organic carbon and especially soil respiration at the grazed pastures can be very heterogeneous. Overgrazing causes an alteration in soil physical, chemical and biological properties at short distances resulting in “patchy” spatial structure, where carbon stocks and fluxes at the neighboring sites are considerably different. Animals tend to graze on areas with the most nutritious plants, whereas they select particular landscape features for resting and ruminating. As a consequence, different types of vegetation develop, which in turn influences the subsequent behavior of animals. Soil nutrients’ contents and availability increase from grazed to resting areas (Badia et al. 2008). For example, the availability of phosphorus increases as a consequence of cattle grazing and defecation (Güsewell et al. 2005). An increase of fresh organic matter and nutrients’ availability may increase microbial decomposition of the native soil carbon stocks by so-called priming effect and therefore intensify $CO_2$ emissions. In mountainous areas, variability in soil C stocks and especially $CO_2$ emissions is further complicated by vertical gradient in abiotic conditions. For example, Komarova et al. (2022) showed that decomposition of organic matter in grazed and nongrazed meadows in the Northwest Caucasus was affected by soil temperature; however, the temperature sensitivity did not depend on vegetation type (Komarova et al. 2022).

Climate mitigation and carbon neutrality goals motivate decision-makers to explore opportunities to decrease carbon emissions and increase carbon sequestration in all principal spheres of the regional economy. In Chechenia Republic, sheep breeding is a traditional sphere of agriculture and regional economy, therefore studying grazing effect on carbon stocks and fluxes in mountainous pastures is highly relevant. Recent research claimed that sustainable practices can have a strong positive effect on carbon balance of the grazed pastures (Gavrichkova et al. 2022). This research aimed to study the effect of grazing on carbon stocks and $CO_2$ emissions in the mountainous pastures in the Northwest Caucasus, Chechenia Republic of Russia.

## 8.2 Materials and Methods

### *8.2.1 Research Area*

The research area (42N, 46E) is located at the mountainous part of Chechenia republic and belongs to the experimental station "WAYCARBON" established in 2022 to monitor the effects of different agricultural practices on carbon stocks and fluxes in mountainous conditions. The research sites are located at the steep and moderate slopes Northwest Caucasus mountains, the elevation ranges between 1800 and 2000 m. The area is dominated by Mollic Leptosols. The vegetation of the area is dominated by subalpine meadow which were historically used as natural pastures. The research sites represent different intensity of grazing: intensive (grazed zone, GZ), light (transit zone, TZ) and reference site, where sheep are not grazed (Fig. 8.1).

### *8.2.2 Measuring Soil C Stocks and Fluxes*

Soil survey in the area was carried out in July 2022. Sampling points were randomly selected in each of the sites, considering the vegetation patterns and elevation. Topsoil (0–10 cm) mixed samples (corners and center) were collected at 30 locations (10 location from each research sites). Additional samples from top 50 cm and top 100 cm were collected at 3 and 1 locations from each site correspondingly to describe the soil profile distribution and to collect. Subsoil samples from the soil horizons. Bulk density was measured in topsoils and further used to estimated soil C stocks. After transporting to the soil-ecological laboratory of RUDN University, soil samples were processed (roots and rock fragments were removed).

In the fresh samples, microbial biomass carbon (MBC) was measured by substrate-induced respiration (SIR) method, based on the registration of the highest initial microbial $CO_2$ production after glucose addition (Anderson and Domsch 1978;

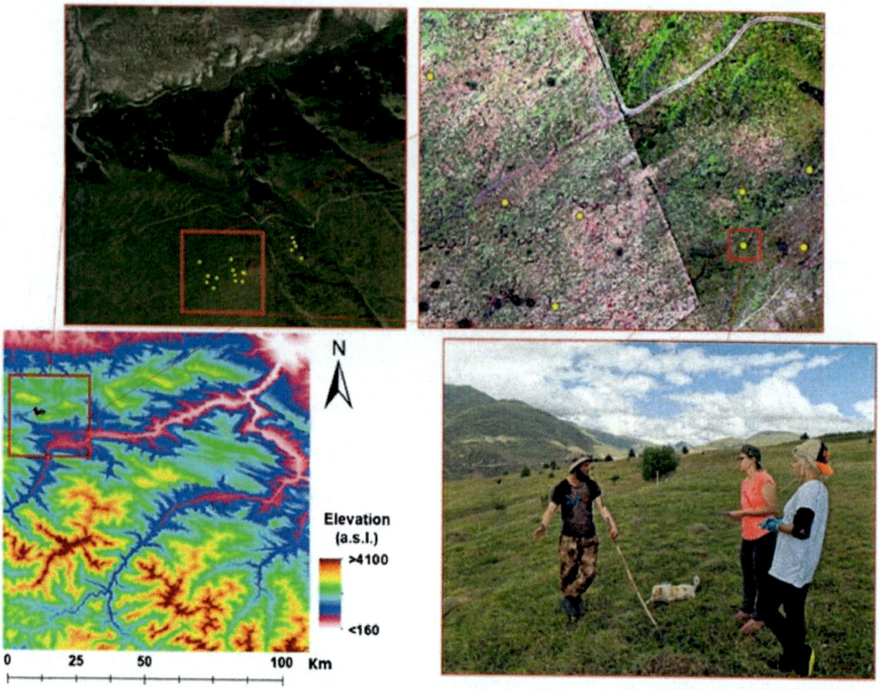


**Fig. 8.1** Research area and soil survey locations (yellow points)

Ananyeva et al. 2008). The subsamples (2.0 g each) were placed in a vial (15 mL volume) and a glucose solution was added dropwise (10 mg glucose $g^{-1}$, volume was 0.1 mL). The vial was tightly closed and incubated at 22 °C during 3.5 h. The measured SIR was converted to MBC units (μg C $g^{-1}$) by the following equation: SIR (μl $CO_2$ $g^{-1}$ $h^{-1}$) × 40.04 + 0.37 (Anderson and Domsch 1978). Basal respiration (BR) was measured in soil samples collected at zero moment, 2 and 14 months after Technosols' construction. In laboratory conditions, soil samples were moistened up to 55–65% water holding capacity (WHC) by distilled water and preincubated during 72 h at 25 °C. After preincubation 2 g subsamples were taken and placed in a vial (15 ml). The vials were tightly closed (time recorded) and incubated at the 22 °C during 24 h (ISO 16072:2002). After incubation of the subsamples, air samples from the vials were taken by syringe and injected into a gas chromatograph (KrystaL 2000 M, 'Meta-Chrom' manufacturer, Yoshkar-Ola, Russia) equipped with a thermal conductivity detector for measuring $CO_2$ concentration. The BR reflected the $CO_2$ produced by microbes in each subsample during the exposition period and was expressed in μg C $g^{-1}$ dry soil $h^{-1}$. After the analysis, soil samples were dried in oven at 105 °C for 8 h to calculate soil moisture. Microbial metabolic quotient (specific respiration, $qCO_2$) was estimated as the ratio of BR to MBC (Anderson and Domsch 1993). In the dried and sieved samples $pH_{H2O}$, total, organic (SOC) and inorganic carbon (SIC) contents were analyzed. Relationships between C content and BR were analyzed to assess the resistance of soil organic matter to biodegradation, based on the biodegradation coefficient k and half-life time $T_{0.5}$. The half-life time is the time period (in years) needed for microbial decomposition of the half of soil organic matter (Smagin et al. 2018).

Soil $CO_2$ emissions were measured in situ at 10 locations by infrared gas analyzer Li-7810SC in the July and August once in each month with the parallel observation of soil temperature and soil moisture. Relationship between soil respiration and temperature were analyzed to study the abiotic drivers of soil $CO_2$ emissions, whereas correlation between C content and microbial parameters (MBC and BR) were analyzed to study the biotic factors of soil $CO_2$ emissions.

## 8.3 Results and Discussions

### *8.3.1 Soil C Stocks: Total, Organic and Inorganic Components*

The descriptions of 100 cm deep soil profiles (RZ, TZ and GZ) reflect the main natural and anthropogenic factors of soil formation and functioning in the area. Soil of the area were formed on carbonated parent materials and inclusions of carbonates are abundant along all the three profiles. In soil profile of the grazed area, the carbonated $C_{Ca}$ horizon is identifies, that can indicate a more intensive erosion compared to the other sites (Fig. 8.2). The highest bulk density was obtained for the transit zone, where it was 10–15% higher than at the other sites. The observed values 1.1–1.2 g $cm^{-3}$ do not indicate over-compaction, that is likely explained by the short exposure time—active grazing started less than one month before the soil survey. All soil had basic pH, which did not differ significantly between the sites or along the profile. The highest content of total carbon was observed at the reference site, whereas the lowest values were obtained for the transit zone. The grazed zone was the most heterogeneous in C content with CV above 60% which was three times higher than at the other sites. This outcome confirms the previously reported effects of grazing on the short-distance spatial variability of soil properties at the natural pastures, when different activities of cattle (e.g., grazing, for resting or ruminating) results in different C inputs and outputs (Badia et al. 2008; Güsewell et al. 2005). Similarly, mean values and variance of N content at the grazed zone was the highest. C/N ratio ranged from 12 to 16 with the highest values observed at the reference site. Topsoil C stocks at the reference site were 10.4 kg $m^{-2}$ which was 10 and 40% higher than at the grazing and transit zones correspondingly. Almost 30% of the total topsoil C stocks at the reference site were distinguished by inorganic C, which was almost two time higher than at the grazed zone (Fig. 8.3).

Fig. 8.2 Soil profiles exposed to different grazing intensity

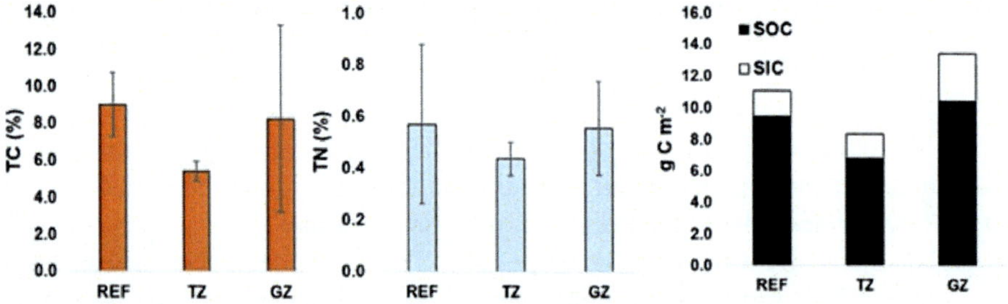


Fig. 8.3 Soil carbon and nitrogen contents and topsoil carbon stocks (organic and inorganic components)

## 8.3.2 Soil Microbial Properties and $CO_2$ Emissions

Microbial activity expressed basal respiration in reference areas was 20% lower compared to the sites affected by grazing. Likely, input of fresh organic matter with dropping could activate development of soil microbial community. A similar pattern was reported for the microbial biomass carbon, whereas $qCO_2$ were similar for all the sites. The magnitude of $qCO_2$ close to 1 indicate favorable conditions for microbial community development. This is quite unexpected for the grazed area, since physical disturbance is known to increase $qCO_2$ (Kooch et al. 2020; Qi et al. 2011; Vasenev et al. 2013). Possible reason could be the short period for which the areas was continuously grazed before soil sampling, or the positive effect of the implemented grazing scheme based on the rotation of the grazing sites. Previously the different impacts of high and low grazing intensity on soil microbial properties were reported for Altai mountains (Goenster-Jordan et al. 2021) or Mediterranean pastures (Gavrichkova et al. 2008). The sustainability of soil C stocks expressed by T0.5 and T0.95 were similar for the grazed and reference sites, whereas soil C stocks

**Table 8.1** Soil mirobial properties at the research sites

| Site | MBC $\mu$g C $g^{-1}$ | BR $\mu$g C $g^{-1}$ dry soil $h^{-1}$ | $qCO_2$ | k $years^{-1}$ | T0.5 years | T0.95 years |
|---|---|---|---|---|---|---|
| REF | 2327 | 2.3 | 0.96 | 0.11 | 7 | 29 |
| TZ | 2897 | 3.1 | 1.06 | 0.09 | 2 | 10 |
| GZ | 2838 | 3.0 | 1.03 | 0.14 | 7 | 30 |

at the transit area were almost three times more vulnerable to biodegradation (Table 8.1). Soil $CO_2$ emissions were 2.8 ± 0.9, 1.7 ± 0.3 and 2.4 ± 0.9 g C $m^{-2}$ $day^{-1}$ at the grazed, transit and reference site correspondingly. Soil temperature effect was not significant, likely due to the short observation period.

## 8.4 Conclusion

Sheep grazing at the natural mountainous pastures is historically a highly important agricultural practice in Chechnya which makes a considerable contribution to the republican economy. Monitoring soil carbons stocks and fluxes and the WAYCARBON polygon started in 2022 aim to assess the grazing effect on carbon balance and improve understanding sheep breeding as a sustainable agricultural practice. The first outcomes obtained in summer 2022 do not show negative effects of grazing on soil C stocks. Microbial biomass and activity at the grazed areas was also similar to the undisturbed reference site. The most likely reason is a short exposure period the area was continuously grazed for less than two months before the soil survey was organized; however, the effect of the light grazing intensity is also a possible option. The latter outcome would be an optimistic message for land-use planners in the region. This message shall be further tested and verified based on the results obtained in the research which can be considered a baseline for further investigation.

**Acknowledgements** Estimation of C stocks was supported by Russian Scientific Foundation Project # 19-77-30012. Monitoring soil emissions was supported by Kadyrov Chechen state university development program 2021–2030.

## References

Ananyeva ND, Susyan EA, Chernova OV, Wirth S (2008) Microbial respiration activities of soils from different climatic regions of European Russia. Eur J Soil Biol 44(2):147–157. https://www.scopus.com/inward/record.uri?eid=2-s2.0-40849136046&doi=10.1016%2fj.ejsobi.2007.05.002&partnerID=40&md5=963216e11538faf38834d60ff14534e8. https://doi.org/10.1016/j.ejsobi.2007.05.002

Anderson JPE, Domsch KH (1978) A physiological method for the quantitative measurement of microbial biomass in soils. Soil Biol Biochem 10(3). https://doi.org/10.1016/0038-0717(78)90099-8

Badia D, Martí C, Sánchez JR, Fillat F, Aguirre J, Gómez D (2008) Influence of livestock soil eutrophication on floral composition in the Pyrenees mountains. J Mt Sci 5:63–72

Chen Y, Lee P, Lee G, Mariko S, Oikawa T (2006) Simulating root responses to grazing of a Mongolian grassland ecosystem. Plant Ecol 183:265–275

Chapin FS, Woodwell GM, Randerson JT, Rastetter EB, Lovett GM, Baldocchi DD, Clark DA, Harmon ME, Schimel DS, Valentini R, Wirth C, Aber JD, Cole JJ, Goulden ML, Harden JW, Heimann M, Howarth RW, Matson PA, McGuire AD, … Schulze ED (2006) Reconciling carbon-cycle concepts, terminology, and methods. Ecosystems 9(7). https://doi.org/10.1007/s10021-005-0105-7

Elmore AJ, Asner GP (2006) Effects of grazing intensity on soil carbon stocks following deforestation of a Hawaiian dry tropical forest. Glob Change Biol 12(9). https://doi.org/10.1111/j.1365-2486.2006.01198.x

Ferraro DO, Oesterheld M (2002) Effect of defoliation on grass growth: a quantitative review. Oikos 98:602–606

Garcia-Pausas J, Romanyà J, Montané F, Rios AI, Taull M, Rovira P, Casals P (2017) Are soil carbon stocks in mountain grasslands compromised by land-use changes? Adv Global Change Res 62:207–230. https://doi.org/10.1007/978-3-319-55982-7_9

Gavrichkova O, Moscatelli MC, Grego S, Valentini R (2008) Soil carbon mineralization in a Mediterranean pasture: effect of grazing and mowing management practices. Agrochimica 52(5):285–296

Gavrichkova O, Pretto G, Brugnoli E, Chiti T, Ivashchenko KV, Mattioni M, Moscatelli MC, Scartazza A, Calfapietra C (2022) Consequences of grazing cessation for soil environment and vegetation in a subalpine grassland ecosystem. Plants 11(16):2121

Goenster-Jordan S, Ingold M, Jannoura R, Buerkert A, Joergensen RG (2021) Soil microbial properties of subalpine steppe soils at different grazing intensities in the Chinese Altai Mountains. Sci Rep 11(1):1653

Güsewell S, Jewell PL, Edwards PJ (2005) Effects of heterogeneous habitat use by cattle on nutrient availability and litter decomposition in soils of an Alpine pasture. Plant Soil 268:135–149

Klumpp K, Fontaine S, Attard E, Le Roux X, Gleixner G, Soussana JF (2009) Grazing triggers soil carbon loss by altering plant roots and their control on soil microbial community. J Ecol 97:876–885

Komarova A, Ivashchenko K, Sushko S, Zhuravleva A, Vasenev V, Blagodatsky S (2022) Temperature sensitivity of topsoil organic matter decomposition does not depend on vegetation types in mountains. Plants 11(20):2765. https://doi.org/10.3390/plants11202765

Kooch Y, Moghimian N, Wirth S, Noghre N (2020) Effects of grazing management on leaf litter decomposition and soil microbial activities in northern Iranian rangeland. Geoderma 361:114100

MacDonald D, Crabtree JR, Wiesinger G, Dax T, Stamou N, Fleury P, Gutierrez Lazpita J, Gibon A (2000) Agricultural abandonment in mountain areas of Europe: environmental consequences and policy response. J Environ Manage 59:47–69

Meusburger K, Alewell C (2014) Soil erosion in the Alps. Experience gained from case studies (2006–2013). Federal Office for the Environment, Bern. Environmental Studies No 1408, 116 pp

Qi S, Zheng H, Lin Q, Li G, Xi Z, Zhao X (2011) Effects of livestock grazing intensity on soil biota in a semiarid steppe of Inner Mongolia. Plant Soil 340(1):117–126

Smagin AV, Sadovnikova NB, Vasenev VI, Smagina MV (2018) Biodegradation of some organic materials in soils and soil constructions: experiments, modeling and prevention. Materials 11(10):1–22. https://doi.org/10.3390/ma11101889

Vasenev VI, Anan'eva ND, Ivashchenko KV (2013) The effect of pollutants (heavy metals and diesel fuel) on the respiratory activity of constructozems (artificial soils). Russ J Ecol 44(6): 475–483. https://doi.org/10.1134/S1067413613060118

Xie ZB, Zhu JG, Liu G, Cadisch G, Hasegawa T, Chen CM, Sun HF, Tang HY, Zheng Q (2007) Soil organic carbon stocks in China and changes from 1980s to 2000s. Glob Change Biol 13:1989–2007

Zhao W, Chen SP, Lin GH (2008) Compensatory growth responses to clipping defoliation in *Leymus chinensis* (Poaceae) under nutrient addition and water deficiency conditions. Plant Ecol 196:85–99

# Chapter 9
# Ecosystem Services and Ecological State of Postpyrogenic Soils in Urban Forest Environments

**Ekaterina Chebykina, Evgeny Abakumov, and Anna Shchepeleva**

**Abstract** Forest fires are among the most significant types of disturbances on a global scale, affecting biodiversity and biogeochemical cycles with an essential contribution to global atmospheric chemistry. Direct damage from forest fires is estimated in billions of rubles annually; however, it can be even higher if indirect losses of environmental and natural values of forests and soils are considered. A novel approach is proposed to assess the ecosystem services of postpyrogenic soils in urban forests of Tolyatti city. A complex indicator to assess the ecological soils state ($P_e$) was used as a quantitative evaluation parameter. It can both serve as an indicator of the degradation of a separate soil property and be used in the economic assessment of individual environmental services of urban forests. Devaluation of ecosystem services by forest ecosystems and soils in result of wildfires shall be transformed into a cost valuation in future. The cost of ecosystem services restoration provided by the soil is very high and determines the need for maximum conservation of the soil cover in the implementation of economic activities.

**Keywords** Wildfires · Ecosystem services · Ecological soil state · Forest ecosystem · Urban landscape

## 9.1 Introduction

The Russian Federation has vast forest resources that provide numerous ecosystem services that are vital to society and the national economy. They contribute to climate mitigation by carbon sequestration, supply a high-quality timber and other forest products, prevent and mitigate damage from catastrophic weather events (storms, droughts, floods) (Schulze et al. 2022; Reitz et al. 2021). Forest tree stands regulate

E. Chebykina (✉) · E. Abakumov
Saint-Petersburg State University, Saint-Petersburg, Russian Federation
e-mail: e.chebykina@spbu.ru

A. Shchepeleva
RUDN University, Moscow, Russian Federation



M. Korneykova et al. (eds.), *Smart and Sustainable Urban Ecosystems: Challenges and Solutions*, Springer Geography, https://doi.org/10.1007/978-3-031-37216-2_9

the hydrological regime at various scales, purify water, contribute to the creation and maintenance of water reserves in rivers, lakes and groundwater reservoirs (Asgharinia et al. 2022; Efrosinin et al. 2021). Forests also play an important role in the processes of pollination and biological plants protection from diseases, serve as a habitat for more than half of the known plant and animal species, form the soil and protect it from destruction (Bolund and Hunhammar 1999; Gamfeldt et al. 2013). The capacity of the forests to supply vital ecosystem services is affected by natural disturbances, among which wildfires are the most dangerous (Dobrovolsky 2002).

Fire risk is controlled by the interplay of climate, fuel (vegetation) and ignitions, with high risk periods occurring during summer droughts when fuels are abundant and ignitions are typically human-caused (Ganteaume et al. 2013). Under the climate change these risks can increase dramatically (Shvidenko and Schepaschenko 2013). The fire season has become virtually year-round. There are more and more forest fires over the years, and they are more widespread and larger and more catastrophic than before. Plants ignite more easily due to rising temperatures, heatwaves and dry weather.

The negative environmental impacts of forest fires are substantial. As sources of additional greenhouse gases emissions, forest fires have a clear negative impact on global climate (Shvidenko and Schepaschenko 2013). The total amount of annual carbon dioxide emissions from wildfires is about a quarter of those from burning oil, coal and gas. Another major contributor to global warming is black carbon produced and emitted by forest fires. On the other hand, forests accumulate and store the carbon, which has a beneficial effect on the climate. A forest damaged by fire weakens, regenerates more slowly, absorbs less carbon dioxide and releases less oxygen, and therefore does not help to slow climate change (Ilina 2011; Boisramé et al. 2017).

The area of natural wildfires in Russia over the past 20 years has averaged 8.9 million hectares per year, according to Information system for remote monitoring of the Federal Forestry Agency (Rosleskhoz). The area of forest fires in Russia in 2021 amounted to 18.81 million ha. This is maximum value since the beginning of the twenty-first century (Fig. 9.1).

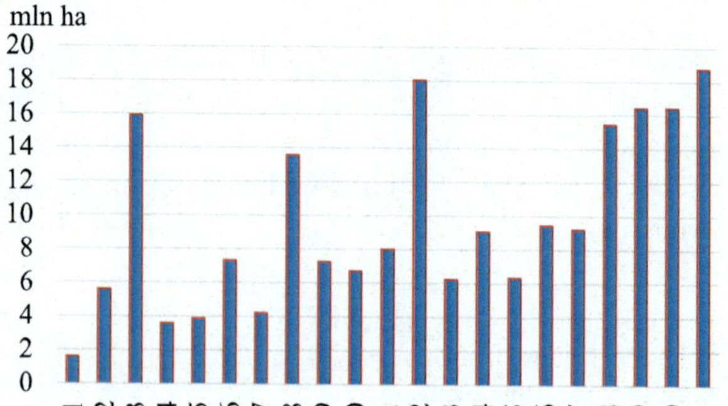


**Fig. 9.1** The area of forest fires in Russia during 2001–2021 according to Information system for remote monitoring of the Federal Forestry Agency (Rosleskhoz) (Instructions for determining the damage caused by forest fires 1998)

The average temperature in the European part of Russia in June 2021 was the highest for more than 130 years of meteorological observations. There has been a drought since May in the Urals, the middle Volga and in many regions of Siberia. Yakutia region suffered the most from forest fires. In the conditions of the extreme heats, urban forests become very vulnerable to fire events and the potential health risks and damages from urban forest fires is especially high. A forest fire is fraught with great danger, since in addition to destroying large material values and killing animals and plants, the fire can spread to settlements, industrial enterprises and disable power lines. For example, the forest fires caused by extreme summer heat 2010 caused considerable human victims (Grigorieva and Revich 2021) and total damage of 85.5 billion rubles according to the Federal Forestry Agency. These data are likely considerably underestimated due to the fact that they take into account only the direct loss of wood resources, but do not take into account the loss of environmental and natural values by forests, and even more so do not take into account the indirect damage caused by fire and smoke to human life and health. In particular, a study showed an increase in the number of deaths among the population in the Volga basin (in the Samara region) and the proportion of deaths from diseases of the circulatory system was conducted (Lifirenko and Lifirenko 2010).

The Instructions for determining damage caused by forest fires (Instructions for determining the damage caused by forest fires 1998) states that it is necessary to take into account the damage from pollution of the environment by combustion products, but it does not take into account the fact of an increase in mortality due to fires, which was presented by the study described above (Lifirenko and Lifirenko 2010). Moreover, the Instructions does not take into account the damage from changes in ecological functions and properties of soils as a result of the burnt plant residues and other products income into the soil. It must be remembered that the forest is not only trees, but a whole ecosystem where everything is interconnected: apart from vegetation, fire destroys soils, animal communities, pollutes the atmosphere and much more. All these limitations contribute to the considerable underestimation of the forest fire damages by existing approaches, especially in urban forests, where connection between nature and society is very strong.

The identification of ecosystem services performed by green spaces, the development of approaches to their assessment and payment mechanisms are being conducted for urban landscapes (Baranova et al. 2020; Potapova 2016). Usually, only services for air purification from pollution and recreation are evaluated (Bukvareva and Sviridova 2020). Despite the accumulated significant information on soils, the assessment of ecosystem services of the soil component is not given sufficient attention. Soil transformations affect their functioning and fulfillment of their ecological functions, which should be manifested in ecological services by entire landscapes. Therefore, approaches to assessing the ecosystem services of soils, which are a valuable and hard-to-renew resource, deserve special attention. A natural soil profile formation is measured over centuries in contrast to revegetation.

The soil, as an integral part of the forest community, is affected by wildfires in many ways. Pyrogenic soil changes are the result of the direct fire impact, as well as indirect postfire changes in the biogeocenosis, which are being much more widespread. Frequent forest fires damage beneficial microorganisms in the soil and destroy nutrients. The forest does not fully recover, and if fires occur constantly, then it stops growing altogether. Therefore, soil erosion occurs and desertification begins. Such processes threaten not only biodiversity, but also the climate: these lands do not absorb carbon dioxide, it becomes more in the atmosphere, which means that global warming continues. The pyrogenesis processes are a widespread phenomenon that has a huge impact on the soil formation processes, which makes scientists pay special attention to them in case of natural ecosystems study. Therefore, at present, despite fundamentally different technologies in economic activity, the pyrogenic factor remains an acute problem for forest countries, which requires solutions both in connection with global climate change and a number of economic issues—the loss of ecosystem services provided by forests and soils, the loss of forests as an important component in the terms of decarbonization of the economy.

The research aimed to develop a novel approach to assess the ecosystem services of postpyrogenic soils in urban forests of Tolyatti city.

## 9.2 Materials and Methods

*Research area.* The Tolyatti city is a large settlement (more than 700 thousand inhabitants) of a diffuse type. Its areas are separated by forest islands—massifs, which are mostly old pine forests and open spaces (fields, meadows, wastelands). Tolyatti city is a city not only of the Russian automobile industry, but also of a large Russian chemistry (KuibyshevAzot, TolyattiAzot, Sintezkauchuk, Tolyattisintez). Therefore, urban forests are considered a "pine filter" for urban atmosphere. The abnormally dry year of 2010 provoked forest fires on the territory of Tolyatti city as elsewhere in the European part of Russia. This became a local disaster, as it completely changed the functioning of forest ecosystems. Ten destroyed large forest areas. The urban forests of Tolyatti city lost more than 2000 ha of old pine forests as a result of crown fires.

The tree layer survived in forests affected by the surface fire, but the fire of various intensity degree damaged trees trunks, which weakened their vitality and will lead to gradual loss in the future. However, it is woody plants that form communities and will determine the course of restore successions in the future. The grass and shrub layers were damaged to a greater extent. Tree and shrub plantations burned out completely in forest communities affected by the crown fire, as well as the grass and shrub layers, and spontaneous postpyrogenic successions began here. The processes of natural reforestation began already in the following year after the fire. Almost all burnt areas were gradually overgrown with trees and shrubs (Rakov et al. 2011).

The forest grows for a long time. Therefore, the problem of providing city residents with green spaces continues to be very acute, and it should be addressed immediately. According to the "Regional standards for urban design of the Samara region", the total area of green areas for common use (parks, squares, boulevards, etc.) should be at least 9.0 $m^2$ per person. In fact, one inhabitant of Tolyatti accounts for 2.9 $m^2$/person in residential areas, and 5.0 $m^2$/person in forests areas (before the fires in 2010); the average is 7.9 $m^2$/person. Moreover, within the framework of the activities of the Program, "Forestry Development for 2014–2018 and for the period up to 2022", work to clear illiquid forest areas affected by the consequences of forest fires have being carried out on the territory of the Tolyatti forestry. However, unfortunately, the situation is aggravated by the fact that forest fires in the urban forests of Tolyatti city are repeated, as it was in the summer of 2021.

The forests and green spaces of Tolyatti city perform extremely important aesthetic, landscape-forming, environmental, sanitary and architectural and planning roles. They create conditions for the life support of the city and its population, satisfy the needs for fresh air, places for recreation and communication with nature, are a source of forest resources of various nature and a green filter that reduces the degree of environmental pollution. The great achievement of foresters and population of Tolyatti city is the presence (where industrial structures and transport networks developed in the city) and the preservation (even after the large-scale construction in the end of the twentieth century) of large forest massifs (more than 8 thousand hectares) with their inherent forest environment and uniqueness of nature. Urban forests are characteristic of the cities in Siberia and the Urals to a greater extent. There are not such large forests in the Volga basin except for the urban forests of Moscow (for example, the Losiny Ostrov massif), comparable to the urban forests of Krasnoyarsk. Therefore, the assessment of the loss of an ecosystem service—the soil formation service—after forest fires in urban forests, using the example of Tolyatti city, is an urgent task, which is associated with the uniqueness and difficulty of soil replenishment.

*Methods.* The quality of ecosystem services in urban landscapes is undoubtedly affected by the ecological soils state. Methods of assessing the ecological state of the city soils have been developed and tested (Stoma and Romanova 2019; Kurbatov and Bashkin 2004). A complex indicator for assessing the ecological soils state ($P_e$) was used as a quantitative evaluation parameter. The ecological soils state was assessed according to the method of Stroganova et al. (2003). The ecological soils state is a complex of soil properties that determines the degree of their correspondence with the natural and climatic conditions of soil formation and suitability for the sustainable functioning of natural and anthropogenic ecosystems. Eleven parameters were selected for the assessment: thickness of the humus stratum, humus stochs, $pH_{H2O}$, particle size distribution, bulk density, trace metals content (total pollution index), soil respiration (biological activity), number and biomass of soil biota in the upper layer. Depending on the level of change in the soil property relative to its optimal values or natural analogues, an assessment was made on a 5-point scale ($B_i$ is the assessment of individual *i*-diagnostic indicators of soil properties in points (5 is the optimal situation, 1 is an almost irreversible violation of soil property)) and multiplied by the weight significance coefficient of soil properties diagnostic indicators ($K_i$ = 0.5–1.5). Taking into account the weight significance of each property, a complex indicator for assessing the ecological soils state ($P_e$) was calculated, which can serve

a b c

**Fig. 9.2** **a** Surface forest fire revegetation plot; **b** crown forest fire revegetation plot; **c** control

both as an indicator of the degradation of a separate soil property and be used in the economic assessment of individual environmental services of urban forests.

Gray-humus sandy loamy soils with features of spodic processes but without the formation of an independent podzolic horizon formed on ancient alluvial Volga sands—Psamment Entisols were studied for assessment the ecological soils state after forest fires. Soil diagnostics were carried out according to the "Classification and diagnostics of soils of Russia" (Shishov et al. 2004) and the World Reference Base for Soil Resources, FAO 2015 (IUSS Working Group WRB 2015). The impact of two types of wildfires—surface and crown—was studied in comparison with the control variant. Figure 9.2 shows the landscape dynamics, the case immediately after wildfire is presented at (Maksimova and Abakumov 2015a).

## 9.3 Results and Discussion

Data on the main chemical parameters of soils were published in previous works (Maksimova et al. 2019, 2014; Maksimova and Abakumov 2015b; Abakumov et al. 2017). Studies have shown that forest fires alter the morphological and physicochemical properties of soils (Maksimova et al. 2019, 2014; Maksimova and Abakumov 2015b), and lead to complete or partial degradation of organic horizons and the formation of so called pyrogenic horizons (Abakumov et al. 2017).

Diagnosed postpyrogenic soils had the following soil horizons: in the case of postfire areas Qpyr-AY-AC-C; in the case of control area, O-AY-AC-C. The ash was a loose crumbling crust of a dirty gray color of small thickness (1–2 cm), with a significant admixture of small pieces of charcoal and soil particles (the ash seemed to be washed into the mineral horizons). The water-soluble components of the plant litter pyrolysis products together with precipitations will penetrate into the soil stratum and serve as the main agents affecting the organo-mineral part of the soil.

Fires lead to serious changes within the soil profile. The processes of humus loss are especially active when the litter and the upper humus horizon burn out. Changes in soil morphology are most noticeable in the upper horizons (widespread of coals, preservation of ocher tones in the horizons color).

The burning of the litter changes the occurrence and subsequence of soil horizons. As the vegetation develops, a new litter begins to form, but the former pyrogenic horizons persist for a long time. The soils acidity changes as a result of fires, usually toward alkalization. The loss of the most mobile nutrients (potassium, sodium, magnesium and manganese) is noted in the ash formed after fires.

The soil performs many ecological functions in ecosystems, and individual soil properties or their complex are involved in the implementation of certain ecosystem services provided. Transformation of soil indicators under the influence of various factors can lead to a change in the quality of ecosystem services. An assessment of the ecological soil properties state taking into account the degree of their transformation (or difference) relative to the natural potential or optimal values is presented in Table 9.1. Eleven soil properties that can ensure the performance of certain ecosystem services by the urban forest were selected.

The providing ecosystem services are determined by the number and biomass of soil biota from the selected soil indicators: bacteria, fungi, and actinomycetes. Regulatory ecosystem service "filtration and accumulation of chemical elements in ecosystem" correlates with a wide range of soil properties (bulk density, pH, particle size distribution, humus stocks, trace metal pollution), "weathering and soil formation processes" with only two soil properties—a decrease in the thickness of humus stratum and basal respiration as a parameter of biological activity.

The ecological state of postfire soils deteriorates compared to the control: the average $P_e$ value significantly decreases from 5.8 to 4.1 (Table 9.1), and there are no differences between the soils of crown and surface forest fires. It varies significantly for individual soil properties from 2.0 to 7.5 and is below the control by 20–60%. A decrease in the ecological state of individual soil properties relative to the optimum is manifested in postpyrogenic soils in almost 80% of cases. The most negative contribution to the decrease in $P_e$ is made by a decrease in the thickness of the humus horizon and the number and biomass of soil biota. Due to changed habitat conditions and direct burning of soil biota, its condition is significantly deteriorating. Individual indicators of the degradation level of its number and biomass are reduced from 5 to 2–3 points. Moreover, there is a significant deterioration in the ecological state of such diagnostic soil parameters as humus stocks and the thickness of the humus layer (from 7.5 to 3.0 points). Other soil properties do not change significantly as a result of fires.

Devaluation of ecosystem services by forest ecosystems and soils as a result of wildfires shall be transformed into a cost valuation in the future. This is especially true for urban forests in the context of forest fire damage calculations. Similar cost estimates were previously successfully carried out for urban soils (Stoma and Romanova 2019; Semenyuk et al. 2019, 2021). Undoubtedly, the valuation of ecosystem services

**Table 9.1** Ecosystem services and indicators of the ecological soil properties state of urban forests in Tolyatti city

| Ecosystem services (Vasenev et al. 2018; Morel et al. 2015) | Soil properties | Value | | | $B_i$, points | | | $K_i$ | $P_e$ | | |
|---|---|---|---|---|---|---|---|---|---|---|---|
| | | 1 | 2 | 3 | 1 | 2 | 3 | | 1 | 2 | 3 |
| The genetic material of the biota | $\frac{\text{Number (bln cells/g)}}{\text{biomass (mkg/g)}}$ of bacteria | $\frac{0.64}{12.8}$ | $\frac{0.46}{9.2}$ | $\frac{2.04}{20.4}$ | 3 | 2 | 5 | 1.0 | 3 | 2 | 5 |
| | $\frac{\text{Mycelium length (m/g)}}{\text{biomass (mkg/g)}}$ of fungi and actinomycetes | $\frac{144.8}{204.9}$ | $\frac{39.9}{63.38}$ | $\frac{602.9}{832.81}$ | 3 | 2 | 5 | 1.0 | 3 | 2 | 5 |
| Filtration and accumulation of chemical elements in ecosystem | Total pollution index Zc | 32 | 7 | 12 | 4 | 5 | 5 | 1.5 | 6 | 7.5 | 7.5 |
| | Physical clay content, % | 14.9 | 14.1 | 15.8 | 4 | 4 | 4 | 1.0 | 4 | 4 | 4 |
| | Bulk density, g/sm$^3$ | 0.78 | 0.85 | 1.03 | 5 | 5 | 5 | 1.0 | 5 | 5 | 5 |
| | $pH_{H2O}$ | 8.0 | 7.9 | 6.5 | 3 | 3 | 4 | 1.5 | 4.5 | 4.5 | 6 |
| | Humus stocks, t/ha | 36 | 54 | 81 | 3 | 4 | 5 | 1.5 | 4.5 | 6 | 7.5 |
| Weathering and soil formation processes | humus horizon thickness, sm | 14 | 10 | 23 | 3 | 2 | 5 | 1.5 | 4.5 | 3 | 7.5 |
| | Basal respiration (biological activity), mg $CO_2$ per 100 g per day | 75 | 77 | 258 | 3 | 3 | 5 | 1.0 | 3 | 3 | 5 |
| $P_e$ average | | | | | | | | | 4.2 | 4.1 | 5.8 |
| $P_e$ total | | | | | | | | | 37.5 | 37 | 52.5 |

*Note* 1—surface forest fire, 2—crown forest fire, 3—control of urban forests affected by forest fires is of great practical importance in the framework of fire safety, forest fire risk assessment and urban planning and should be used as a tool for organizing the urban environment that regulates economic activity in urban forests and the ecological state of urbanized forests territories.

## 9.4 Conclusions

Forest fires remain one of the powerful natural factors that determine the large-scale environmental changes. The damage from forest fires is estimated in billions of rubles annually, however they take into account only the direct loss of wood resources, but do not take into account the loss of environmental and natural values by forests, and even more so do not take into account the damage caused by fire and smoke to people's lives and health. The ability of the forest and soils to support and supply ecosystem services is lost as a result of forest fires. The cost of ecosystem services restoration provided by the soil is very high and determines the need for maximum conservation of the soil cover in the implementation of economic activities. In order to assess the cost of damage from forest fires, it is necessary, first of all, to take into account the cost of ecosystem services delivered by soil, since their price is comparable to or an order of magnitude higher than the price of ecosystem services of green spaces, and some soil ecosystem services are invaluable due to the complexity of the renewal of this natural resource and the inability to fully compensate for their ecological functions.

**Acknowledgements** The work was supported by the Grant of the President of the Russian Federation for young PhDs No. MK-4596.2022.1.4. This work is dedicated to the 300th anniversary of Saint Petersburg State University.

## References

Abakumov E, Maksimova E, Tsibart A (2017) Assessment of postfire soils degradation dynamics: stability and molecular composition of humic acids with use of spectroscopy methods. Land Degrad Dev 29:2092–2101. https://doi.org/10.1002/ldr.2872

Asgharinia S, Leberecht M, Marchesini LB, Friess N, Gianelle D, Nauss T, Opgenoorth L, Yates J, Valentini R (2022) Towards continuous stem water content and sap flux density monitoring: IoT-based solution for detecting changes in stem water dynamics. Forests 13(7). Manuscript No. 1040. https://doi.org/10.3390/f13071040

Baranova OYu, Semenuk OV, Bodrov KS, Stoma GV (2020) Ecosystem services and regulatory and methodological approach as tools for environmental and economic assessment of a natural area. Akademicheskij vestnik URALNIIPROEKT RAASN 45(2):21–27. https://doi.org/10.25628/UNIIP.2020.45.2.004

Boisramé GFS, Thompson SE, Kelly M, Cavalli J, Kate MW, Stephens SL (2017) Vegetation change during 40 years of repeated managed wildfires in the Sierra Nevada, California. Forest Ecol Manage 402:241–252. https://doi.org/10.1016/j.foreco.2017.07.034

Bolund P, Hunhammar S (1999) Ecosystem services in urban areas. Ecol Econ 29(2):293–301. https://doi.org/10.1016/S0921-8009(99)00013-0

Bukvareva EN, Sviridova TV (eds) (2020) Ecosystem services in Russia: prototype of the national report. Biodiversity and ecosystem services: accounting principles in Russia, vol 2. Publishing House of the Center for Wildlife Conservation, Moscow, p 252

Dobrovolsky GV (2002) Soil degradation and protection. Publishing House of Moscow State University, Moscow

Efrosinin D, Kochetkova I, Stepanova N, Yarovslavtsev A, Samouylov K, Valentini R (2021) Trees classification based on Fourier coefficients of the Sapflow density flux. Ann Math Inform 53:109–123. https://doi.org/10.33039/ami.2021.03.002

Gamfeldt L, Snäll T, Bagchi R, Jonsson M, Gustafsson L, Kjellander P, Ruiz-Jaen MC, Fröberg M, Stendahl J, Philipson CD, Mikusiński G, Andersson E, Westerlund B, Andrén H, Moberg F, Moen J, Bengtsson J (2013) Higher levels of multiple ecosystem services are found in forests with more tree species. Nature Commun 4. Manuscript No. 1340. https://doi.org/10.1038/ncomms2328

Ganteaume A, Camia A, Jappiot M, San-Miguel-Ayanz J, Long-Fournel M, Lampin C (2013) A review of the main driving factors of forest fire ignition over Europe. Environ Manage 51:651–662

Grigorieva EA, Revich BA (2021) Health risks to the Russian population from temperature extremes at the beginning of the XXI century. Atmosphere 12(10). Manuscript No. 1331

Ilina VN (2011) Pyrogenic impact on vegetation cover. Samarskaya Luka Probl Reg Glob Ecol 20(2):4–30

Instructions for determining the damage caused by forest fires. The decree of Federal Forestry Agency (Rosleskhoz) dated 03 of April, 1998, No. 53

IUSS Working Group WRB (2015) World Reference Base for Soil Resources International Soil Classification System for Naming Soils and Creating Legends for Soil Maps, World Soil Resources Reports No. 106, FAO, Rome, Italy

Kurbatov AS, Bashkin VN (eds) (2004) Ecological functions of urban soils. Magenta, Moscow–Smolensk, p 232

Lifirenko DV, Lifirenko NG (2010) The increase in mortality of the population of the Volga basin as a result of the abnormal heat in the summer of 2010. Acad J Izvestia Samara Sci Center Russ Acad Sci 14(1):272–275

Maksimova E, Abakumov E (2015a) Soil organic matter alterations after catastrophic wildfires in forest-steppe ecosystem. FLAMMA 6(1):10–12

Maksimova E, Abakumov E (2015b) Wildfire effects on ash composition and biological properties of soils in forest-steppe ecosystems of Russia. Environ Earth Sci 74:4395–4405. https://doi.org/10.1007/s12665-015-4497-1

Maksimova E, Abakumov E, Shamilishvili G (2018) Sustainable development of forest ecosystems in urbanized territories as a way of wildfire control in Russia. In: Vasenev V, Dovletyarova E, Cheng Z, Prokof'eva T, Morel J, Ananyeva N (eds) Urbanization: challenge and opportunity for soil functions and ecosystem services, SUITMA. Springer Geography. Springer, Cham, pp 279–288. https://doi.org/10.1007/978-3-319-89602-1_33

Maksimova EYu, Tsibart AS, Abakumov EV (2014) Soil properties in the Tol'yatti pine forest after the 2010 catastrophic wildfires. Eurasian Soil Sci 47(9):940–951

Morel JL, Chenu C, Lorenz K (2015) Ecosystem services provided by soils of urban, industrial, traffic, mining, and military areas (SUITMAs). J Soils Sediments 15(8):1659–1666. https://doi.org/10.1007/s11368-014-0926-0

Potapova EV (2016) Ecosystem services of green areas of settlements. Bull Sci Prac 9(10):36–41

Rakov NS, Saxonov SV, Senator SA (2011) Initial stages of pyrogenic succession in the urban forests of Tolyatti (floristic aspect). In: Proceedings of the III international ecological congress (V international scientific and technical conference) "Ecology and life safety of industrial and transport complexes", vol 2. Tolyatti, pp 196–200

Reitz O, Yaroslavtsev A, Meshalkina JL, Vasenev II, Vasenev V, Valentini R, Leuchner M (2021) Carbon dioxide fluxes of an urban forest in Moscow. Springer Geography, pp 40–50. https://doi.org/10.1007/978-3-030-75285-9_5

Schulze ED, Bouriaud O, Irslinger R, Valentini R (2022) The role of wood harvest from sustainably managed forests in the carbon cycle. Ann Forest Sci 79(1). Manuscript No. 17 (2022). https://doi.org/10.1186/s13595-022-01127-x

Semenyuk OV, Bodrov KS, Stoma GV, Yakovlev AS (2019) Estimation of the value of ecosystem services of the natural park "Bitzevsky forest". Vestnik Moskovskogo universiteta. Seriya 17. Soil Sci 3:23–29

Semenyuk OV, Stoma GV, Bodrov KS (2021) Evaluation of the cost of ecosystem services of urban landscapes (using the example of Moscow). Eurasian Soil Sci 12:1–14

Shishov LL, Tonkonogov VD, Lebedeva II, Gerasimova MI (2004) Classification and diagnostics of soils in Russia. Ojkumena, Smolensk, p 342

Shvidenko AZ, Schepaschenko DG (2013) Climate change and wildfires in Russia. Contemp Probl Ecol 6(7):683–692

Stoma GV, Romanova LV (2019) Ecological state of soils and tree vegetation in urban park—recreational landscapes (tha case of the Catherine park of Moscow). Vestnik Moskovskogo universiteta. Seriya 17. Soil Sci 4:11–18

Stroganova MN, Prokofieva TV, Prokhorov AN, Lysak LV, Sizov AP, Yakovlev AS (2003) Ecological state of urban soils and land valuation. Eurasian Soil Sci 7:867–875

Vasenev VI, Van Oudenhoven APE, Romzaykina ON, Hajiaghaeva RA (2018) The ecological functions and ecosystem services of urban and technogenic soils: from theory to practice (a review). Eurasian Soil Sci 51(10):1119–1132. https://doi.org/10.1134/S1064229318100137

# Chapter 10
# The Dynamics Peculiarities of the Xylem Sap of Coniferous Woody Plants on the Example of Picea Abies in Rostov-on-Don

**R. N. Ospischev, S. N. Gorbov, and A. M. Yaroslavtsev**

**Abstract** The intensive process of urbanization increases the demand for urban green building. In order for the development of green infrastructure to correspond to the rate of urban growth and reduce its negative impact on human health, it is necessary to use methods of modern and continuous monitoring. Through innovative methods, it is possible to optimize the greening of the city, which will increase the efficiency of the functioning of green spaces. Monitoring was carried out with the help of TreeTalker+device in this work. The main parameter for the study chosen rate of xylem sap of woody plants. Representatives of common spruce Picea Abies were used as an object of study. When analyzing the data obtained during monitoring, it was possible to establish a rank correlation between xylem sap velocity and temperature, as well as to establish a relationship with phenological phases.

**Keywords** Urban landscaping · Express monitoring · TreeTalker+ · *Picea abies*

R. N. Ospischev (✉) · S. N. Gorbov
Academy of Biology and Biotechnology of D.I. Ivanovsky, Southern Federal University, Rostov-on-Don, Russian Federation
e-mail: rostislav.ospishev.96@mail.ru

R. N. Ospischev
Federal Research Centre the Southern Scientific Centre of the Russian Academy of Sciences, Moscow, Russian Federation

A. M. Yaroslavtsev
Peoples Friendship University of Russia, Moscow, Russian Federation

Russian Agrarian University, Moscow, Russian Federation

 
M. Korneykova et al. (eds.), *Smart and Sustainable Urban Ecosystems: Challenges and Solutions*, Springer Geography, https://doi.org/10.1007/978-3-031-37216-2_10

## 10.1 Introduction

Currently, most of the world's population lives in cities, and the proportion of urban dwellers is steadily increasing, outpacing the population of rural areas in terms of growth rate. According to the forecasts of several specialists, the gap in this ratio of inhabitants will only increase (Dye 2008). As a consequence, the man-made load on the green framework of urban ecosystems, which, in turn, provides a neutralization of the negative factors of the urban environment, increases (Matasov 2020).

Urban green infrastructure, as the main compensating component of the city, affects both the formation of a comfortable microclimate for living and reduces the concentration of pollutants in the soil and the surrounding environment. At the same time, modern cities are characterized by spatial expansion due to the occupation of adjacent areas, and green infrastructure should keep up with this growth. However, there are several problems that prevent the formation of a balanced combination of park and recreational and industrial-settlement areas associated with the specific nature and climatic conditions, anthropogenic pressures, and limited space (Lisikov 2015).

In the context of Rostov-on-Don, these problems are exacerbated by the steppe type of biocenosis. Special climatic conditions with a large amplitude of temperature differences between summer and winter make it difficult to grow woody plants in this area. In addition, woody plants play an important role in the formation of the microclimate in the city of Rostov-on-Don. A decrease in the number of trees leads to dust storms and droughts. The solution to this problem is the introduction of the most appropriate species for the area, which requires a lot of time resources. At the same time, the introduced plants require special timely care and control of their condition, and consequently, the use of modern monitoring methods is necessary to optimize this process as much as possible. These methods make it possible to quickly give a quantitative assessment of the state of plants, namely trees growing in certain soil and climatic conditions. Such an approach will probably make it possible to leave the extensive development of urban greening to the intensive one, increasing its quality and efficiency (Gudzenko and Gudzenko 2012; Kozlovskiy et al. 2013).

The main problem of growing woody plants in the city of Rostov-on-Don is regular watering against the background of extremely high summer temperatures and dry winds, especially in the early years of rooting. In this case, the question of watering doses and maximum seasonal peaks in moisture consumption can be indirectly determined with TreeTalker+, as the device allows you to track the rate of xylem sap and establish in what period of the year the plant needs abundant watering for comfortable growth.

Thus, the quality of green infrastructure will be higher, and therefore, woody plants will perform their functions better. Thus, it is possible to move from extensive development of urban landscaping to intensive.

In this study, it is expected to establish a correlation between the rate of xylem sap and the temperature of the near-trunk part of the tree. It is also assumed that the phenological phases directly depend on the flow rate.

## 10.2 Objects and Methods

The territory of Rostov-on-Don is characterized by an arid climate. The amount of precipitation in the spring–summer–autumn period is 270–300 mm. In summer, the average temperature is + 22– + 23 °C. Also 180–190 days with a temperature above 0° C. Winter is moderately mild: the average temperature in January is −5 °C, the average absolute minimum of air temperature during the winter is −20–25 °C (Kozlovskiy).

Monitoring of Picea Abies was carried out in the nursery of the botanical garden of Southern Federal University. This decision was due to the growth of single-aged specimens of Picea Abies, formed in similar soil and microclimatic conditions, at the monitoring point. Picea Abies is actively used in the landscaping of the city of Rostov-on-Don, as it is resistant to the specific climatic conditions of the steppe zone. Covering an area of 165 ha and located in the central part of the city, the botanical garden is a unique reference monitoring site where it is possible to approximate the results obtained with TreeTalker+ to trees growing in watershed areas throughout the city. This relates to the fact that one of the advantageous features of the SFU botanical garden is the presence of all the main components of mesorelief characteristic of Rostov-on-Don city, including the flat upland part of the right bank of the Don River (Kozlovskiy et al. 2011).

The main factor in choosing a tree crop for monitoring was the frequency of its use in urban landscaping and its range. R. abies is widespread in the north-east of Europe, there it forms solid forested areas, in Russia the northern border of the range often coincides with the border of forests, and the southern reaches the chernozem zone (Novikov et al. 2008). P. Abies has a high drought and frost tolerance, which is one of the main qualities in its integration into the green frame of Rostov-on-Don. High shade tolerance of this tree allows its use in city parks in conjunction with broad-leaved trees.

We used three specimens of P. Abies. All selected trees were of the same age, identical habitus, and formed near each other (Fig. 10.1).

The TreeTalker+ device contains external and internal sensors. With the help of external sensors, the device is able to measure the intensity of sap flow (temperature and humidity of the xylem of the trunk) and an IR distance sensor (for measuring point radial growth). Internal sensors measure the temperature and humidity of the air, the intensity of solar radiation, and the light reflected in the crown in various wavelength ranges (12 ranges, from blue to near infrared), and the angular deviation of the trunk from the normal along three coordinate axes, using a gyroscopic sensor.

The monitoring was carried out from March 15 to September 15, 2021. The TreeTalker+ device was used to record parameter sap velocity (according to the Granier method) and air temperature in the rootstock part of the tree. In this method, two sensors are embedded in a tree trunk to a depth of 3 cm, one of which measures the temperature of a linear heat source (which it is itself) implanted in the sapwood part of the tree, in relation to the temperature of unheated wood, measured by the

**Fig. 10.1** Map of TreeTalker devices in the nursery of the SFU Botanical Garden where: 1–3 Picea Abies

second sensor. When the flow rate is zero or minimum, the temperature difference between the two sensors will be maximum. As the flow accelerates, this temperature difference decreases.

During the growing season, the visual method (according to V.A. Alekseev) recorded the phenophases of Picea Abies. The presence of a meteorological station in the immediate vicinity of the nursery allowed recording microclimatic conditions for the entire monitoring period.

The rank correlation between xylem sap rate (variable) and tree rootstock air temperature (factor) was established using the STATISTICA program. In most cases, the rank correlation was found according to Pearson; the non-normal distribution method (according to Spearman) was used when searching for the dependence during the whole month. Scale diagrams were used to establish the reliability of the differences in juice flow (variable) from the time of day (grouping).

## 10.3 Results and Discussions

The analysis of daily TreeTalker+ data taken from Picea Abies rank correlation of xylem sap rate with temperature was traced daily in most cases in all investigated individuals, and the distribution was normal and the dependence was direct. The degree of reliability ranged from strong to weak. The correlation was strong ($p = 0.79–0.87$) in the spring months, moderate ($r = 0.36–0.71$) in the summer months, and weak ($r = 0.11–0.21$) in September (Fig. 10.2).

There was only rarely recorded an inverse dependence of xylem sap rate on temperature in individual specimens ($p = -0.41–0.62$), these anomalies occurred in the spring, and less frequently in the summer period. A similar inverse correlation appeared sporodically for each tested tree and was unaffected by weather conditions (Fig. 10.3).

A correlation was detected sporadically and not for all specimens during analysis of data of monthly fluctuations of sap flow, and the distribution was non-normal. The correlation was most frequent in the spring months, and it was weak ($r = 0.11–0.21$).

Mann–Whitney U-criterion analysis showed practically no reliability of differences in the rate of sap flow from the time of day. It was noted only on some summer days, with the highest rate of xylem sap occurring at night. However, the low occurrence of such cases casts doubt on the objectivity and representativeness of the results obtained.

A certain dependence can be traced when comparing phenological phases of Picea Abies and dynamics of xylem sap. Swelling of vegetative buds (PB1) in spruce begins

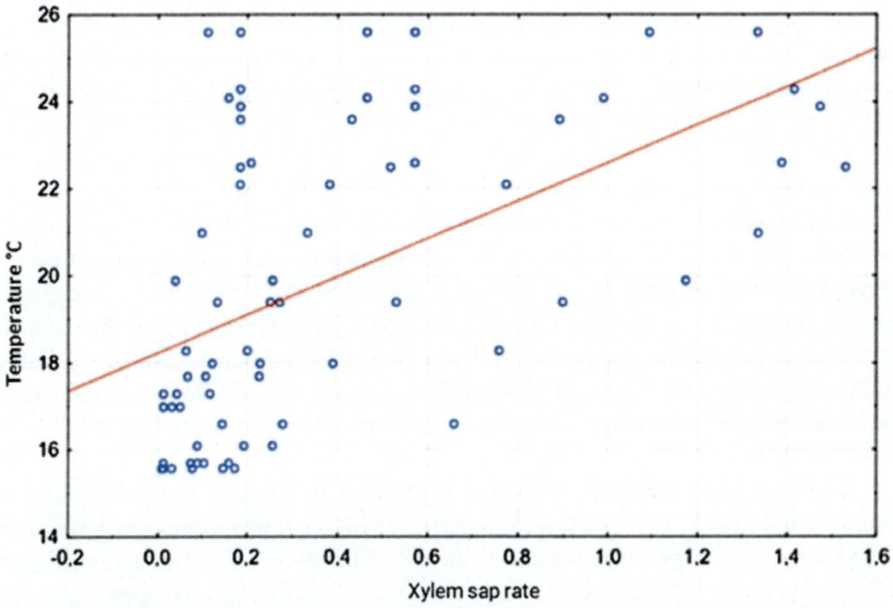


**Fig. 10.2** Scatter diagram of xylem sap velocity and temperature of Picea Abies (16.05.2021)

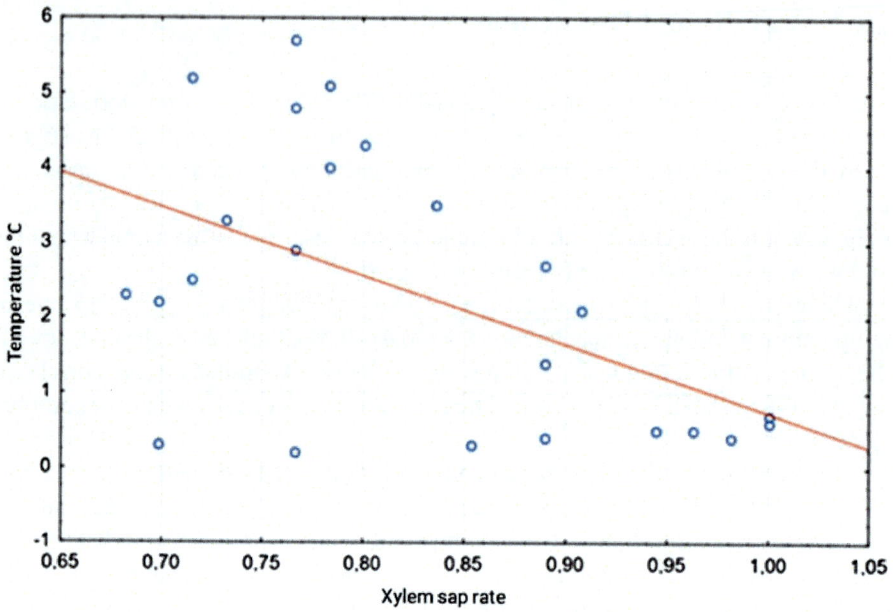


**Fig. 10.3** Scatter diagram of xylem sap velocity and temperature of a single specimen of Picea Abies (26.03.2021)

in the middle of April, and from that moment a sharp increase of sap flow rate until the beginning of May is noted. In the described time interval, cutting of vegetative buds (PB2) and the beginning of linear growth of shoots (PB3) also take place. All these processes require a large amount of moisture, which is shown in Fig. 10.4. Further, from the middle of May, needles detachment (L1), flowering (C1), and growth shoots capsizing (PB4) start. These processes are also reflected in the form of response on the graph.

Nevertheless, the correlation observed cannot indicate that the rate of xylem sap depends only on the ambient temperature and phenological phase. The xylem construction and the structure of the root system of this or that plant species play an important role in the studied process (Chernishenko 1998). A significant role is played by spring and summer precipitation, the influence of which has been partially recorded. Thus, the analysis of weather archives data allowed us to trace the increase of sap flow rate on the days when precipitation of more than 15 mm and the day following it. However, it should be noted that such peaks of sap flow associated with atmospheric precipitation were not always tracked and were characteristic of heavy rainfall in summer (Gismeteo Weather forecast 2022).

A characteristic feature of holosemous plants is two phases of fluid saturation: summer and winter. In winter, there is much more moisture in the tree than in summer, which can also be partially traced by the TreeTalker+ results (Fig. 10.5). During summer periods, there is a decrease in the rate of xylem sap movement, with the maximum decrease occurring in August, when dry winds dominate the region. It

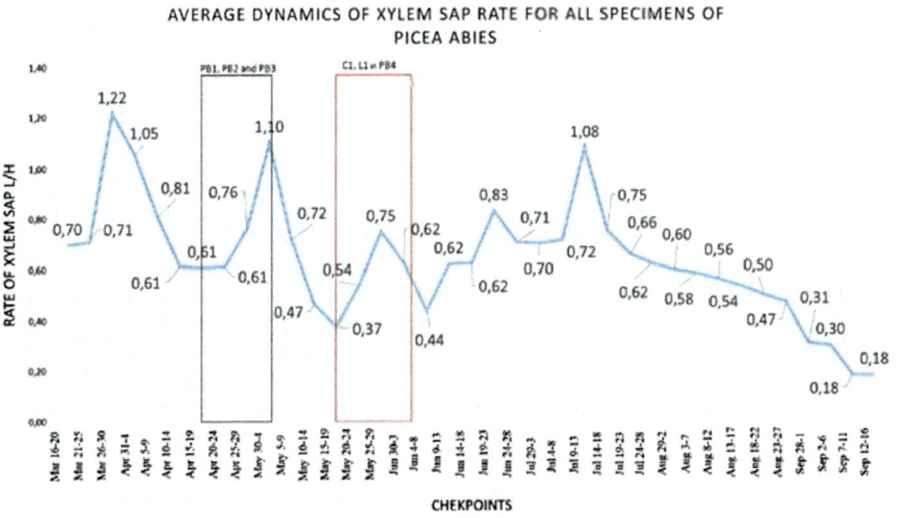


**Fig. 10.4** Seasonal dynamics of the sap flow rate of Picea Abies from March 15 to September 15

should be noted that Picea Abies is characterized by a difference in dynamics and correlation between sap flow rate and temperature relative to other conifers of the region, such as Pínus sylvéstris. Thus P. Abies has a clear correlation throughout the season, and it is characterized by a clear dependence of individual phenological phases on the nature of xylem sap dynamics (Kramer and Kozlovskiy 1983; Ospischev et al. 2022).

## 10.4 Conclusions

1. As a result of IoT monitoring using TreeTalker+, a daily and monthly correlation between sap flow rate and temperature for Picea Abies was established. The dependence is observed throughout the monitoring period. However, the Mann–Whitney U-criterion analysis showed no significant difference between the sap flow rate and time of day.
2. The analysis of xylem sap dynamics during the whole vegetation season recorded the correlation of the studied index with the phenological phases of Picea Abies. It is most clearly traced at the beginning of the tree's vegetation. The first peak of sap flow occurs during the period of vegetative bud swelling and cutting, as well as during linear growth of shoots. The second peak of sap flow is associated with detachment of needles, flowering and capsizing of growth shoots.

**Acknowledgements** The study was carried out at financial support by of the Russian Science Foundation (RSF) № 19-77-300-12. The study was supported by the Strategic Academic Leadership Program of Southern Federal University ("Priority 2030").

## References

Chernishenko OV (1998) Ekofiziologicheskie aspekti vodnogo obmena rastuschego dereva (Ecophysiological aspects of water exchange in a growing tree). Vestnik MGUL – Lesnoy vestnik, vol 1 (in Russian)

Dye C (2008) Health and urban living. Science 319:766

Gismeteo Weather forecast (2022) https://www.gismeteo.ru/diary/5110/2021/6/. Date of the application 20 Oct 2022

Gudzenko EO, Gudzenko IIO (2012) problemah ozeleneniya goroda Rostova-na-Donu (On the problems of landscaping the city of Rostov-on-Don). Strategiya ustoichivogo razvitiya regionov Rossii (Strategy for sustainable development of Russian regions), vol 11. (in Russian)

Kozlovskiy BL, Assortiment drevesnih rasteniy dlya zelyonogo stroitelstva v Rostovskoy oblasti: monographiya (Assortment of woody plants for green building in the Rostov region: monograph). In: Kozlovskiy BL, Ogorodnikova TK, Kuropyatnikov MV, Fedorinova OI. Publishing house SFU, Rostov-na-Donu (in Russian)

Kozlovskiy BL, Kuropyatnikova MV, Fedorinova OI (2011) Ekologo-biologicheskaya haracteristika drevesnih rasteniy urbanoflori Rostova-na-Donu (Ecological and biological characteristics of woody plants of the urban flora of Rostov-on-Don). Biol Ecol T.4(2 C):38–43 (in Russian)

Kozlovskiy BL, Kuropyatnikov MV, Fedorinova OI (2013) Osnovnoy i dopolnitelniy assortiment drevesnih rasteniy dlya zelyonogo stroitelstva na yugo-zapade Rostovskoy oblasti (The main and additional range of woody plants for green building in the south-west of the Rostov region). IVD 2(25) (in Russian)

Kramer P, Kozlovskiy T (1983) Fiziologiya drevesnih rasteniy (Physiology of woody plants). Lesnaya promishlennost, Moscov, 464 s (in Russian)

Lisikov AB (2015) Aktualnie voprosi gorodskogo ozelenenia (Current issues of urban gardening). Aktualnie problemi lesnogo kompleksa 43 (in Russian)

Matasov V, Marchesini LB (2020) IoT monitoring of urban tree ecosystem services: possibilities and challenges. Forests 11:755

Novikov VS, Gubanov IA (2008) Rod El' (Picea) [Genus Spruce (Picea)] // Populyarniy atlas-opredelitel'. Dikorastuschie rasteniya. — 5-e izdanie, stereotip. Drofa, Moscow, S. 65—66. — 415 s (in Russia)

Ospischev RN, Gorbov SN, Vasenyov VI (2022) Sezonnie I sutochnie izmereniya skorosti transpiracii zhidkosti drevesnih rasteniy na primer sosni obiknovennoy Pinus sylvestris (Seasonal and daily changes in the rate of transpiration of the liquid of woody plants on the example of Scots pine Pinus sylvestris). Biologicheskoe raznoobrazie I bioresursi stepnoy zoni v usloviyah izmenyayuschegosya klimata: materiali Mezhdunarodnoy nauchnoy konferencii, posvyaschyennoy 95-letiyu Botanicheskogo sada Yuzhnogo federalnogo universiteta, Rostov-on-Don – Taganrog, 24–29 may 2022 year. Izdatelstvo Yuzhnogo federalnogo universiteta, vol S, pp 448–454 (in Russia)

# Chapter 11
# Daily and Seasonal Dynamics of Mixed Forest Biodiversity in the Moscow Region According to Acoustic Monitoring Data

**M. Goretskaya, I. Beme, J. Betancourt, M. Perov, and V. Matasov**

**Abstract** The Moscow agglomeration is one of the largest in the world and the largest in Russia. Natural landscapes here influenced by the strongest anthropogenic impact, manifested in the increase of landscape fragmentation and reduction of natural habitats. Monitoring of biodiversity and study of species behavior patterns under anthropogenic pressure is one of the important goals of ecoacoustics—a scientific discipline at the interface of geography and ecology, which is only beginning to develop in Russia. In this work, the main subject of research is the diurnal and seasonal dynamics of avifauna of the temperate forests of European Russia. The key study sites were the mixed forests of the western Moscow Region ecopark "Nachiniye," the park "Korsar" and the preserve "Zvenigorod Biological Station". Audio data were collected by passive acoustic monitoring using Song Meter SM4 devices between January 2020 and October 2021. Five major acoustic indices were calculated using the R "seewave" package. Statistical analysis showed the relationship of bird vocalizations with the time of day, the season of the year, and the place of recording. The hourly changes in acoustic indices by seasons of the year for the temperate forests of the Moscow region were identified. Our results slightly differ from the previously identified regularities.

**Keywords** Ecoacoustics · Biodiversity · Song Meter · Temperate forest · Temporal dynamics

M. Goretskaya · I. Beme · M. Perov
Lomonosov Moscow State University, Leninskie gory—1, 119991 Moscow, Russia

J. Betancourt · V. Matasov (✉)
RUDN University, Miklukho-Maklaya St., 6, 117198 Moscow, Russia
e-mail: matasov_vm@pfur.ru



M. Korneykova et al. (eds.), *Smart and Sustainable Urban Ecosystems: Challenges and Solutions*, Springer Geography, https://doi.org/10.1007/978-3-031-37216-2_11

## 11.1 Introduction

Numerous global initiatives aim to conserve biodiversity, but it would be successful only if biodiversity can be measured and its rate of change quantified (Buckland et al. 2005). Rapid changes in landscape use and the impact of climate change have a lot of consequences (Foley 2005; Stocker 2013), such as growing fragmentation of natural landscapes globally and further habitat loss with biodiversity drop, changes in behavioral patterns and temporal activity of different species, and transformation in ecosystems functioning in general (Crooks et al. 2017). Thus, the development of cost-effective methods for biodiversity monitoring at scale is an urgent global imperative (Newbold et al. 2015; Campos et al. 2021).

The sounds produced in the environment by geophysical (geophonies), biological (biophonies), and anthropogenic (technophonies) processes have been found to be an important indicator of the quality of a system (Schafer 1977). Recent developments in soundscape analysis have demonstrated the existence of a strict connection between ecology and sounds (Farina et al. 2014a). Acoustic diversity has been considered to be a good proxy for overall vocal animal richness (Depraetere et al. 2012). Different from bioacoustics which mainly investigates sound as a signal that transfers information between individuals, sound is regarded as an indicator of ecological processes for ecoacoustics (Xie et al. 2020; Doser et al. 2020). In particular, ecoacoustics aims to investigate natural and anthropogenic sounds and their relationship with the environment including populations and communities (Sueur et al. 2014; Gasc et al. 2015). Acoustic patterns may be used for assessing the complexity of ecosystems (Matsinos et al. 2008; Servick 2014).

In the literature, there is a long-standing idea of a biphasic, with peaks in the morning and evening hours, distribution of the daily activity of birds, including singing activity (Kasatkina, 1963; Ptushenko and Inozemtsev, 1968; Sheldon, 1994). As a rule, singing activity was measured in several (2–3) individuals of one species, and then, the results were extended to other representatives of the species studied (Podarueva, 1977). The use of automatic monitoring methods makes it possible to fill this gap, and such work is currently being carried out in different regions and on different continents (Sugai et al. 2019a). In the temperate forests of European Russia, such work is carried out for the first time, and it will allow us to describe in detail the nature of bird vocalization in different periods of the year. Therefore, the aim of the work is to analyze the seasonal and daily variability of biodiversity acoustic activity in the temperate forests of European Russia. So, our tasks were to collect acoustic data at key sites, calculate acoustic indices, and identify temporal variability.

## 11.2 Materials and Methods

### *11.2.1 Study Area*

The research was conducted at three key sites: eco-park "Nachinanie," park "Corsair," Zvenigorod Biological Station. All three key sites are located in the Moscow region. The eco-park "Nachinanie" and park "Corsair" are located on the opposite banks of the Istrinsky reservoir, 38 km northwest of Moscow. Both key sites are located within the moraine-outwash plains of the Istrinsky landscape, which is characterized by a slightly undulating relief (Annenskaya et al. 1997). Zonal spruce and spruce-broadleaved forests due to early development of the territory have not been preserved—today within the ecopark, there are not only spruce and oak, but also larch and small-leaved forests due to reforestation after bark beetle damage.

Zvenigorod Biological Station is located 41 km west of Moscow on the right bank of the Moscow River. This key site is located within the Zvenigorod landscape, which is characterized by undulating outwash plains on interfluves and flat sandy terraces in the Moscow valley (Annenskaya et al. 1997). Spruce and pine-broadleaved forests are predominantly distributed within the study area.

### *11.2.2 Data Collection*

The data collection method used in this work is passive acoustic monitoring. This method makes it possible to collect large acoustic data arrays without active human participation using special sound recording devices (Sugai et al. 2019b). The sound recording device used at all key sites is the Song Meter SM4. Sound recordings were made during the first half of each hour during the observation period. Thus, we recorded 12 h of sound per day. The sampling rate used was 48 kHz, with 16 bits quantization. Recording points were chosen to minimize anthropogenic influence on the collected audio data. For this purpose, the points were installed away from roads, paths, or other potential sources of anthropogenic noise. At each point of each key site, audio was recorded for at least one full day, after which the recorder was usually moved to another point or to another key site.

At the Zvenigorod Biological Station, sound recording was conducted at two points in May, June, August, and October 2020 and in April, May, June, and October 2021. In the Ecopark "Nachinanie," sound was recorded at four locations in January, March, April, May, June, and September 2020 and in April, May, June, August, and October 2021. Corsair Park was recorded at four locations in January, February, March, June, and September 2020 and in April, May, June, August, and October 2021. In result, more than 500 gigabytes of audio data (1440 records in 2020 and 1603 records in 2021) were collected.

### 11.2.3 Data Processing

First, to clean the data, each recording was checked (listening and examining frequency plots) for extreme values and interference using the Avisoft-pro software, since weather conditions, mainly rain or very strong wind, external or artificial noise, or explosive vocal behavior of animals (frogs, insects, etc.) create erroneous recordings. When these inconsistencies in extreme conditions were checked, the corresponding wav files were deleted. In addition, only files with a recording duration of 30 min were selected to calculate the acoustic indices. All calculations of the acoustic indices were performed in the R environment using the "*seewave*" package (R Core Team 2014). For this work, five acoustic indices were selected to describe the daily and seasonal dynamics of the sound environments of selected key sites: ACI, BI, ADI, AEI, and NDSI.

The Acoustic Complexity Index (ACI) is based on the hypothesis derived from the observation that natural biotic sounds (e.g., birdsong or wolf howls) are characterized by intrinsic variability in intensity. Anthropogenic noises (e.g., the sound of passing cars or airplanes taking off), on the contrary, are often monotonous, even if not always. Thus, ACI should theoretically "extract" its natural component from an audio recording and quantify changes in behavior or biophonic composition from recording to recording (Pieretti et al. 2011). In this work, to reduce the influence of anthropogenic noise on the results obtained, only the frequency range of 500–12,000 Hz was used in the calculation of the ACI.

The Bioacoustic Index (BIO) is based on the hypothesis that most of the biophonic activity of the sound environment falls within a specific frequency range, which can be adjusted depending on the research objectives (Farina and Pieretti 2014). In this work, to reduce the influence of anthropogenic noise on the results obtained, only the frequency range of 500–12,000 Hz was used in the calculation of BI.

Acoustic Diversity Index (ADI) and Acoustic Evenness Index (AEI) are a pair of indices, which evaluate the variability of signals within an audio recording. Similarly to ACI, these indices use hypothesis about high internal variability of biophony. ADI is a Shannon index for signals louder than a certain threshold in various frequency bands. AEI similarly represents the Gini index. Thus, the presence of a single signal in multiple frequency bands marks the natural origin of the signal (Kasten et al. 2012). In this paper, to reduce the influence of anthropogenic noise on the results obtained, only the frequency range from 0 to 12,000 Hz with a threshold of 50 decibels was used in calculating the ADI and AEI.

Normalized Difference Soundscape Index (NDSI) is an index, the purpose of which is to determine the level of influence of anthropogenic impact on the soundscape by identifying the ratio of biophony to anthropony in the collected sound recordings (Farina and Pieretti 2014; Gasc et al. 2015). Thus, index values from −1 to 0 indicate the predominance of anthrophony in a particular sound recording, and from 0 to 1—biophony. In this work, the frequency range from 0 to 1600 Hz was chosen for anthrophony, and from 1600 to 12,000 kHz for biophony.

Statistical processing was also performed in the R environment using the "nlme" package. In this work, we used Linear Mixed Effects (LME) models, which are effectively used to estimate fixed and random variables as predictors of the model. The fixed variables were the time of day (hour of sound recording), the period of sound recording, and the key plot. Thus, these three characteristics of the sound recording were chosen as the primary determinants of the values of the acoustic indices. The recording device and the number of the point within the key section were chosen as random variables. The obtained values of acoustic indices showed a significant relationship with the fixed variables with 95% reliability. Thus, it was concluded that the statistical significance of bird vocalizations from the time of day, season, and place of sound recording.

To plot the daily and seasonal dynamics, the values of acoustic indices in a particular hour were averaged over the values of indices in the same hour at other points and other key sites. Thus, the average value of the acoustic index for a particular hour within all key sites was obtained. This method allows minimizing the influence of weather, human, and other interference in specific recordings. As a disadvantage of the chosen method, we should mention ignoring the spatial differentiation of the sound environment both at the level of one key site and at the regional landscape level.

The periods of the year covered with acoustic data were divided into seasons as follows: Period 1—January and February—Winter; Period 2—March—Spring; Period 3—April and early May—Spring; Period 4—End of May and beginning of June—Early Summer; Period 5—End of June and early July—Summer; Period 6—August—Late Summer; Period 7—September and October—Autumn.

## 11.3 Results

### *11.3.1 ACI Daily and Seasonal Dynamics*

During the seven selected periods, changes of daily dynamics were observed, which are manifested primarily in the number and time of recorded peaks of ACI (Fig. 11.1).

In January–February, the latest hour of the morning dawn peak is at 9 a.m. This peak smoothly passes into a plateau of similar index values, which by 19 h compares with the night values. The March period of sounding is characterized by a similar daily dynamic: a morning peak by 8 a.m. and a rapid return to the night values of the index around 4 p.m. It should be noted here that the highest ACI values are recorded at 1 p.m. In the records from April and early May, the morning peak is shifted a few more hours back to 4–5 h in the morning. After a slight decline of the index during the day, a small evening peak at 8 p.m. is observed. On the ACI chart for period 4 during the dawn peak at 4 a.m., the maximum values for the day are reached; the evening peak is more pronounced than in May and early June, following a dip in the chart during the hot afternoon hours. In late June and early July, the maximum

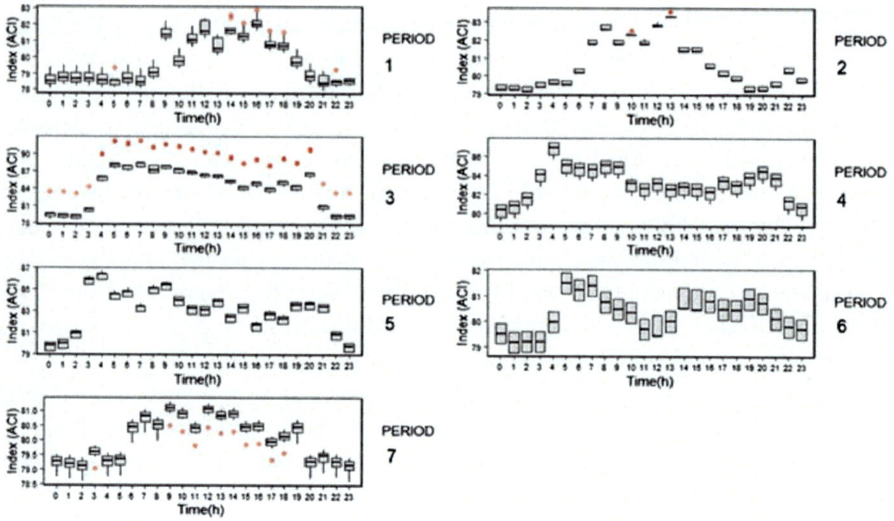


**Fig. 11.1** Daily dynamics by seasons of the year of ACI index

ACI values are also recorded during the morning; similar ACI values are recorded throughout the day until darkness at 22 h, without a pronounced evening ACI. In August, the morning peak again begins to shift in the later hours, after which, as in other summer months, the index values fall in the hottest hours. The evening peak is not pronounced, comparing with the afternoon index values. In September and October, it is difficult to distinguish even the morning peak—we can simply distinguish a daytime period of increased ACI and a quieter nighttime period, which lasts almost the same amount of time.

Thus, the daily dynamics by seasons of the year has significant differences. In addition, seasonal differences are also manifested in the absolute values of ACI. From May to July, the peak values of the index are 86–87, which is 4–5 higher than at any other time. The night values of the index vary little throughout the year, ranging from 78 to 80.

### *11.3.2 BIO Daily and Seasonal Dynamics*

The bioacoustic index shows similar ACI dynamics, with peak occurring in periods 3–5, i.e., from May to July (Fig. 11.2). Similarly, in January and October, peaks are not pronounced, but rather long periods of relatively high BIO during the day and silence at night. Similarly, the morning and evening peaks are most pronounced in May–July, with generally high daytime index values. A rare noticeable difference is the August period of sounding, during which the peaks, like in October, are not expressed, and the graph is divided into day and night periods.

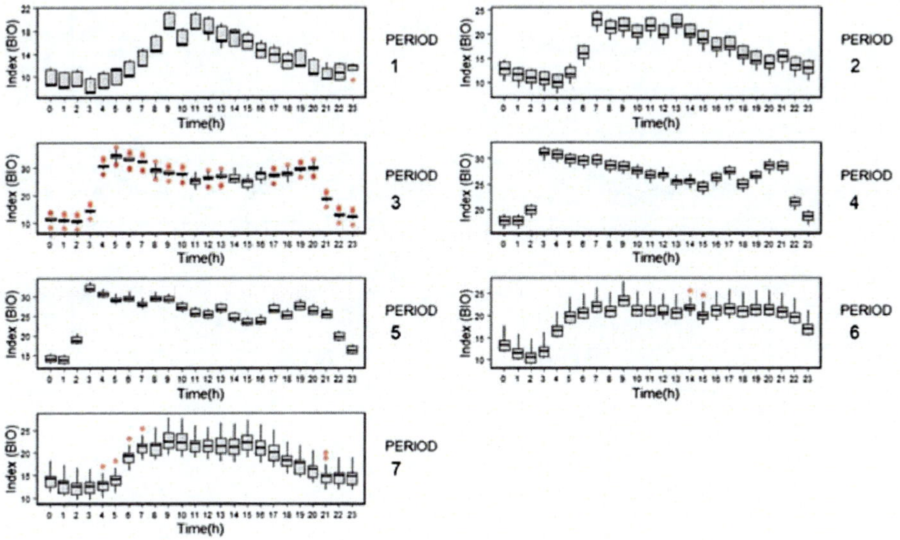


**Fig. 11.2** Daily dynamics by seasons of the year of BIO index

### *11.3.3 ADI and AEI Daily and Seasonal Dynamics*

The ADI and AEI indices, as evaluating the same phenomenon from opposite sides, will be considered in pairs (Fig. 11.3). The ADI index by higher values highlights periods of variable acoustic environment, and the AEI highlights the same periods by lower values. In general, the graphs of these indices repeat the previously described graphs of ACI and BIO with some peculiarities. Thus, ADI highlights the morning peaks brighter than the evening ones, unlike ACI and BIO, whose peaks in the morning and in the evening had approximately the same values. The exception is the August recording period, during which the maximum values are recorded during the sunset peak and smoothly pass into the nighttime. In addition, it is possible to distinguish morning and evening peaks ADI (and, hence, AEI) plots even in October soundings-the above-described ACI and BIO indices represented daytime soundings of this period as a plateau with approximately equal values.

### *11.3.4 NDSI Daily and Seasonal Dynamics*

During the winter period of audio recording, the NDSI extreme values are recorded at 3, 9, 12, and 15 o'clock. At the same time, the difference in absolute values varies within 0.04. In general, the NDSI values in January and February have practically minimal values approximately all the time. In the March recording period, the night values of the morning peak at 6–8 a.m. clearly stand out against the background of the daytime values of the index, which do not differ much from the nighttime

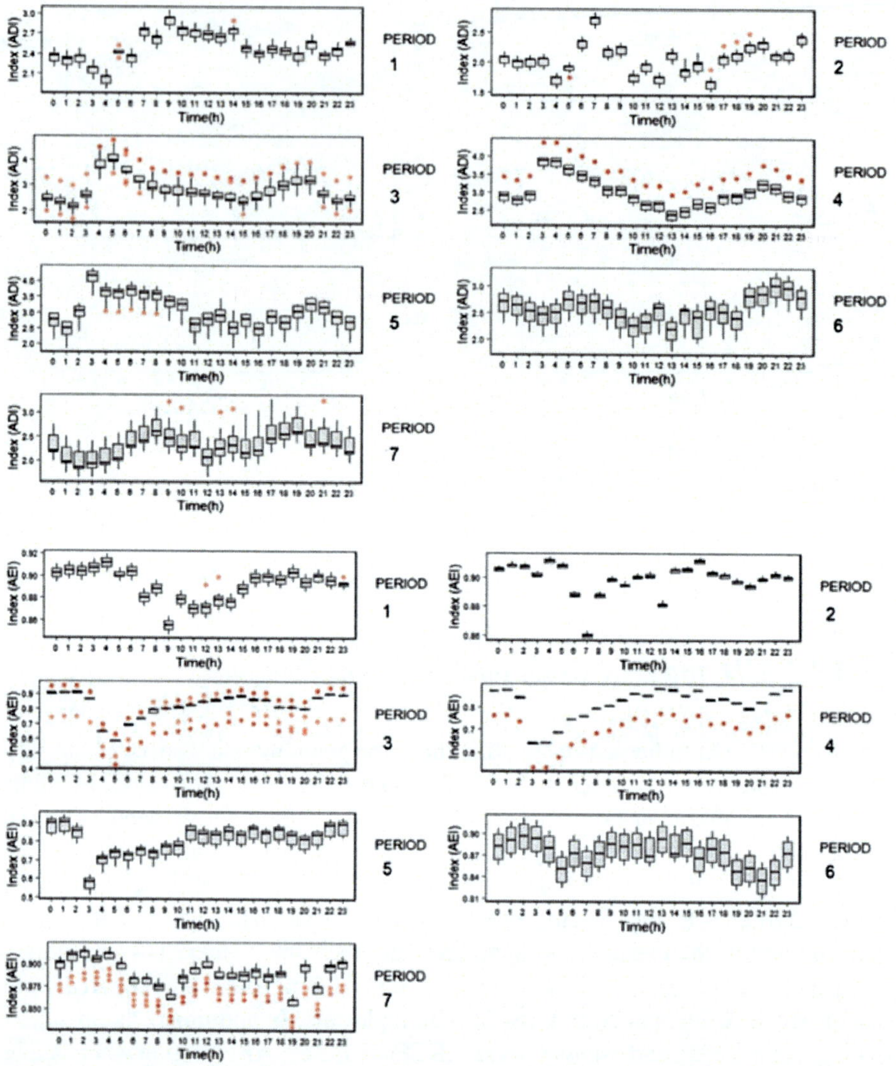


**Fig. 11.3** Daily dynamics by seasons of the year of ADI (top) and AEI (bottom) indices

values—not more than 0.05. Nevertheless, even the sound recordings with the highest NDSI values show a clear predominance of low frequencies in the sound recordings (Fig. 11.4).

The graphs of NDSI in periods 3, 4, and 5 are very similar to each other. At night, almost the minimum possible values of the index are fixed, after which the index values rise sharply above zero at 3–5 a.m. in the morning. The daytime values of the index still indicate the predominance of nominal anthropony, ranging from −0.5 to −0.8. However, these values are significantly (by 0.2–0.5) lower than during similar hours in early spring, autumn, and winter. In August, there is a sharp drop in the absolute values of NDSI—the maximum values do not exceed −0.9 even in the

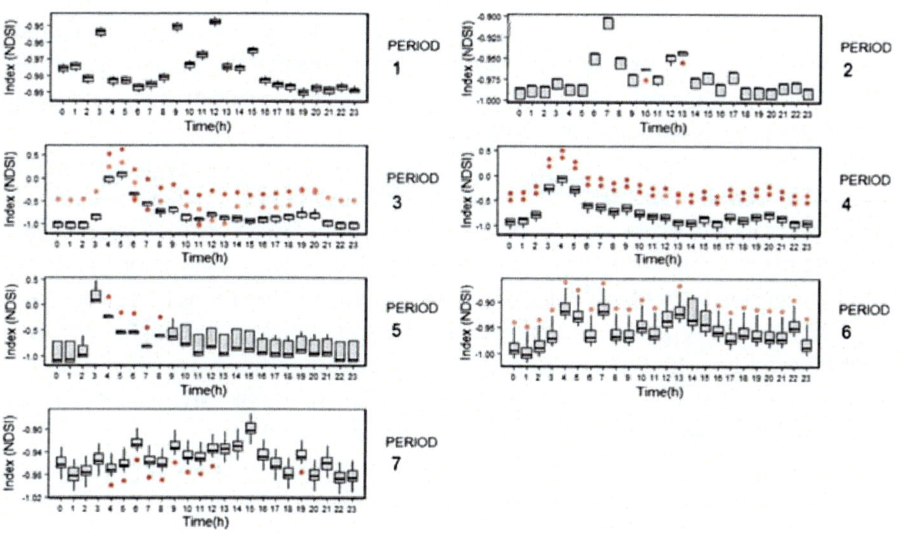


**Fig. 11.4** Daily dynamics by seasons of the year of NDSI index

morning, and the “summer” appearance of the graph with a clear peak is lost. In the fall, the NDSI takes on the appearance of a set of extremely low values, which are only slightly higher than in January and February.

## 11.4 Discussion

Our data showed that, in general, in the temperate forests the acoustic indices reflecting the complexity of the acoustic environment and biodiversity have a pronounced seasonal variability. In the cold season, vocalization begins shortly before dawn and remains at a high level until 2 p.m., then it decreases slightly and disappears completely with sunset. In March, the period of vocalization is longer, but also limited by daylight hours. In spring and summer period, vocalization begins long before dawn (at 4 o’clock in April–early May and at 3 o’clock in June) and continues after sunset; the biotic component is absent only in the period from 23 to 2 o’clock. Interestingly, the maximum of vocal activity during the breeding season falls in the pre-dawn hours, the pre-dawn chorus characteristic of birds (Catchpole and Slater 2008), and then the level of vocalization declines somewhat, but remains high until 10–11 h. A slight decrease in vocalization activity occurs during the daytime only from 12 to 15 o’clock and then increases. Thus, the automatic monitoring revealed a much higher level of vocalization during the day than previously assumed (Depraetere et al. 2012; Eldridge et al. 2018).

We were limited by the number of sensors, so different points were recorded at different times within a season. Perhaps a preliminary comparison with AudioMoth (Farina et al. 2014b; Hill et al. 2019) and their use could help reduce costs and thus get more points recorded in parallel. The influence of the weather was also eliminated by averaging, but this factor should be considered separately in a further detailed analysis (Sueur et al. 2014; Borker et al. 2020). Verification with traditional biodiversity assessment methods is also needed.

With the development of acoustic monitoring systems, it is possible to incorporate them into decision-making systems. Since for the system of nature reserves and national parks one of the tasks is the monitoring of biodiversity, the introduction of such algorithms can significantly improve our understanding of the natural processes. In addition, analysis of vocalization is also important for the urban environment biodiversity assessment, including recreational infrastructure development in urban forests, as it allows to assess the impact of anthropogenic influence (Cohen et al. 2014; Klingberg et al. 2017). In addition, the combination of noninvasive monitoring with crowdsourced observations by city residents (Mydlarz et al. 2019) seems particularly promising for biodiversity and noise pollution assessment in urban conditions (Fang and Ling 2003; Dowling et al. 2012).

## 11.5 Conclusions

- For the first time, long-term acoustic monitoring studies in the Moscow region have been carried out.
- Representative series of daily and seasonal dynamics of acoustic indices in background mixed forests of temperate climate are obtained.
- According to the results of the analysis, the duration of the morning activity hours during the day (from predawn to noon) was longer than in the previous assumptions.
- The period of minimum activity of biota in summer months was only three hours (from 23 to 2 o'clock)

**Acknowledgements** Data collection was done with support of "RUDN University Strategic Academic Leadership Program." Data analysis was done under the Russian Scientific Foundation Project # 19-77-300-12.

**Conflict of Interest** The authors declare no conflict of interest.

## References

Annenskaya GN, Zhuchkova VK, Kalinina VR et al (1997) Landscapes of Moscow Region and their modern State. S: Publishing house of Moscow State University, 298p

Borker AL, Buxton RT, Jones IL et al (2020) Do soundscape indices predict landscape-scale restoration outcomes? A comparative study of restored seabird island soundscapes. Restor Ecol 28:252–260. https://doi.org/10.1111/rec.13038

Buckland ST, Magurran AE, Green RE, Fewster RM (2005) Monitoring change in biodiversity through composite indices. Phil Trans R Soc B 360:243–254. https://doi.org/10.1098/rstb.2004.1589

Campos IB, Fewster R, Truskinger A et al (2021) Assessing the potential of acoustic indices for protected area monitoring in the Serra do Cipó National Park, Brazil. Ecol Ind 120:106953. https://doi.org/10.1016/j.ecolind.2020.106953

Catchpole CK, Slater PJB (2008) Bird song. Biological themes and variations. 2nd edn. Cambridge University Press

Cohen P, Potchter O, Schnell I (2014) The impact of an urban park on air pollution and noise levels in the Mediterranean city of Tel-Aviv, Israel. Environ Pollut 195:73–83. https://doi.org/10.1016/j.envpol.2014.08.015

Crooks KR, Burdett CL, Theobald DM et al (2017) Quantification of habitat fragmentation reveals extinction risk in terrestrial mammals. Proc Natl Acad Sci USA 114:7635–7640. https://doi.org/10.1073/pnas.1705769114

Depraetere M, Pavoine S, Jiguet F et al (2012) Monitoring animal diversity using acoustic indices: implementation in a temperate woodland. Ecol Ind 13:46–54. https://doi.org/10.1016/j.ecolind.2011.05.006

Doser JW, Finley AO, Kasten EP, Gage SH (2020) Assessing soundscape disturbance through hierarchical models and acoustic indices: a case study on a shelterwood logged northern Michigan forest. Ecol Ind 113:106244. https://doi.org/10.1016/j.ecolind.2020.106244

Dowling JL, Luther DA, Marra PP (2012) Comparative effects of urban development and anthropogenic noise on bird songs. Behav Ecol 23:201–209. https://doi.org/10.1093/beheco/arr176

Eldridge A, Guyot P, Moscoso P et al (2018) Sounding out ecoacoustic metrics: avian species richness is predicted by acoustic indices in temperate but not tropical habitats. Ecol Ind 95:939–952. https://doi.org/10.1016/j.ecolind.2018.06.012

Fang C-F, Ling D-L (2003) Investigation of the noise reduction provided by tree belts. Landsc Urban Plan 63:187–195. https://doi.org/10.1016/S0169-2046(02)00190-1

Farina A, Buscaino G, Ceraulo M, Pieretti N (2014a) The soundscape approach for the assessment and conservation of Mediterranean landscapes: principles and case studies. J Landsc Ecol 7:10–22. https://doi.org/10.2478/jlecol-2014-0007

Farina A, James P, Bobryk C et al (2014b) Low cost (audio) recording (LCR) for advancing soundscape ecology towards the conservation of sonic complexity and biodiversity in natural and urban landscapes. Urban Ecosyst 17:923–944. https://doi.org/10.1007/s11252-014-0365-0

Farina A, Pieretti N (2014) Acoustic codes in action in a soundscape context. Biosemiotics 7:321–328. https://doi.org/10.1007/s12304-014-9213-0

Foley JA (2005) Global consequences of land use. Science 309:570–574. https://doi.org/10.1126/science.1111772

Gasc A, Pavoine S, Lellouch L et al (2015) Acoustic indices for biodiversity assessments: analyses of bias based on simulated bird assemblages and recommendations for field surveys. Biol Cons 191:306–312. https://doi.org/10.1016/j.biocon.2015.06.018

Hill AP, Prince P, Snaddon JL et al (2019) AudioMoth: a low-cost acoustic device for monitoring biodiversity and the environment. HardwareX 6:e00073. https://doi.org/10.1016/j.ohx.2019.e00073

Kasatkina VI (1963) On daily activity of birds during nesting period in polar regions. J Zool 42(2):303–306

Kasten EP, Gage SH, Fox J et al (2012) The remote environmental assessment laboratory's acoustic library: an archive for studying soundscape ecology. Ecol Inform 12:50–67. https://doi.org/10.1016/j.ecoinf.2012.08.001

Klingberg J, Broberg M, Strandberg B et al (2017) Influence of urban vegetation on air pollution and noise exposure—a case study in Gothenburg, Sweden. Sci Total Environ 599–600:1728–1739. https://doi.org/10.1016/j.scitotenv.2017.05.051

Matsinos YG, Mazaris AD, Papadimitriou KD, et al (2008) Spatio-temporal variability in human and natural sounds in a rural landscape. Landsc Ecol. https://doi.org/10.1007/s10980-008-9250-7

Mydlarz C, Sharma M, Lockerman Y et al (2019) The life of a New York city noise sensor network. Sensors 19:1415. https://doi.org/10.3390/s19061415

Newbold T, Hudson LN, Hill SLL et al (2015) Global effects of land use on local terrestrial biodiversity. Nature 520:45–50. https://doi.org/10.1038/nature14324

Pieretti N, Farina A, Morri D (2011) A new methodology to infer the singing activity of an avian community: the acoustic complexity index (ACI). Ecol Ind 11:868–873. https://doi.org/10.1016/j.ecolind.2010.11.005

Podarueva VI (1977) Review of studies on daily and seasonal activity of birdsong. Geographical variability of some ecological and physiological characters in birds. pp 3–33 (in Krupskaya MEPI collection)

Ptushenko ES, Inozemtsev AA (1968) Biology and economic significance of birds of the Moscow Region and adjacent territories. Moscow State University Press, Moscow, 461p

R Core Team (2014) R: a language and environment for statistical computing. R Foundation for Statistical Computing, Vienna, Austria

Schafer RM (1977) The soundscape: our sonic environment and the tuning of the world. Destiny-Books, Rochester, NY

Servick K (2014) Eavesdropping on ecosystems. Science 343:834–837. https://doi.org/10.1126/science.343.6173.834

Sheldon BC (1994) Song rate and fertility in the chaffinch. Anim Behav 47(4):986–987

Stocker TF (ed) (2013) Climate change 2013—IPCC. Intergovernmental Panel on Climate Change, New York

Sueur J, Farina A, Gasc A et al (2014) Acoustic indices for biodiversity assessment and landscape investigation. Acta Acust Acust 100:772–781. https://doi.org/10.3813/AAA.918757

Sugai LSM, Desjonquères C, Silva TSF, Llusia D (2019a) A roadmap for survey designs in terrestrial acoustic monitoring. Remote Sens Ecol Conserv. https://doi.org/10.1002/rse2.131

Sugai LSM, Silva TSF, Ribeiro JW, Llusia D (2019b) Terrestrial passive acoustic monitoring: review and perspectives. Bioscience 69:15–25

Xie J, Hu K, Zhu M, Guo Y (2020) Data-driven analysis of global research trends in bioacoustics and ecoacoustics from 1991 to 2018. Eco Inform 57:101068. https://doi.org/10.1016/j.ecoinf.2020.101068

# Chapter 12
# A Tremendous Green Roof or Biodiversity Museum? First Outcomes from Soil Survey in Zaryadye Park

**V. Vasenev, M. Korneykova, I. Krokhmal, E. Kozlova, A. Robert, A. Losev, D. Sarzhanov, Y. Sotnikova, K. Makhinya, D. Gosse, E. Dovletyarova, and M. Nakhaev**

**Abstract** Soils play a crucial role in functionality, sustainability and biodiversity of urban green infrastructures. Constructed Technosols not only contribute to biodiversity by creating conditions for vegetation growth, but also are niches for microbial communities. This effect is especially important in botanical gardens and urban parks, where Technosols aim to reconstruct the soil conditions typical for the plants from different biomes. In this research, Technosols properties were studied in Zaryadye Park—a unique green area in the centre of Moscow city, where the diversity of bioclimatic zones ranging from tundra to subtropics was reproduced. Soil survey was carried out in summer 2022. In total, 36 locations from 9 aggregated biomes were visited: tundra (*T*), coniferous forest (CF), mixed forest (MF), deciduous forest (DF), steppe (*S*), floodplain (*F*), meadow (*M*), subtropical forest (SF) and urban greening (UG). At each location, both disturbed and non-disturbed sires were examined. Soil organic matter (SOM) in Technosols was significantly different between the biomes, whereas soil pH, *P* and *K* contents were similar. Physical disturbance

V. Vasenev (✉)
Soil Geography and Landscape, Wageningen University, Wageningen, The Netherlands
e-mail: Slava.vasenev@wur.nl

M. Korneykova · E. Kozlova · A. Robert · A. Losev · D. Sarzhanov · Y. Sotnikova · K. Makhinya · E. Dovletyarova
Smart Urban Nature Research Center, RUDN University, Moscow, Russia

I. Krokhmal
Zaryadye Park, Moscow, Russia

D. Gosse
Soil Science Faculty, Lomonosov Moscow State University, Moscow, Russia

M. Nakhaev
Kadyrov Chechen State University, Grozny, Russia

I. Krokhmal
V.L. Talrose Institute for Energy Problems of Chemical Physics, N.N. Semenov Federal Research Center for Chemical Physics Russian Academy of Sciences, Moscow, Russia



M. Korneykova et al. (eds.), *Smart and Sustainable Urban Ecosystems: Challenges and Solutions*, Springer Geography, https://doi.org/10.1007/978-3-031-37216-2_12

from recreational load resulted in topsoil over-compaction with the highest bulk density above 1.5 g $cm^{-3}$. Anthropogenic factors (e.g. urban heat island, aerial dust deposition and implementation of mineral fertilizers) resulted in mineralization of organic matters, shift in soil pH and phosphorous inputs decreased initial variability of soil properties, decreased their functionality and microbial diversity. Site-specific precise soil management followed by regular soil monitoring is needed to preserve biodiversity of the Zaryadye Park.

**Keywords** Technosols · Urban green infrastructures · Soil organic matter · Microbial diversity · MicroResp

## 12.1 Introduction

Although urban areas cover around only 2% of the global land area, in some countries, their extent reaches 20% and continues growing fast (Sharma et al. 2016). Urbanization has a strong and diverse impact on ecosystems, resulting in considerable and as a rule irreversible changes in vegetation and soils. Deforestation, pollution and soil sealing are the main negative consequence of urbanization which deplete urban ecosystem services (Dvornikov et al. 2021; Romzaykina et al. 2021; Vasenev et al. 2019). Development of green infrastructures, in contrast, enhance urban ecosystem services (Elmqvist et al. 2015; Gómez-Baggethun and Barton 2013). Urban green infrastructures provide cooling effect and contribute to climate mitigation and adaptation (Yan et al. 2018), support carbon accumulation and storage (Deeb et al. 2020; Vasenev et al. 2017), preserve and enhance biodiversity (Filazzola et al. 2019; Hostetler et al. 2011; Savvas 2016). In fact, urban green infrastructures are new ecological niches with unique environmental conditions for living organisms (Korneykova et al. 2021; McPhearson et al. 2016). Traditionally, urban sustainable development strategies and nature-based solutions focused on biodiversity preservation and control focus on the plant diversity and abundance as the main indicator (Basnou et al. 2020; Hoyle et al. 2017; Pierce et al. 2020). However, the role of urban soils remains overlooked or completely ignored.

Urban soils play a crucial role in functionality and sustainable development of urban ecosystems by delivering and supporting important ecosystem services, e.g. carbon accumulation, nutrients supply and water balance (Morel et al. 2014; Vasenev et al. 2018). Urban soils are exposed to direct and indirect anthropogenic influences, and they are very heterogeneous and dynamic. This variability is just partly considered by international soil classifications (e.g. Technosols in World Reference Base for Soil Resources, (Rossiter 2007; Schad 2018)), whereas much more detailed classification schemes were developed at the country and city levels (Charzyński et al. 2018; Kabała et al. 2020; Prokofyeva et al. 2011). Technosols are man-made soil constructions which “properties and functions are dominated by technical human activity” (Rossiter 2007). Technosols’ properties and functions are usually studied from the perspective of a particular service they deliver, including reclamation of degraded lands (Moreno-Barriga et al. 2017; Slukovskaya et al. 2019), capturing surface run-off (Mueller and Thompson 2009; Sun et al. 2009) or supporting green infrastructures for sports and recreation (Ignatieva et al. 2017; Monteiro 2017). In contrast to long-term formation and evolution of natural soils, Technosols are constructed at a single moment in time and can be considered as a model for understanding the early dynamics of urban soil formation. At the zero moment, Technosols' properties are mainly distinguished by materials used for their construction (Smagin et al. 2012; Neina et al. 2016). Physicochemical and biological properties of peat materials, organic wastes, composts, biosolids, dredge sediments and other materials establish a baseline for Technosols' functions and ecosystem services (Ivashchenko et al. 2021). Their further evolution is driven by soil-forming factors, including climate, parent material and living organisms (Deeb et al. 2020; Séré et al. 2010).

Technosols have direct and indirect impact on urban biodiversity. Directly, Technosols under green infrastructures are niches for microbial communities, which functional and taxonomic diversity can be much higher compared to more disturbed urban areas or even to natural ecosystems (Burrow 2018; Hedde et al. 2019; Korneykova et al. 2021). Indirectly, Technosols contribute to biodiversity by creating conditions for vegetation growth. This indirect effect is especially important in botanical gardens and urban parks where diversity of plants from different biomes and bioclimatic zones is exhibited. In this case, Technosols aim to reconstruct the soil conditions typical for the plants from different biomes. At the same time, they are exposed to the effect of plants via, e.g. carbon input or metabolic exudate. These mutual relationships between Technosols and vegetation drive sustainability of urban green infrastructures.

The research aimed to investigate diversity of Technosols properties in Zaryadye Park—a unique green area in the centre of Moscow city, where the diversity of bioclimatic zones ranging from tundra to subtropics was reproduced. The concept of Zaryadye Park was to create analogue landscapes that reflect the natural zones of Russia: tundra, coniferous and broad-leaved forest, and steppe. A zonal vegetation is also represented in birch forests as an element of secondary vegetation of forest zones and meadows. Artificial phytocenoses have also been created in the park. Plants are placed on artificial relief and soil, mostly on the roof. These factors necessitate the study of the regularities of the dynamics of multispecies communities and the development of the regulations for their maintenance.

We hypothesized that Technosols properties will differ between the plant biomes. We also assume that dynamics in plant distribution and soil properties in comparison with the initial state will be observed.

## 12.2 Materials and Methods

### 12.2.1 Research Area

Zaryadye Park locates in the historical centre of Moscow megapolis (55.75 N; 37.63 E). Created in 2014 in the area of the demolished "Rossia" hotel, Zaryadye is one of the most famous places in Moscow, attracting more than 10,000 visitors every day (Kasyanov 2017). The principal aim of the park was exhibiting natural and cultural diversity of Russia. In result, biomes from tundra, taiga, deciduous and mixed forests, steppe, meadows, floodplain and even subtropical forest co-exist on the 10.5 ha area (Fig. 12.1). The collection of plants of Zaryadye includes 38 varieties, forms, varieties of 226 species, with 34 species listed in the Red Data Book of Moscow and/or the Moscow region and 7 species in the Red Book of the Russian Federation. The collection includes 29 varieties of trees and shrubs of 74 species belonging to 39 genera and 21 families (22 tree species and 52 shrub species); 9 varieties of herbaceous perennials, varieties forms of 137 species belonging to 101 genera 34 families; 9 species of aquatic and coastal aquatic plants belonging to 9 genera and 7 families; 6 species of ferns belonging to 5 genera and 6 families. Family Rosaceae Juss. is the most widely represented among tree and shrub species. Among the herbaceous perennials, the most common are the Asteraceae Bercht. and J. Presl, Poaceae Barnhart and Rosaceae Juss.

To support such a diverse vegetation, Technosols were constructed with consideration of the natural soil properties for the corresponding biomes, e.g. high-moor peat was recommended for tundra zones to create acid conditions, whereas valley peat was used to construct Technosols of the floodplain biomes (Rappoport et al. 2019). Soil functionality in the park is constrained by a very limited soil depth, which is less than 50 cm for the major part of the area. Technosols are underlain by

**Fig. 12.1** Soil sampling locations (red points) at different biomes of Zaryadye Park

concrete constructions and engineering facilities and pipelines. In fact, Zaryadye Park can be considered a tremendous green roof, and therefore soil formation and functioning processes occur in a relatively shallow layer and are very intensive. Besides, these processes occur under permanent anthropogenic pressure. Thousands of visitors affect soils by over-compaction and physical disturbance. Located in the centre of Moscow and surrounded by roads with intensive traffic, Zaryadye ecosystem is also exposed to aerial deposition of dusts and heavy metals (Romzaykina et al. 2021) as well as to a strong urban heat island effect (Vasenev et al. 2021).

### *12.2.2 Soil Survey and Laboratory Analysis*

To study variability of soil properties in Zaryadye Park, soil survey was carried out in summer 2022. In total, 36 locations from 9 aggregated biomes were visited: tundra (*T*), coniferous forest (CF), mixed forest (MF), deciduous forest (DF), steppe (*S*), floodplain (*F*), meadow (*M*), subtropical forest (SF) and urban greening (UG). At each location, both disturbed and non-disturbed sires were examined. Composite soil samples were collected from 0 to 10 cm by Edelman auger and transported to laboratory for physical, chemical, and microbial analysis. Bulk density samples were collected by Edelman and estimated after drying in the oven at 105 °C for 8 h. The contents of organic carbon ($C_{org}$, dichromate oxidation), available phosphorus and potassium (extract 0.5 M $CH_3COOH$, colorimetry and flame photometry, respectively) were determined in the soil samples. The $pH_{KCl}$ was determined by potentiometry (soil:solution = 1:2.5).

The plate count method was used to determine the number of viable microbial cells. The next day after sampling, a series of soil dilution were prepared and plated on nutrient agar mediums: meat-peptone agar (MPA) to determine the total number of microorganisms and Czapek medium, acidified with lactic acid (4 ml/l), for micromycetes. After the incubation time, the grown colonies were counted in the form of colony forming units (CFU), calculation per gram of dry soil. The species composition of micromycetes was determined based on cultural and morphological characteristics using the following identification keys (Domsch et al. 2007; Raper and Thom 1968) and subsequent refinement using Index Fungorum database (http://www.indexfungorum.org).

Community level physiological profile (*CLPP*) was assessed by microbial respiration response on addition of different low molecular weight organic substrates, measured by MicroResp™ technique (Marinari et al. 2013; Moscatelli et al. 2018; Campbell et al. 2003). The substrates represent C sources of different quality: amino acids (glycine, L-arginine, L-leucine, α-aminobutyric and L-aspartic acids), carbohydrates (D-fructose, D-galactose and D-glucose), carboxylic acids (L-ascorbic, citric and oxalic acids) and phenolic acids (vanillic and syringic). The response of microbial community was detected by $CO_2$ production using colorimetric method after 6 h of incubation with detection gel at 25 °C. The absorbance by the detection gel was analysed at 595 nm wave length (microplate spectrophotometer FilterMax F5, US) before and after incubation. The absorption units were converted to μg C $g^{-1}$ $h^{-1}$ according to Moscatelli et al. (2018). Microbial functional diversity was assessed using the Shannon–Wiener index: $H = -\Sigma Pi \times \ln Pi$, where $Pi$ is the ratio of respiration response to $i$ substrate addition to total respiration response for all studied substrates.

### 12.2.3 *Data Processing and Analysis*

The primary outcomes were analysed by basic descriptive statistical tools and presented as mean and standard error for each biome. The significance of difference between the biomes was checked by analysis of variance (one-way ANOVA). The effect of disturbance was checked by *T*-test. Cluster analysis was implemented to distinguish microbial respiration response to different substrates for the CLPP analysis. To analyse temporal dynamics in soil properties, the research outcomes were compared to the initial values recommended for each biome when Technosols were constructed.

## 12.3 Results and Discussions

### 12.3.1 *Variability of Soil Physical and Chemical Properties*

All Technosols had neutral pH ranging from 6.0 to 7.2. The lowest values were obtained for tundra biomes, whereas soils under steppe vegetation had the highest pH. Neutral or slightly alkaline pH is typical for urban soils even in the areas where natural soils have acid reaction. It is usually explained by the aerial deposition of dust containing concrete particles and other substances (Brianskaia et al. 2020; Prokof'eva TV, Kiryushin AV, Shishkov VA, Ivannikov FA 2017). The investigated biomes varied in SOM content considerably. Technosols under tundra biome contained more than 14% of organic matter which was significantly higher than in any other biome. The lowest SOM contents were observed in steppe and meadow biomes. An average SOM content was 5.9% which is comparable to the values previously reported for urban soils in Moscow (Ivashchenko et al. 2014; Vasenev et al. 2013); however, the variance coefficient above 70% indicated high spatial variability of SOM content in Zaryadye Park. Available phosphorous ranged between 200 and 800 mg $kg^{-1}$ and available potassium between 100 and 500 mg $kg^{-1}$. The highest nutrient contents were observed in soils under forest and steppe biomes, whereas tundra soils contained the lowest amount of available *P* and *K* (Fig. 12.2).

The average *P* and *K* contents were considerably higher than in natural references of the corresponding biomes. The maximal *P* concentrations above 800 mg $kg^{-1}$ can be even considered toxic and likely indicate anthropogenic *P* income. High

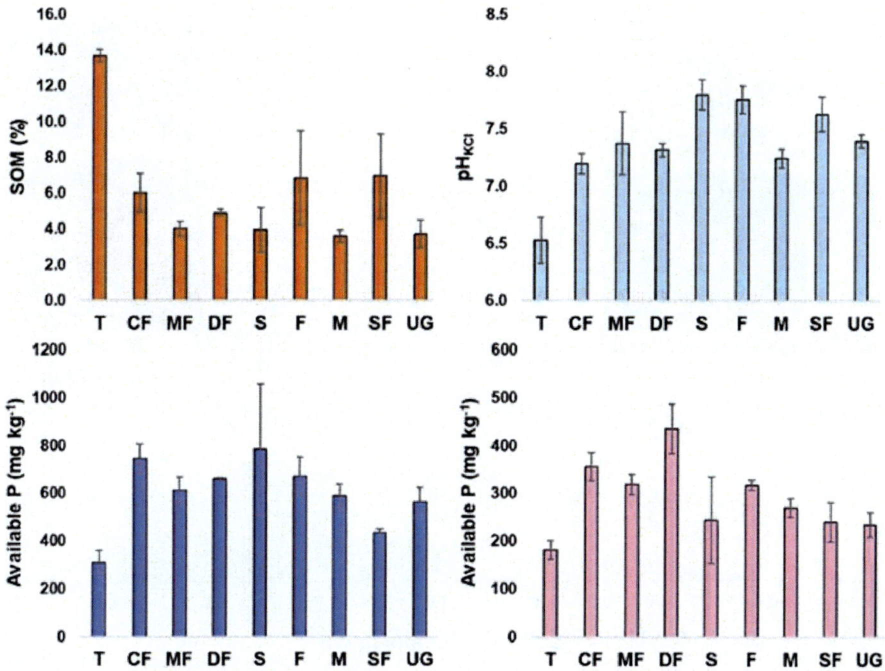


**Fig. 12.2** Variability of soil chemical properties between and within biomes (mean and SD, see Sect. 2.1 for abbreviations)

*P* concentrations in urban soils were previously reported for Beijing (Han et al. 2011; Zhao and Xia 2012), Nanjing (Yang et al. 2021) or Apatity (Korneykova et al. 2022), which is sometimes referred as phosphorous pollution, which coincided with considerable environmental risks. The biome factor explained from 50 to 70% of total variance in soil properties with the highest effect reported for SOM ($R^2 = 0.73$, $p < 0.05$) and the lowest for pH and *P* content.

The effect of disturbance on soil agrochemical propertied was not significant (*T*-test, $p < 0.05$). This is not surprising considering that recreational load coincides with over-compaction and soil cover disturbance that has a stronger impact on soil physical properties. Soil bulk density in the park was normally distributed with the median value close to 1.0 g $cm^{-3}$. The lowest bulk density was observed at the non-disturbed plots of the coniferous forest biome. The highest bulk density observed at the disturbed meadow area was above 1.5 g $cm^{-3}$ that indicates over-compaction with possible negative consequences for plant roots and soil microbiota (Fig. 12.3).

## *12.3.2 Variability of Soil Microbial Properties*

The main integral indicator of the biological activity of the soil is the total microbial number, ranged from a minimum of $12 \times 10^3$ CFU/g soil in disturbed areas

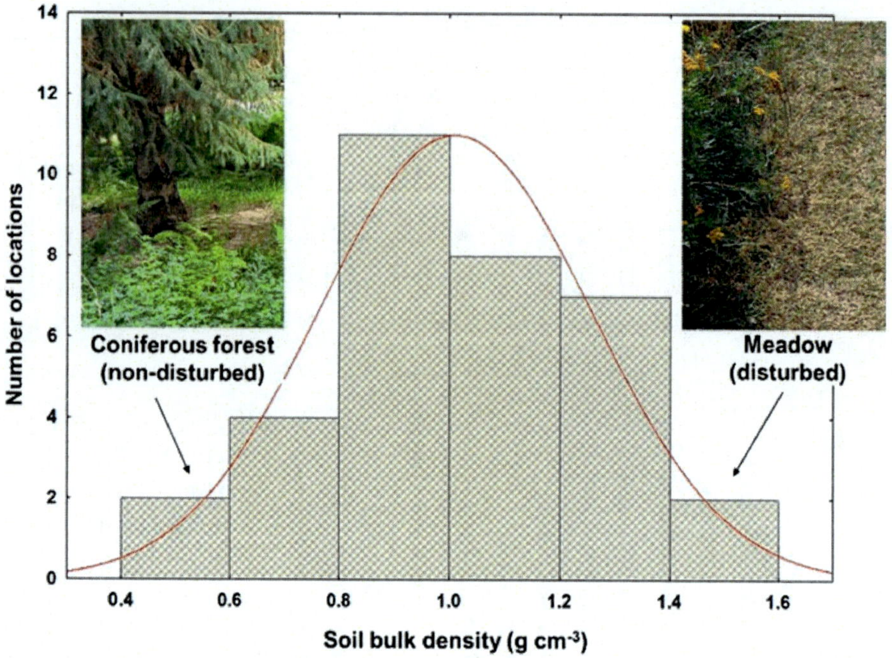


**Fig. 12.3** Distribution of soil bulk density in Zaryadye Park

of the floodplain to a maximum of 1788 × $10^3$ CFU/g in non-disturbed areas of the subtropic forest. The number of cultivated microscopic fungi in different landscape zones varied from 20 CFU/g soil to 2.6 × $10^3$ CFU/g soil. The potential for decomposition of organic matter in soils is low. The number of micromycetes was higher in the soil samples of the "northern" landscapes, alpine and humid tundra, which corresponds to zonal patterns and can be explained by a more acidic reaction of the environment. In the species composition of the cultivated mycobiota, 20 species of micromycetes were identified, excluding the group of fungi with white sterile mycelium. These are mainly cosmopolitan species found on various organic, mainly plant, substrates. Representatives of the phylum *Ascomycota* prevailed (95% of the total species composition) the phylum *Mucoromycota* represented by the only species of *Rhizopus stolonifer*. Species diversity is low compared to zonal analogue, which corresponds to the concept of "dominance concentration" (Terekhova and Ashikhmina 2013) and is a characteristic consequence of the anthropogenic impact on the soil microbiota (Zvyagintsev et al. 2002). In terms of species diversity and occurrence, the genus *Penicillium* prevailed (Fig. 12.4) with the species *Penicillium miczynskii* dominated in various landscapes zones. The highest species diversity observed in the non-disturbed area of tundra landscape. In the landscapes of deciduous forest, monospecific communities have been identified.

The soil physiological profile showed the high level of carbohydrates and carboxylic acids mineralization. In the soils of tundra, high values of mineralization of amino acids (arginine, aspartic acid) were also noted. Physiological profile of

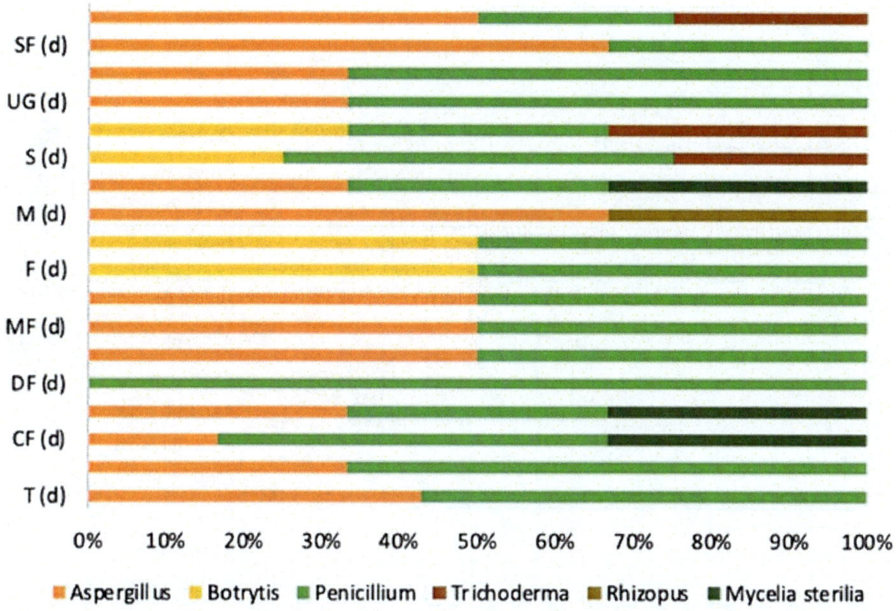


**Fig. 12.4** Genus's taxonomy of micromycetes in soils of disturbed (*d*) and non-disturbed areas of different biomes of Zaryadye Park

microbial communities of non-disturbed zones of urban greening, steppe and subtropical forest turned out to be the most similar, demonstrating the greatest response to the introduction of various substrates (Fig. 12.5).

As a measure of microbial functional diversity of prokaryotic soil communities, the Shannon–Wiener index was used, the values of which varied from a minimum of 1.4 for disturbed mixed forest to a maximum of 2.6 for the disturbed urban greening (Fig. 12.6).

### *12.3.3 Dynamics in Soil Properties*

Soil survey was conducted five years after the park was constructed that allowed analysing dynamics in soil properties by comparison with the initially recommended values. Even though initial data was not available for some of the biomes, the general patterns were clearly identified (Table 12.1). Depletion of SOM was shown for all biomes with the highest losses reported for tundra, steppe, meadows and floodplains where SOM contents decreased more than two times. Likely, depletion of SOM was caused by biodegradation of easily mineralizable compounds of soil mixtures (e.g. peat and compost) used for Technosols' constructions. High emissions from constructed Technosols in Moscow caused by biodegradation of organic compounds and intensified by urban heat island effect was previously reported for Moscow

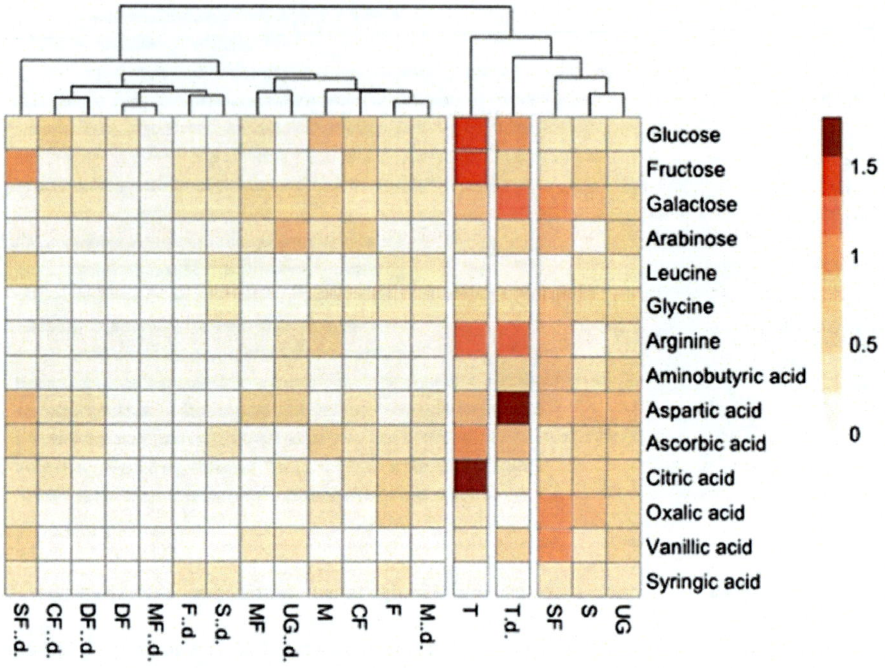


**Fig. 12.5** Physiological profile of the microbial community of soils in disturbed (*d*) and non-disturbed areas of different biomes of Zaryadye Park

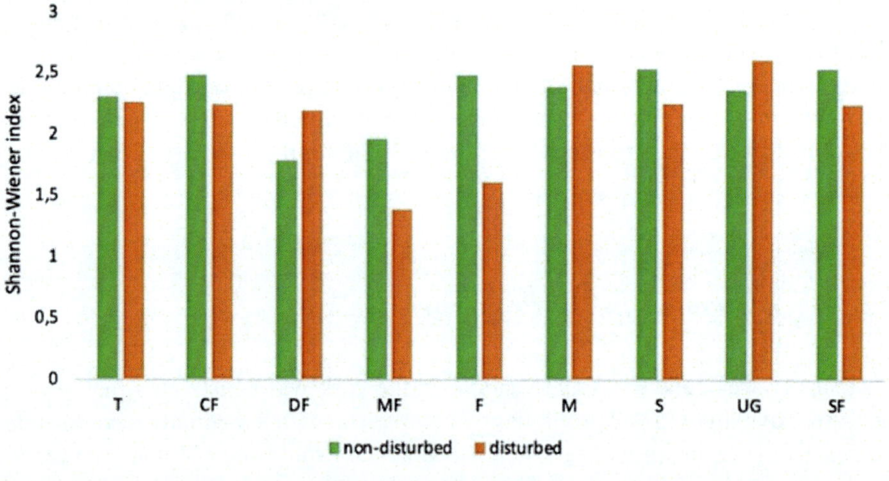


**Fig. 12.6** Functional diversity of soil microbial communities in disturbed and non-disturbed areas of different biomes of Zaryadye Park, calculated using the Shannon–Wiener index

**Table 12.1** Changes in soil properties in comparison with the initially recommended values averaged for biomes

| Biome | SOM (%) | | | $pH_{KCl}$ | | | $P$ (mg kg$^{-1}$) | | | $K$ (mg kg$^{-1}$) | | |
|---|---|---|---|---|---|---|---|---|---|---|---|---|
| | 2014 | 2022 | Δ (%) | 2014 | 2022 | Δ (%) | 2014 | 2022 | Δ (%) | 2014 | 2022 | Δ (%) |
| *T* | 27.5 | 13.7 | – 101 | 4.0 | 6.5 | 39 | 89 | 308 | 71 | 138 | 181 | 24 |
| CF | 8.0 | 6.0 | – 33 | 5.5 | 7.2 | 24 | 152 | 746 | 80 | 225 | 356 | 37 |
| DF | 8.0 | 4.9 | – 63 | 6.0 | 7.3 | 18 | 152 | 659 | 77 | 225 | 436 | 48 |
| *S* | 8.0 | 4.0 | – 103 | 6.9 | 7.8 | 12 | 300 | 783 | 62 | 300 | 245 | – 23 |
| *F* | 15.0 | 6.8 | – 120 | 6.0 | 7.8 | 23 | 250 | 671 | 63 | 250 | 317 | 21 |
| *M* | 8.0 | 3.6 | – 122 | 6.3 | 7.2 | 14 | 225 | 588 | 62 | 225 | 270 | 17 |

(Shchepeleva et al. 2019; Vasenev et al. 2021). An increase of phosphorous was another pattern reported for all biomes. In 2022, soils of all biomes contained 60–70% more *P* than they used to in 2014. Similar pattern in *P* increase reported for all biomes indicates the anthropogenic source of *P* inputs, e.g. with mineral fertilizers. Potassium content also increased 20–40% in all biomes excluding steppe, where *K* content in 2022 was 23% less compared to the initially recommended values. Most remarkable changes were found for soil pH, which shifted 1.5 units in average and 2.5 units in tundra zone. Moreover, in comparison with the initial variability in soil pH, a homogenization patterns was clearly reported—CV in 2022 decreased almost three times compared to 2017.

### 12.3.4 Dynamics in Plants' Distribution and Variability

Based on the analysis of the abundance of species of herbaceous perennials over a five-year period, 4 resistant species were identified for the tundra biome, 4 species for the steppe biome, 9 species for the dry meadow, 8 species for the broad-leaved forest and 2 species for the coniferous forest. It has been established that about 50% of the species have reduced their abundance and 16% of the species have disappeared from the communities. Based on population studies of herbaceous perennials, it was revealed that some forest species (e.g., *Primula elatior* (L.) Hill, *Corydalis solida* (L.) Clairv., *Scilla siberica* Andrews) and steppe species (e.g., *Eryngium planum* L., *Tulipa biebersteiniana* Schult. & Schult. f.) self-renewed by self-seeding in multi-species communities of the park. The formation of sustainable artificial populations of forest herbaceous species is associated with both microclimatic conditions of growth and the possibility of seed or vegetative reproduction. Seed regeneration for many forest species in the broad-leaved forest of Zaryadye is associated with insects, mainly myrmecochores species. The diversification of the *Anemonoides ranunculoides* L. species occurs in two stages: the spread of seeds from the parent clump and the subsequent development of the daughter clump in a vegetative way. As a result, a sporadic spatial structure of the population is created. The species *C. solida* is characterized by vivipary, which ensures rapid seed germination already in the current growing season before the onset of the dry season. The possibility of formation of populations of forest species in the broad-leaved forest community of the park is also explained by the high coefficients of fruit flowering, seedification and real seed productivity. Thus, such species as *Anemone ranunculoides*, *P. elatior* and *S. siberica* are characterized by a high percentage of fruit flowering (87, 82 and 80%, respectively) and a high seeding coefficient (48% for *S. siberica* and 35% for *A. ranunculoides*) in the natural and climatic conditions of Moscow.

A constant migration of species between different communities revealed in the park, mainly characteristic of meadow species: 12 atypical species were recorded in tundra; 14 species in dry meadow; 19 species in steppe; 39 species in broad-leaved forest with 25 species belonging to the meadow plant community; 14 species in pine forest with 10 meadow species; 21 species in spruce forest with 9 meadow species. Migrate mainly such meadow species as *Linaria vulgaris* Mill., *Vicia cracca* L., *Poa pratensis* L., *Tanacetum vulgare* L., *Alopecurus pratensis* L., *Phleum pratense* L., *Melilotus albus* Medik., *Melilotus officinalis* (L.) Lam., *Medicago lupulina* L., *Trifolium pratense* L., *Trifolium repens* L., *Medicago falcata* L., *Dactylis glomerata* L., *Calamagrostis x acutiflora* (Schrad.) Rchb. etc.

## 12.4 Conclusion

Zaryadye Park is a unique ecosystem where plants from different bioclimatic zones co-exist under high recreational load and anthropogenic pressure of Moscow megapolis centre. In this circumstance, Technosols play a crucial role in supporting plants diversity and ecosystem sustainability. In their turn, Technosols are affected by plant communities and are also exposed to anthropogenic impact. In result, SOM in Technosols was significantly different between the biomes, whereas soil pH, *P* and *K* contents did not differ significantly. Physical disturbance from recreational load resulted in topsoil over-compaction with the highest bulk density above 1.5 g $cm^{-3}$. Anthropogenic factors such as urban heat island effect, deposition of alkaline dust and implementation of mineral fertilizers during the seven years of park exploitation resulted in considerable changes in soil properties. Mineralization of organic matters, shift in soil pH and phosphorous inputs decreased initial variability of soil properties, decreased their functionality and microbial diversity. Site-specific precise soil management followed by regular soil monitoring is needed to keep Zaryadye a biodiversity museum and prevent its evolution to a regular urban park. Based on phytocenotic, ecological-geographical and physiognomic principles, an assortment of new species was selected for the communities of the park for their restoration, reconstruction and renewal. Plans have been developed for planting natural species in communities, considering their phytocenotic ratio. These landings are planned for the fall of 2023 after all recommendations, based on the results of soil and environmental monitoring of Zaryadye Park in 2022, which will be implemented.

**Acknowledgements** Analysis of soil chemical propertied was supported by Russian Scientific Foundation Project #19-77-30012. Monitoring soil microbial properties were supported by Kadyrov Chechen State University Development Program 2021–2030. We acknowledge the research and organizational support of park administration and staff members as well as their contribution in discussion and validation of the research outcomes.

## References

Basnou C, Baró F, Langemeyer J, Castell C, Dalmases C, Pino J (2020) Advancing the green infrastructure approach in the Province of Barcelona: integrating biodiversity, ecosystem functions and services into landscape planning. Urban For Urban Green 55.https://doi.org/10.1016/j.ufug.2020.126797

Brianskaia IP, Vasenev VI, Brykova RA, Markelova VN, Ushakova NV, Gosse DD et al (2020) Analysis of volume and properties of imported soils for prediction of carbon stocks in soil constructions in the Moscow metropolis. Eurasian Soil Sci 53. https://doi.org/10.1134/S1064229320120042

Burrow C (2018) Influence of connectivity & topsoil management practices of a constructed Technosol on pedofauna colonization: A field study. Appl Soil Ecol 123. https://doi.org/10.1016/j.apsoil.2017.12.001

Campbell CD, Chapman SJ, Cameron CM, Davidson MS, Potts JM (2003) A rapid microtiter plate method to measure carbon dioxide evolved from carbon substrate amendments so as to determine the physiological profiles of soil microbial communities by using whole soil. Appl Environ Microbio 69(6):3593–3599. https://doi.org/10.1128/AEM.69.6.3593-3599.2003

Charzyński P, Bednarek R, Hudańska P, Świtoniak M (2018) Issues related to classification of garden soils from the urban area of Toruń, Poland. Soil Sci Plant Nutr 64. https://doi.org/10.1080/00380768.2018.1429833

Deeb M, Groffman P, Blouin M, Perl Egendorf S, Vergnes A, Vasenev V et al (2020) Using constructed soils for green infrastructure—challenges and limitations. Soil 6.https://doi.org/10.5194/soil-6-413-2020

Domsch KH, Gams W, Anderson T-H (2007) Compendium of soil Fungi, 2nd edn. IHW Verlag, Ehing

Dvornikov YA, Vasenev VI, Romzaykina ON, Grigorieva VE, Litvinov YA, Gorbov SN et al (2021) Projecting the urbanization effect on soil organic carbon stocks in polar and steppe areas of European Russia by remote sensing. Geoderma 399. https://doi.org/10.1016/j.geoderma.2021.115039

Elmqvist T, Setälä H, Handel SN, van der Ploeg S, Aronson J, Blignaut JN et al (2015) Benefits of restoring ecosystem services in urban areas. Curr Opin Environ Sustain 14. https://doi.org/10.1016/j.cosust.2015.05.001

Filazzola A, Shrestha N, MacIvor JS (2019) The contribution of constructed green infrastructure to urban biodiversity: a synthesis and meta-analysis. J Appl Ecol 56. https://doi.org/10.1111/1365-2664.13475

Gómez-Baggethun E, Barton DN (2013) Classifying and valuing ecosystem services for urban planning. Ecol Econ 86:235–245. https://doi.org/10.1016/j.ecolecon.2012.08.019

Han Y, Li X, Nan Z (2011) Net anthropogenic phosphorus accumulation in the Beijing metropolitan region. Ecosystems 14(3):445–457

Hedde M, Nahmani J, Séré G, Auclerc A, Cortet J (2019) Early colonization of constructed Technosols by macro-invertebrates. J Soils Sediments 19:3193–3203. https://doi.org/10.1007/s11368-018-2142-9

Hostetler M, Allen W, Meurk C (2011) Conserving urban biodiversity? Creating green infrastructure is only the first step. Landsc Urban Plan 100. https://doi.org/10.1016/j.landurbplan.2011.01.011

Hoyle H, Hitchmough J, Jorgensen A (2017) All about the 'wow factor'? The relationships between aesthetics, restorative effect and perceived biodiversity in designed urban planting. Landsc Urban Plan 164. https://doi.org/10.1016/j.landurbplan.2017.03.011

Ignatieva M, Eriksson F, Eriksson T, Berg P, Hedblom M (2017) The lawn as a social and cultural phenomenon in Sweden. Urban for Urban Green 21. https://doi.org/10.1016/j.ufug.2016.12.006

Ivashchenko K, Lepore E, Vasenev V, Ananyeva N, Demina S, Khabibullina F et al (2021) Assessing soil-like materials for ecosystem services provided by constructed Technosols. Land 10. https://doi.org/10.3390/land10111185

Ivashchenko KV, Ananyeva ND, Vasenev VI, Kudeyarov VN, Valentini R (2014) Biomass and respiration activity of soil microorganisms in anthropogenically transformed ecosystems (Moscow region). Eurasian Soil Sci 47. https://doi.org/10.1134/S1064229314090051

Kabała C, Greinert A, Charzyński P, Uzarowicz Ł (2020) Technogenic soils—soils of the year 2020 in Poland. Concept, properties and classification of technogenic soils in Poland. Soil Sci Annu 71. https://doi.org/10.37501/soilsa/131609

Kasyanov NV (2017) New perspectives for landscape architecture in Russia—Park Zaryadye in Moscow. Landsc Archit Art 11(11):7–14. https://doi.org/10.22616/j.landarchart.2017.11.01

Korneykova MV, Vasenev VI, Nikitin DA, Soshina AS, Dolgikh AV, Sotnikova YL (2021) Urbanization affects soil microbiome profile distribution in the Russian arctic region. Int J Environ Res Publ Health 18. https://doi.org/10.3390/ijerph182111665

Korneykova MV, Vasenev VI, Nikitin DA, Dolgikh AV, Soshina AS, Myazin VA, Nakhaev MR (2022) Soil microbial community of urban green infrastructures in a polar city. Urban Ecosyst 25(5):1399–1415. https://doi.org/10.1007/s11252-022-01233-8

Marinari S, Bonifacio E, Moscatelli MC, Falsone G, Antisari LV, Vianello G (2013) Soil development and microbial functional diversity: Proposal for a methodological approach (2013) Geoderma 192(1):437–445. https://doi.org/10.1016/j.geoderma.2012.08.023

McPhearson T, Pickett STA, Grimm NB, Niemelä J, Alberti M, Elmqvist T et al (2016) Advancing urban ecology toward a science of cities. Bioscience 66. https://doi.org/10.1093/biosci/biw002

Monteiro JA (2017) Ecosystem services from turfgrass landscapes. Urban for Urban Green 26. https://doi.org/10.1016/j.ufug.2017.04.001

Morel JL, Chenu C, Lorenz K (2014) Ecosystem services provided by soils of urban, industrial, traffic, mining, and military areas (SUITMAs). J Soils Sediments 15:1659–1666. https://doi.org/10.1007/s11368-014-0926-0

Moreno-Barriga F, Díaz V, Acosta JA, Muñoz MÁ, Faz Á, Zornoza R (2017) Organic matter dynamics, soil aggregation and microbial biomass and activity in Technosols created with metalliferous mine residues, biochar and marble waste. Geoderma 301. https://doi.org/10.1016/j.geoderma.2017.04.017

Moscatelli MC, Secondi L, Marabottini R, Papp R, Stazi SR, Mania E, Marinari S (2018) Assessment of soil microbial functional diversity: land use and soil properties affect CLPP-MicroResp and enzymes responses. Pedobiologia 66:36–42. https://doi.org/10.1016/j.pedobi.2018.01.001

Mueller GD, Thompson AM (2009) The ability of urban residential lawns to disconnect impervious area from municipal sewer systems. J Am Water Resour Assoc 45. https://doi.org/10.1111/j.1752-1688.2009.00347.x

Neina D, Buerkert A, Joergensen RG (2016) Microbial response to the restoration of a Technosol amended with local organic materials. Soil Tillage Res 163. https://doi.org/10.1016/j.still.2016.06.008

Pierce JR, Barton MA, Tan MMJ, Oertel G, Halder MD, Lopez-Guijosa PA et al (2020) Actions, indicators, and outputs in urban biodiversity plans: a multinational analysis of city practice. PLoS ONE 15. https://doi.org/10.1371/journal.pone.0235773

Prokof'eva TV, Kiryushin AV, Shishkov VA, Ivannikov FA (2017) The importance of dust material in urban soil formation: the experience on study of two young Technosols on dust depositions. J Soils Sediments 17:515–524.https://doi.org/10.1007/s11368-016-1546-7

Prokofyeva TV, Martynenko IA, Ivannikov FA (2011) Classification of Moscow soils and parent materials and its possible inclusion in the classification system of Russian soils. Eurasian Soil Sci 44:561–571.https://doi.org/10.1134/S1064229311050127

Rappoport AV, Umarova AB, Butylkina MA, Kutyeva EV (2019) Technology of creating soils of natural zones in "Zaryadye Park" in Moscow. IOP Conf Ser Earth Environ Sci 368(1):012039. https://doi.org/10.1088/1755-1315/368/1/012039

Raper KB, Thom C (1968) A manual of the Penicillia. Hefner Publishing Co., New York

Romzaykina ON, Vasenev VI, Paltseva A, Kuzyakov YV, Neaman A, Dovletyarova EA (2021) Assessing and mapping urban soils as geochemical barriers for contamination by heavy metal(loid)s in Moscow megapolis. J Environ Qual 50. https://doi.org/10.1002/jeq2.20142

Rossiter DG (2007) Classification of urban and industrial soils in the world reference base for soil resources. J Soils Sediments 7:96–100. https://doi.org/10.1065/jss2007.02.208

Savvas W (2016) Green infrastructure and urban biodiversity. Landsc Archit Front 4

Schad P (2018) Technosols in the world reference base for soil resources—history and definitions. Soil Sci Plant Nutr 64. https://doi.org/10.1080/00380768.2018.1432973

Séré G, Schwartz C, Ouvrard S, Renat JC, Watteau F, Villemin G et al (2010) Early pedogenic evolution of constructed Technosols. J Soils Sediments 10. https://doi.org/10.1007/s11368-010-0206-6

Sharma RC, Tateishi R, Hara K, Gharechelou S, Iizuka K (2016) Global mapping of urban built-up areas of year 2014 by combining MODIS multispectral data with VIIRS nighttime light data. Int J Digit Earth 9:1004–1020. https://doi.org/10.1080/17538947.2016.1168879

Shchepeleva AS, Vizirskaya MM, Vasenev VI, Vasenev II (2019) Analysis of carbon stocks and fluxes of urban lawn ecosystems in Moscow megapolis. . https://doi.org/10.1007/978-3-319-89602-1_11

Slukovskaya MV, Vasenev VI, Ivashchenko KV, Morev DV, Drogobuzhskaya SV, Ivanova LA et al (2019) Technosols on mining wastes in the subarctic: efficiency of remediation under Cu-Ni atmospheric pollution. Int Soil Water Conserv Res 7. https://doi.org/10.1016/j.iswcr.2019.04.002

Smagin AV (2012) Column-centrifugation method for determining water retention curves of soils and disperse sediments (2012) Eurasian Soil Sci 45(4):416–422. https://doi.org/10.1134/S1064229312040126

Sun SM, Shan BQ, Peng WJ (2009) Roles of urban lawn system in controlling rainfall-induced runoff nitrogen pollution. Chin J Ecol 28

Terekhova VA, Ashikhmina TY (2013) Biodiagnostics in the environmental assessment of soils and adjacent environments. Theor Appl Ecol 1:107–118

Vasenev VI, Stoorvogel JJ, Dolgikh AV, Ananyeva ND, Ivashchenko KV, Valentini R (2017) Changes in soil organic carbon stocks by urbanization. Urban Soils. https://doi.org/10.1201/9781315154251

Vasenev VI, Stoorvogel JJ, Vasenev II (2013) Urban soil organic carbon and its spatial heterogeneity in comparison with natural and agricultural areas in the Moscow region. CATENA 107. https://doi.org/10.1016/j.catena.2013.02.009

Vasenev VI, Van Oudenhoven APE, Romzaykina ON, Hajiaghaeva RA (2018) The ecological functions and ecosystem services of urban and technogenic soils: from theory to practice (a review). Eurasian Soil Sci 51. https://doi.org/10.1134/S1064229318100137

Vasenev VI, Yaroslavtsev AM, Vasenev II, Demina SA, Dovltetyarova EA (2019) Land-use change in new Moscow: first outcomes after five years of urbanization. Geogr Environ Sustain. https://doi.org/10.24057/2071-9388-2019-89

Vasenev V, Varentsov M, Konstantinov P, Romzaykina O, Kanareykina I, Dvornikov Y et al (2021) Projecting urban heat island effect on the spatial-temporal variation of microbial respiration in urban soils of Moscow megalopolis. Sci Total Environ 786. https://doi.org/10.1016/j.scitotenv.2021.147457

Yan H, Wu F, Dong L (2018) Influence of a large urban park on the local urban thermal environment. Sci Total Environ 622–623. https://doi.org/10.1016/j.scitotenv.2017.11.327

Yang JL, Yuan DG, Zhao YG, He Y, Zhang GL (2021) Stoichiometric relations of C, N, and P in urban top soils in Nanjing, China, and their biogeochemical implications. J Soils Sediments 21. https://doi.org/10.1007/s11368-020-02826-6

Zhao X, Xia X (2012) Total nitrogen and total phosphorous in urban soils used for different purposes in Beijing, China. Proc Environ Sci 3:95–104

Zvyagintsev DG, Umarov MM, Chernov YI (2002) Microbial communities and their functioning in the processes of soil degradation and self-restoration. In: Dobrovolsky GV (ed) Publishing House of Moscow State University, 441–445

# Chapter 13
# Carbon Stocks and Fluxes in Soils of the Urban Park in Grozny City

**K. Ivashchenko, R. Bekmurzaeva, Z. Iriskhanova, V. Vasenev, D. Sarzhanov, M. Korneykova, and M. Nakhaev**

**Abstract** Development of urban green infrastructures (UGI) is considered among the main nature-based solutions for climate mitigation and adaptation in cities; however, the role of urban soils in carbon balance of UGI remains overlooked and likely underestimated. In this research, we observed soil stocks of organic and inorganic carbon and soil carbon dioxide ($CO_2$) emission for the city park "Mothers' Glory" in Grozny, Chechen Republic, Russia. Soil survey in the park was performed in July 2022. Sampling points were randomly selected within the landscape part of the park. Topsoil (0–10 cm) samples were collected at sites of irrigated lawns, non-irrigated lawns and non-irrigated lawns with trees. Additional samples from top 50 cm and top 100 cm were collected. Soil respiration was measured in situ at the sites by infrared gas analyzer Li-7810SC during the period April–October 2022 with an average time step 2–3 weeks. In parallel, soil temperature and moisture were observed. Total soil $CO_2$ emission for the observation period was above 0.5 kg $Cm^{-2}$. Based on the estimated 5.5 years half-life time, C stocks at the managed green areas of Grozny are not resistant to biodegradation, and their vulnerability will likely further increase under ongoing climate change.

**Keywords** Technosols · Urban green infrastructures · Carbon balance · Carbon neutrality · Chechen Republic

---

K. Ivashchenko (✉) · D. Sarzhanov · M. Korneykova
Smart Urban Nature Research Center, RUDN University, Moscow, Russia
e-mail: ivashchenko_kv@pfur.ru

K. Ivashchenko
Institute of Physicochemical and Biological Problems in Soil Science, Russian Academy of Sciences, Pushchino, Russia

R. Bekmurzaeva · Z. Iriskhanova · M. Nakhaev
Kadyrov Chechen State University, Grozny, Russia

V. Vasenev
Soil Geography and Landscape, Wageningen University, Wageningen, The Netherlands



M. Korneykova et al. (eds.), *Smart and Sustainable Urban Ecosystems: Challenges and Solutions*, Springer Geography, https://doi.org/10.1007/978-3-031-37216-2_13

## 13.1 Introduction

Globally, cities consume almost three-fourth of energy resources and produce more than half of greenhouse gas' (GHG) emissions. Climate-mitigation strategies claim reduction in urban GHG emissions among the key priorities. The recent IPCC assessment report highlights cities as global warming hotspots, where climate change is exacerbated by urban heat island (UHI) and urban dryness (UD) effects (i.e., higher temperature and lower humidity compared to the surrounding rural areas). Actions on climate adaptation and carbon (C) neutrality are claimed a priority by more than a hundred cities all over the world (e.g., C40 or Cities Alliance). Development of urban green infrastructures (UGI) is considered among the main nature-based solutions (NBS) for climate mitigation and adaptation in cities (Sturiale and Scuderi 2019). City governments and investors have very high expectations from C sequestration by UGI and promote such initiatives as Million Trees in New York (https://www.milliontreesnyc.org/) or My Street in Moscow (https://www.mos.ru/city/projects/moscow2020/) as a universal recipe to reach C neutrality. However, none of these or similar projects consider C stocks in and emission from urban soils. In comparison with natural soils, which are considered in C accounting as achieved stocks (Mikhailova et al. 2021), urban soils shall be also considered as potential C sources due to artificial origin (i.e., importing C with materials for urban soil's construction, Vasenev and Kuzyakov 2018) and high soil respiration further intensified by UHI (Vasenev et al. 2021).

The role of urban soils in C balance of UGI remains overlooked, and this *knowledge gap* increases uncertainty in C accounting and probably overestimates the real potential of UGI for C sequestration and therefore their role in climate mitigation and adaptation measures. Although high net primary production (NPP) of urban trees was reported (Nowak and Crane 2002; Elmqvist et al. 2015), high $CO_2$ emissions from soils will likely change some UGI ecosystems from sinks to sources, like it was shown for managed lawns (Shchepeleva et al. 2016). Under the climate change, the role of urban soils in UGI C balance will likely increase considering high sensitivity of soil C emissions to changes in temperature and moisture regimes. Risks of further intensification of soil $CO_2$ emissions are especially high in cities with hot and arid climate where high soil temperature results in intensive heterotrophic respiration during a considerable part of the season. Projecting C dynamics in urban soils and UGI under climate change shall also consider the effects of land cover, soil management, and maintenance of UGI (e.g., mowing, pruning, or irrigation), e.g., Zirkle et al. (2011) showed higher C stocks and potential C sequestration in unmanaged lawns in comparison with intensively managed lawns due to limited C input (grass clips were removed) and higher temperature contributing to intensive soil respiration. Higher soil temperature and moisture were the main abiotic factors, which determined an increased biodegradation coefficient (ratio between $CO_2$ emissions and topsoil C stocks) in urban soils of Kursk in comparison with the virgin steppe (Sarzhanov et al. 2015, 2017). Preliminary studies in Moscow in summer 2019 also showed that higher average soil temperature (+4.5 °C) and moisture (+20%) under lawns compared to soils under green stands increased soil $CO_2$ emissions by 40% (Vasenev et al. 2021).

Grozny city is an interesting case study to observe C stocks and fluxes in soil of UGI under hot and dry summer conditions. An average air temperature is above 15 C° from April to October, reaching + 35 C° in July. Based on the remote sensing data, green areas including trees, shrubs, and lawns cover up to 70% of the city area. A considerable part of these green areas comprises natural or semi-natural meadows and forests, where the management effect is minimal. However, several urban parks in Grozny are well-managed that allows studying microclimatic and management effect on C stocks and fluxes in UGI. In 2021, Chechen Republic was selected among the pilot regions in Russia, where experimental stations for C monitoring (C polygons) were set-up. However, so far C stocks and fluxes in urban soils were not observed (Markova et al. 2022). Our research aimed to fill this knowledge gap by investigating C stocks and $CO_2$ emissions in soils of the urban park "Mothers' Glory" in the center of Grozny city.

## 13.2 Materials and Methods

### *13.2.1 Research Area*

Grozny city (43.3°N, 45.7°E) is the administrative center of Chechen Republic and of the biggest city in the North Caucuses region of Russia. Although a considerable part of Chechen Republic is dominated by mountainous landscapes, Grozny city is located on a relatively flat area on the banks of Sunzha River. Chechen Republic exhibits a huge diversity of bioclimatic conditions ranging from semi-desert landscape on the North to alpine meadows on the South. Grozny city locates in steppe conditions, and therefore, natural vegetation is dominated by grasses, and most of the tree species used in urban greening are introduced. Natural soils in the area are dominated by Calcic Chernozems and Kastanozems; however, in the city, Anthrosols and Technosols are widely spread. Grozny city was almost completely redeveloped in early 2000-s that resulted in establishment of new green areas or reorganization of the existing ones. One of them is an urban park "Mothers' Glory" (hereinafter MG Park) located in center of Grozny. The park contains of the two areas: area of attractions and a landscape area. Area of attractions is almost completely sealed, and therefore, its contribution to biogenic C balance is minimal (or difficult to assess). The landscape area, in contrast, is more than 90% unsealed and includes lawn areas, ornamental trees, and shrubs. The area is maintained, but maintenance practices vary between the plots (e.g., not all green lawns are irrigated) resulting in a patchy structure of land cover (Fig. 13.1). This patchiness is likely reflected in the spatial patterns on soil C stocks and $CO_2$ emissions.

**Fig. 13.1** MG Park as a part of urban green infrastructure of Grozny and location of the observation locations (yellow points are the soil sampling locations and red points are locations where soil samples were collected and $CO_2$ was monitored)

### *13.2.2 Measuring Soil C Stocks and $CO_2$ Fluxes*

Soil survey in the park was performed in July 2022. Sampling points were randomly selected within the landscape part of the park. Topsoil (0–10 cm) samples were collected at irrigated lawns (MG-1), non-irrigated lawns (MG-2) and non-irrigated lawns with trees (MG-3) sites. Additional samples from top 50 cm and top 100 cm were collected at the sites. All the subsoil samples were collected based on the distinguished horizons that allowed describing profile distribution of soil properties. Bulk density was measured and further used to estimate soil C stocks. All the collected samples were processed in the laboratory (roots and rock fragments were removed), and the following chemical properties were analyzed: $pH_{H2O}$, total and inorganic carbon (TC and SIC, respectively) contents. Soil organic carbon (SOC, %) was calculated by following: TC - SIC. Soil C stocks (kg $m^{-2}$) were calculated by the formula BD × h × C / 10, where BD is the soil bulk density (g $cm^{-3}$); h is the thickness of the soil layer (cm); C is the content of carbon in the soil (%). Soil respiration was measured in situ by infrared gas analyzer Li-7810SC during the period April–October 2022 with an average time step 2–3 weeks. In parallel, soil temperature and moisture were observed. Total C emission for the season was estimated based on extrapolating the daily measurements for the observation period. Ratio between C emissions extrapolated for the season and topsoil C stocks were estimated to approximate the biodegradation coefficient and resistance on soil C stocks. Relationship between soil respiration and temperature was analyzed by multiple regression.

## 13.3 Results and Discussions

### 13.3.1 *Soil C Stocks: Total, Organic, and Inorganic Components*

Based on the description of 100 cm deep soil profiles (MG-1, MG-2, and MG-3), soils of the park have anthropogenic genesis or have experienced considerable anthropogenic effect. Profile MG-1 is a constructed Technosols with a clear RAT horizon comprised from the introduced organic material. The underlying UR horizon also exposes features of urban pedogenesis such as inclusions of bricks and other artifacts. Profiles MG-2 and MG-3 also contain anthropogenic subsoils horizons (TCH and UR); however, organic material was not added to topsoil artificially (at least, it is not evident in contrast to MG-1 profile). The identified AY horizon was likely formed in result of a gradual carbon input with leaves, grass clips, and fine roots residuals. All the profiles include buried horizons of natural Calcic Chernozems (Fig. 13.2). All soil had basic pH, which did not differ significantly along the profile. Land cover and management (mainly, irrigation) were the main factors which determined topsoil C stocks. Topsoils' SOC stocks under irrigated lawns (MG-1) were 35% higher than under non-irrigated sites (data not shown). Among non-irrigated site, topsoil C stock in MG-2 (3.04 kg $Cm^{-2}$) was slightly higher than in MG-3 (3.00 kg $Cm^{-2}$) that can be likely explained by more intensive C input with above- and belowground biomass residuals from lawns. An opposite pattern was shown for topsoil inorganic C stocks, which were 0.4, 0.5, and 0.7 kg $Cm^{-2}$ for MG-1, MG-2, and MG-3, correspondingly. Profile distribution for SIC showed a gradual increase of SIC content in deeper horizons. High SIC stocks in subsoil could be partly inherited from the carbonate soil formations of the buried horizons. Dust deposition with further migration down the profile is another possible source of SIC in subsoil. An opposite profile distribution was shown for SOC content, which were more than two times higher in top 50 cm in comparison with the depth 50–100 cm (Table 13.1). TC stocks in 0–100 cm ranged from 18 kg $Cm^{-2}$ for MG-2 to 32 kg $Cm^{-2}$ in MG-1. This outcome is comparable to C stocks under green lawns in cities located in Chernozemic zone, such as Kursk (Sarzhanov et al. 2017) or Rostov-on-Don (Dvornikov et al. 2021) and considerably higher than in Moscow (Vasenev et al. 2013) or northern cities (Korneykova et al. 2021). Fraction of SIC in total C stocks ranged from 11 to 64% that is significantly higher than for Moscow. Such a considerable SIC fraction in total C stocks could contribute to a higher resistance to biodegradation as it was proposed by (Lorenz and Lal 2015). However, monitoring and assessment of $CO_2$ emissions are needed to check this assumption.

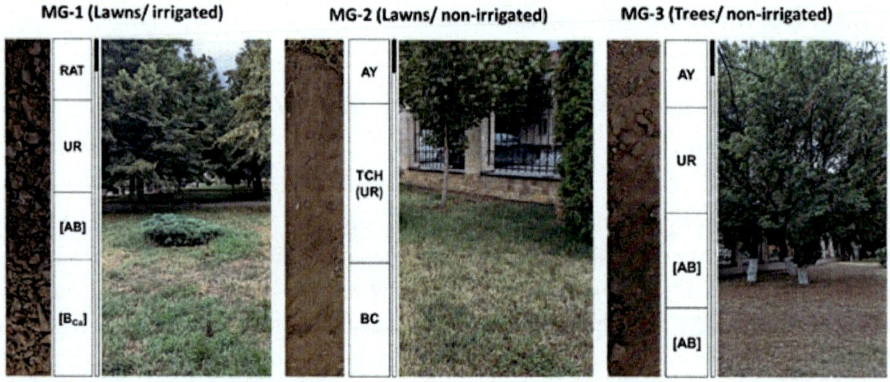


Fig. 13.2 Soil profiles under different land cover types and management regimes

**Table 13.1** Profile distribution of soil properties (BD, bulk density; TC, total carbon; SOC, soil organic carbon; SIC, soil inorganic carbon content)

| Horizon | Depth | Texture | BD (g $cm^{-3}$) | $pH_{H2O}$ | TC (%) | SOC (%) | SIC (%) |
|---|---|---|---|---|---|---|---|
| *MG-1 (lawns/irrigated)* | | | | | | | |
| RAT | 0–20 | Loam | 1.2 | 8.5 | 3.9 | 3.5 | 0.43 |
| Ur | 20–50 | Clay loam | 1.2 | 8.7 | 3.9 | 3.4 | 0.55 |
| [AB] | 50–75 | Clay loam | 1.1 | 8.8 | 1.3 | 1.0 | 0.29 |
| [Bca] | 75–100 | Clay loam | 1.1 | 8.8 | 1.9 | 1.0 | 0.95 |
| *MG-2 (lawns/non-irrigated)* | | | | | | | |
| AY | 0–20 | Sandy loam | 1.0 | 8.6 | 3.4 | 2.8 | 0.62 |
| Ur | 20–50 | Loam | 1.0 | 8.4 | 2.7 | 1.8 | 0.92 |
| TCH (Ur) | 20–50 | Sandy loam | 1.0 | 8.8 | 2.3 | 1.8 | 0.53 |
| BC | 50–80 | Loam | 1.0 | 8.8 | 1.6 | 0.6 | 1.03 |
| *MG-3 (trees/non-irrigated)* | | | | | | | |
| AY | 0–20 | Loam | 1.1 | 8.8 | 2.9 | 2.1 | 0.85 |
| [AB] | 50–80 | Clay loam | 1.2 | 8.6 | 3.6 | 2.9 | 0.68 |
| [B] | 80–100 | Clay loam | 1.2 | 8.7 | 1.8 | 0.8 | 0.98 |

### 13.3.2 Seasonal Dynamics in Soil Temperature and $CO_2$ Emissions

During the observation period, soil temperature varied between 12 and 30 °C with an average + 20.2 °C. For more than six months of observation, temperature never dropped below 10 °C, which is much warmer than in many other cities where soil respiration was previously measured, e.g., Moscow (Goncharova et al. 2019), Kursk (Sarzhanov et al. 2015), or Boston (Decina et al. 2016). Therefore, favorable climatic conditions for soil microbial activity were observed for at least 210 days during the

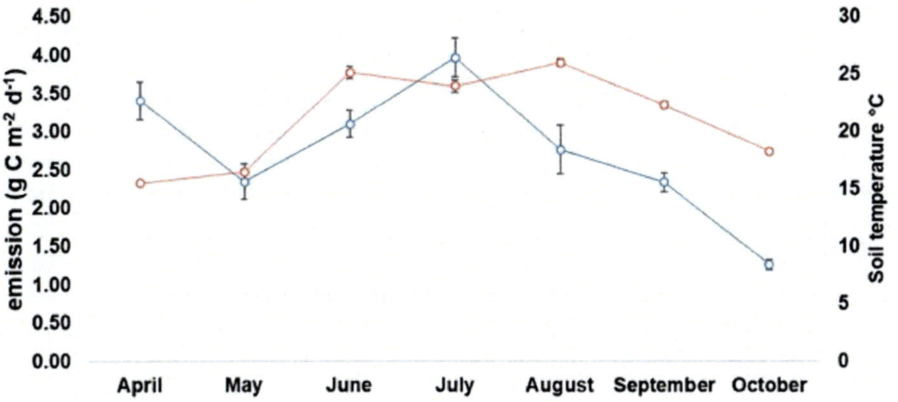


**Fig. 13.3** Seasonal dynamics in soil temperature (red line) and $CO_2$ emissions (blue line) at the irrigated site (MG-1)

season that could result in intensive heterotrophic soil respiration. June and August were the hottest months with soil temperature above 25 °C, whereas the lowest monthly average temperature was reported in October. Soil respiration exposed a similar seasonal dynamic with the highest monthly values in July and the lowest in October (Fig. 13.3). However, we did not find significant positive linear or exponential relationships between soil temperature and soil respiration which is usually reported for urban and natural soils in temperature climate (Shchepeleva et al. 2016; Sushko et al. 2019; Kurganova et al. 2020). Moreover, during the summer months, relationship between soil respiration and temperature was negative, indicating limitation of microbial activity by high temperatures under conditions of the water scarcity. Apparently, even at the irrigated sites, water content in summer is not enough to stimulate gradual growth of soil respiration by increasing temperature. Indirectly, this outcome is confirmed by high soil respiration in April, when soil temperature was almost 10 °C lower than in July, but water content was abundant. In result, observed average emissions were 20 to 50% lower than in green areas of Moscow or Kursk and comparable to green lawns in Rostov-on-Don, where climatic conditions are comparable and water scarcity remains the main limitation factor.

### *13.3.3 Comparison Between C Stocks and Fluxes*

Extrapolation of monthly average $CO_2$ emissions results in total outflow of 573 g $Cm^{-2}$ for the period between the beginning of April and end of October 2022. This outcome can be considered a minimal estimate of the annual emission, since respiration in November–March is highly likely more than zero. This emission is comparable to the estimates obtained for urban lawns in Moscow and Kursk (Shchepeleva et al. 2016) and Kursk (Sarzhanov et al. 2015; Sushko et al. 2019).

The estimated soil SOC stocks at the MG-1 site where soil respiration was monitored were 4.6 kg $Cm^{-2}$ in top 10 cm and 26 kg $Cm^{-2}$ in top 100 cm. Based on the approach proposed by (Sarzhanov et al. 2017) for Kursk, almost 15% of the total topsoil C stocks can be emitted annually by heterotrophic soil respiration. This value is considerably lower than at the industrial site in Kursk, where the ratio was above 50%. When a more advanced approach to estimate biodegradation coefficient ($k$) and half-life time ($T_{0.5}$) proposed by (Smagin et al. 2018) is used, an average half-life time of 5.5 years is obtained. This outcome is comparable to the average values obtained for Moscow (Vasenev et al. 2021) but considerably lower than for most of the natural biomes. A short half-life time indicates low resistance of soil C stocks to biodegradation, which can further decrease with ongoing climate change.

## 13.4 Conclusion

For the first time, soil C stocks and $CO_2$ emission were estimated for UGI in Grozny city, Chechen Republic. C stocks under managed (irrigated) lawns were considerably higher than under non-irrigated lawns and trees, and therefore, management and maintenance are important factors contribution to C balance in urban soils under hot and arid climatic conditions. The contribution of inorganic C to total C stocks was high and increased down the profile. Although C stocks in irrigates sites were considerable, at least 15% of the topsoil stocks could be released annually by heterotrophic respiration. The estimated half-life time of topsoil C stocks was just 5.5 years that indicate low resistance of C stocks to biodegradation and shall be considered in soil management strategies and maintenance practices.

**Acknowledgements** Estimation of C stocks was supported by Russian Scientific Foundation Project # 19-77-30012. Soil respiration measurements were supported by Kadyrov Chechen State University Development Program 2021–2030.

## References

Decina SM, Hutyra LR, Gately CK, Getson JM, Reinmann AB, Short Gianotti AG et al. (2016) Soil respiration contributes substantially to urban carbon fluxes in the greater Boston area. Environ Pollut 212:433–439. https://doi.org/10.1016/j.envpol.2016.01.012

Dvornikov YA, Vasenev VI, Romzaykina ON, Grigorieva VE, Litvinov YA, Gorbov SN et al. (2021) Projecting the urbanization effect on soil organic carbon stocks in polar and steppe areas of European Russia by remote sensing. Geoderma 399. https://doi.org/10.1016/j.geoderma.2021.115039

Elmqvist T, Setälä H, Handel SN, van der Ploeg S, Aronson J, Blignaut JN et al. (2015) Benefits of restoring ecosystem services in urban areas. Curr Opin Environ Sustain 14. https://doi.org/10.1016/j.cosust.2015.05.001

Goncharova OY, Matyshak GV, Udovenko MM, Bobrik AA, Semenyuk OV (2019). Seasonal and annual variations in soil respiration of the artificial landscapes (Moscow Botanical Garden). Springer Geogr. https://doi.org/10.1007/978-3-319-89602-1_15

Korneykova MV, Vasenev VI, Nikitin DA, Soshina AS, Dolgikh AV, Sotnikova YL (2021) Urbanization affects soil microbiome profile distribution in the Russian arctic region. Int J Environ Res Publ Health 18. https://doi.org/10.3390/ijerph182111665

Kurganova IN, Lopes de Gerenyu VO, Khoroshaev DA, Myakshina TN, Sapronov DV, Zhmurin VA et al (2020) Analysis of the long-term soil respiration dynamics in the forest and meadow cenoses of the prioksko-terrasny biosphere reserve in the perspective of current climate trends. Eurasian Soil Sci. https://doi.org/10.1134/S1064229320100117

Lorenz K, Lal R (2015) Managing soil carbon stocks to enhance the resilience of urban ecosystems. Carbon Manag 6:35–50. https://doi.org/10.1080/17583004.2015.1071182

Markarova MY, Makhmudova LS, Kerimov IA, Elgaev AS, Mamadiev NA (2022) Analysis of greenhouse gases in oil-controlled areas and development of methods for their biorecultivation (for example of Grozny) (2022) Geol i Geofiz Yuga Rossii 12(4):153–168. https://doi.org/10.46698/VNC.2022.19.77.012

Mikhailova EA, Zurqani HA, Post CJ, Schlautman MA, Post GC, Lin L et al (2021) Soil carbon regulating ecosystem services in the state of South Carolina, USA. Land 10. https://doi.org/10.3390/land10030309

Nowak DJ, Crane DE (2002) Carbon storage and sequestration by urban trees in the USA. Environ Pollut 116:381–389. https://doi.org/10.1016/S0269-7491(01)00214-7

Sarzhanov DA, Vasenev VI, Sotnikova YL, Tembo A, Vasenev II, Valentini R (2015) Short-term dynamics and spatial heterogeneity of $CO_2$ emission from the soils of natural and urban ecosystems in the Central Chernozemic Region. Eurasian Soil Sci 48:416–424. https://doi.org/10.1134/S1064229315040092

Sarzhanov DA, Vasenev VI, Vasenev II, Sotnikova YL, Ryzhkov OV, Morin T (2017) Carbon stocks and $CO_2$ emissions of urban and natural soils in Central Chernozemic region of Russia. Catena 158. https://doi.org/10.1016/j.catena.2017.06.021

Shchepeleva AS, Vasenev VI, Mazirov IM, Vasenev II, Prokhorov IS, Gosse DD (2016) Changes of soil organic carbon stocks and $CO_2$ emissions at the early stages of urban turf grasses' development. Urban Ecosyst 1–13. https://doi.org/10.1007/s11252-016-0594-5

Smagin AV, Sadovnikova NB, Vasenev VI, Smagina MV (2018) Biodegradation of some organic materials in soils and soil constructions: experiments, modeling and prevention. Materials 11:1–22. https://doi.org/10.3390/ma11101889

Sturiale L, Scuderi A (2019) The role of green infrastructures in urban planning for climate change adaptation. Climate 7. https://doi.org/10.3390/cli7100119

Sushko S, Ananyeva N, Ivashchenko K, Vasenev V, Kudeyarov V (2019) Soil $CO_2$ emission, microbial biomass, and microbial respiration of woody and grassy areas in Moscow (Russia). J Soils Sediments 19. https://doi.org/10.1007/s11368-018-2151-8

Vasenev VI, Stoorvogel JJ, Vasenev II (2013) Urban soil organic carbon and its spatial heterogeneity in comparison with natural and agricultural areas in the Moscow region. Catena 107. https://doi.org/10.1016/j.catena.2013.02.009

Vasenev V, Varentsov M, Konstantinov P, Romzaykina O, Kanareykina I, Dvornikov Y et al (2021) Projecting urban heat island effect on the spatial-temporal variation of microbial respiration in urban soils of Moscow megalopolis. Sci Total Environ 786. https://doi.org/10.1016/j.scitotenv.2021.147457

Vasenev V, Kuzyakov Y (2018) Urban soils as hot spots of anthropogenic carbon accumulation: review of stocks, mechanisms and driving factors. L Degrad Dev. https://doi.org/10.1002/ldr.2944

Zirkle G, Lal R, Augustin B (2011) Modeling carbon sequestration in home lawns. HortScience 46:808–814

# Chapter 14
# Water Table and Dissolved Organic Carbon Seasonal Dynamic at the Different Ecosystems of the Ombrotrophic Bog (Mukhrino, West Siberia)

**Evgeny A. Zarov, Anastasiia V. Meshcheryakova, Viktoria S. Shanyova, and Vitaliy Yu Khoroshavin**

**Abstract** Peatland streams have repeatedly been shown to be responsible for the high amounts of produced in the soil ecosystems dissolved organic carbon (DOC) exported to the Arctic rivers. They contribute to the global carbon budget and remove carbon amounts equal to 50% of the net ecosystem exchange at the peatland. This study aimed to examine the patterns of DOC concentration dynamics for the different vegetation types and to estimate the total amount of DOC loss over the warm season for the typical ombrotrophic peatland in western Siberia. The greatest seasonal averaged DOC concentration (72.0 mg $l^{-1}$) was found at the driest ecosystem (ryam), while the smallest seasonal averaged concentrations (31.3 and 37.8 mg $l^{-1}$) were found at the wettest ecosystems (pool and stream). The highest seasonal DOC variation was found for the ridge (range 31–154 mg $l^{-1}$) and ryam (range 38–107 mg $l^{-1}$) sites; the DOC concentration at the pool site did not change significantly during the whole season (range 27–53 mg $l^{-1}$). A seasonal DOC concentration dynamic was discovered, increasing from the beginning of June until the end of August with a subsequent decrease to the initial values. Most of the DOC losses (42.5%) during the spring freshet with the subsequent decrease in the water discharge and DOC loss. The amount of 9.8 tons of carbon has been lost from the catchment area for the spring–autumn season 2021, i.e., ~ 11.14 g C $m^{-2}$ $year^{-1}$.

**Keywords** Dissolved organic carbon · Peatland · West Siberia · Seasonal dynamic · Carbon loss

E. A. Zarov (✉) · A. V. Meshcheryakova · V. S. Shanyova · V. Y. Khoroshavin
Yugra State University, Khanty-Mansiysk, Russia
e-mail: zarov.evgen@yandex.ru
URL: http://www.ugrasu.ru; http://www.utmn.ru

V. Y. Khoroshavin
Univerisity of Tyumen, Tyumen, Russia



M. Korneykova et al. (eds.), *Smart and Sustainable Urban Ecosystems: Challenges and Solutions*, Springer Geography, https://doi.org/10.1007/978-3-031-37216-2_14

## 14.1 Introduction

Numerous studies highlight the importance of the dissolved organic carbon (DOC) transfer between soil and freshwater systems (Shirokova et al. 2013). Arctic rivers featured the highest concentrations of organic matter (Dittmar and Kattner 2003) coming from the carbon-rich soils spread on the watershed area. In the peatland carbon cycle, the DOC is accounted for 30–50% of net ecosystem exchange fixed carbon (Roulet et al. 2007; Nilsson et al. 2008; Dinsmore et al. 2010). Increasing export of the DOC from the peatlands to the streams causes an increased flux of organic carbon to the oceans (Freeman et al. 2004; Frey and Smith 2005), where it turns to carbon dioxide and methane, which returns to the atmosphere that could contribute to significant positive feedbacks to climate warming (Field et al. 2007).

The West Siberian plain is the most waterlogged area in the world, where peatlands occupy ~ 600,000 $km^2$, i.e., ~ 25% of the whole area (Sheng et al. 2004; Smith et al. 2004). Thus, the catchment area of two main West Siberian rivers mostly covers the carbon-rich soils storing at least 70.2 Pg C (Sheng et al. 2004; Smith et al. 2004) where during decomposition under waterlogged conditions, part of the litter and peat converts to a dissolved form and is flushed out by surface runoff. It was estimated that the average stream DOC concentration of ~ 16 mg $l^{-1}$ in West Siberia will possibly rise up to ~ 21–24 mg $l^{-1}$ by 2100 (Frey and Smith 2005).

Numerous global drivers affecting the DOC concentration have been suggested, such as air temperature (Freeman et al. 2001; Clark et al. 2009), elevated carbon dioxide concentration in atmosphere (Freeman et al. 2004; Fenner et al. 2007a, b) and hydrological regime (Tranvik and Jansson 2002; Clark et al. 2009). Additionally, the number of local factors controlling the DOC export from the peatlands to the streams are presented: the catchment area size (Olefeldt et al. 2013; Pokrovsky et al. 2015), water table depth (Strack et al. 2015), air and soil temperature (Bonnett et al. 2006), and vegetation type (Armstrong et al. 2012).

Thus, the DOC export might be expressed as a function of DOC production and hydrological pathways by which water flows in the peatland (Waddington and Roulet 1997). In addition, the peatland vegetation reflects the integrated properties of the habitat, such as the geochemistry of soil water (Kuhry et al. 1993), the peat physical properties (McNamara et al. 2008; Dyukarev et al. 2021a, b), the micro-biota, the quality and quantity of plant and litter input (Wickland et al. 2007), amount of DOC (Armstrong et al. 2012). Summarizing all the above, the vegetation type might be taken into account for predicting changes in DOC loss. Finally, it might be used to estimate the DOC export using remote access data, such as a landuse thematic map.

In this paper, we aim to study the patterns of DOC concentration dynamics for the vegetation types of the typical ombrotrophic bog in West Siberia and to estimate the total amount of DOC loss over the warm season.

## 14.2 Materials and Methods

### *14.2.1 Mukhrino Location Map and Measurement Points*

Mukhrino field station is located in West Siberia belonging to the middle taiga zone (55–65 N latitude, the Russian Federation). The station occupies the left terrace of the Irtysh River, about 20 km to the west from Khanty-Mansiysk city and about 20 km to the south from the confluence point of the Irtysh and Ob' Rivers.

Most measurements are concentrated at the Mukhrino peatland, a vast pristine ombrotrophic (rain-fed) bog. The most dominant ecosystems are ridge-hollow complexes and ryams. The vegetation consists of a mix of sedges (*Carex limosa*, *Eriophorum vaginatum*) covering a *Sphagnum* mosses mat in hollows, which may become submerged after a snow-melting period. Ridges (30–40 cm height, 10–100 m length) are elongated across water flow, dominated by a mix of *Pinus sylvestris*, dwarf shrubs (*Chamaedaphne calyculata*, *Ledum palustre*, *Andromeda polifolia*) and *Sphagnum fuscum*. Ryams are dense treed by *Pinus sylvestris* together with the dwarf shrubs similar to the ridge vegetation diversity (Filippov and Lapshina 2008). The peatland was initiated ~ 12 kyr ago as a minerotrophic fen with brown mosses, sedges, and wood dominance. The average thickness of the peat deposits is ~ 3.3 m mostly composed of oligotrophic peat (dominant remains are *Sphagnum* mosses and shrubs) (Zarov et al. 2021).

The environmental monitoring system is organized at the station. The main measured properties are the air pressure, radiation balance, precipitation, water level, air and soil temperatures, greenhouse gas fluxes (methane and carbon dioxide), as well as geobotanical surveys (vegetation and fungus) and hydrochemistry (major ions and DOC). The environmental monitoring data are available since 2010 (Dyukarev et al. 2019, 2021a, b).

### *14.2.2 Precipitation (Rain, Snow)*

Precipitation was monitored using a tipping-bucket rain gauge (Hobo RG3-M, resolution 0.2 mm) located in the two sites—ryam and ridge-hollow complex. First, the amount of snow was measured at the end of the snow season (end of March–beginning of April). Then, the snow depth was measured using a measuring rod, and the amount of water stored in the snow was estimated using a metal tube (24 cm diameter, 1.2 m height) and a weighing scale. For this, the exact volume of the snow was cut off by the metal tube and weighted in the field. Totally, forty points was visited during the snow-measurement session.

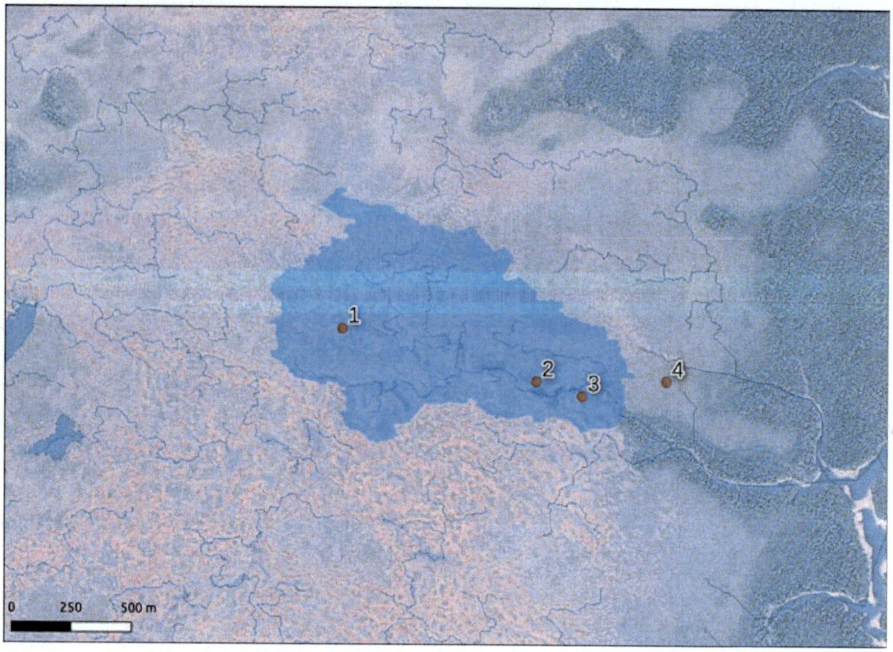


**Fig. 14.1** Stream catchment area (blue shade) and piezometer location (1—hollow, 2—ridge-hollow ecotone, 3—pool, 4—ryam)

### 14.2.3 *Water Levels and Air Temperature*

A pressure logger (Mini-Diver D1501, accuracy ± 5 mm of water, frequency measurement 30–180 min) was installed in groundwater observation tubes (5 cm diameter) with a filter at 100–250 cm below the peat surface. The groundwater wells were fixed in the mineral soil (at a depth of 350–550 cm below the peat surface, depending on the location) to have a fixed reference datum and avoid vertical changes of position due to peat volume expansion or compression. For barometric compensation, the air pressure and temperature were recorded with a pressure sensor (Baro-Diver, accuracy ± 5 mm of water, frequency measurement 30–180 min) placed in the center of the mire 2 m above the surface. Location of the loggers is shown in Fig. 14.1.

### 14.2.4 *Mapping (DEM and Landunit)*

A digital elevation model (DEM) was created based on the Global Multi-resolution Terrain Elevation Data 2010 (GMTED2010). The initial model was converted to the

ASCII-formatted file and processed via Python environment (version 3.9.7) integrated with PcRaster (Karssenberg et al. 2010). The flow direction was calculated using the function «*lddcreate*» on its base, and a catchment area was calculated using the function «*catchment*».

Six land units (hollow, ridge, ryam, pool, forest, and stream) (Filippov and Lapshina 2008) were delineated by visual classification of ESRI Satellite (2 m spatial resolution), Resource P (0.5 m spatial resolution) and field survey data. Classification was preformed within the QGIS software (v. 3.24).

### 14.2.5 DOC Analysis

Water was sampled once a month in three replications from the most typical ecosystems (hollow, ridge, ryam, pool and stream) located at a distance of 100–200 m from each other to cover a spatial variation. The water samples were collected from the plastic pipes (perforated from the surface to the depth 20 cm for the hollows and pools, to the depth 50 cm for the ridges and to the depth 80 cm for the ryams) installed at each point and directly from the stream mouth. The water were sampled into a glass bottle (made of black glass, 70 ml) at the upper water level (10–25 cm depending on the season). All bottles were transported to the lab (Yugra State University), stored in the fridge (no more than 1 day, temperature 4 °C).

In the laboratory, all water samples were filtered through the membrane filter (Whatman©, 0.45 $\mu$m pore-size, polyethersulfone) and immediately analyzed on the elemental analyzer Thermo Scientific (Flash© 2000, Germany) equipped with the autosampler for the liquid substrates (AS1310) to define DOC concentration. Calibration was done using urea standard ($C = 22.56\%$, $N = 7.84\%$, $H = 15.01\%$, $O = 18.59\%$).

### 14.2.6 Discharge and DOC Loss Calculation

Discharge from the study area was measured at the beginning of the stream, where the underground flow turns to the visible stream and forms a small waterfall, by collecting overflow in a bucket of exact volume (12 l) and measuring a filling time. Additionally, the APIK (atmosphere-soil measurement complex, produced by Institute of Monitoring of Climatic and Ecological Systems, Tomsk city) device was installed downstream to measure the water level and electroconductivity every 30 min. Finally, the water levels for the APIK were converted to the discharge values by matching the integrated function "time vs water level" to the annual water discharge volume (catchment area multiplied by the sum of snow and rain precipitation).

Loss of DOC was calculated as the daily water discharge (l $day^{-1}$) multiplied by the monthly average DOC concentration in stream water (kg $l^{-1}$). The cumulative DOC loss was calculated as a sum of daily DOC losses over the season (from April 15, 2021 to October 31, 2021) and is the estimated value.

## 14.3 Results

### *14.3.1 General Hydrology*

The amount precipitation for the study period (April–October) was 429.4 mm, whereas average annual precipitation is 470 ± 68 mm (¼ is snow) (Dyukarev et al. 2021a, b). Precipitation events were not evenly distributed during the study period: most rain falls in May–June and September. Short-term and light rainfalls characterized the mid-summertime.

Snow melting starting at the beginning of April caused a rapid and dramatic water level increase. During this period, water is kept on the surface creating ponds since the peatland is frozen with the lack of infiltration. When the surface thaws, all the water rapidly infiltrates into the peatland profile, increasing the water level. The wet period lasts up to 2.5 months (April–the first decade of June), and then the water level drops. The subsequent temporal variation in precipitation is reflected in the position of the water table, causing a short-term rise. Due to autumn rainfalls and a decrease in evapotranspiration, the water level again increases in September. Discharge stops completely at the end of October when the peatland freezes.

The water table fluctuation range was featured exactly to the studied ecosystems (Figs. 14.1 and 14.2). At the ridge-hollow ecotone located at the central part of the catchment area, the water level was sustained through the summer, varying at a depth of ~ 10 cm and rising above the surface only at the end of May because of snow-melting water. The highest fluctuation was found for the ryam ecosystem where water flow had been blocked by hummocks during the snow-melting period and stayed 30 cm above the surface. The water level was continuously decreasing during the summer and autumn to the lowest value of ~ 90 cm below the surface in September. Pools were covered by water until the first decade of July; further, the water level dropped down the surface and stayed 5–7 cm below until the mid of September when the active rains occurred. The hollow site, located at the far edge of the catchment area, showed around-surface water table position until the mid of June, then the water level dropped until the end of September, with the short-term rising as a response to the heavy rain event.

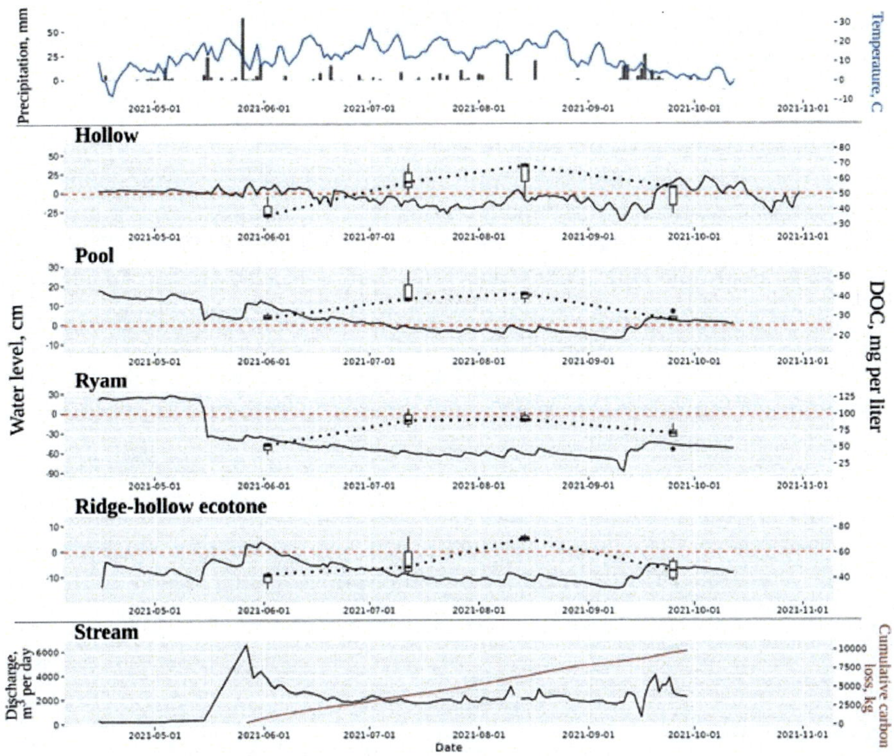


**Fig. 14.2** Seasonal DOC concentration dynamics for the different ecosystems (boxes with whiskers), air temperature, precipitation amount (bars), stream water discharge, and cumulative DOC loss

### *14.3.2 The Seasonal DOC Dynamic*

Concentrations of DOC are generally increasing in the summer because of higher evapotranspiration rates and plant tissue decomposition (Moore 1987). However, pore-water dilution occurs during the snow-melting period (April–May) and autumn rainfall events, which cause the DOC concentration to decline.

DOC concentrations in the peatland ranged from 23.5 to 153.5 mg $l^{-1}$ for the whole season (Fig. 14.2), with the lowest seasonal average concentrations for the pool (31.3 mg $l^{-1}$) and stream (37.8 mg $l^{-1}$) sites, and the highest seasonal average concentration for the ryam (72 mg $l^{-1}$) site. During the lowest water level period in mid-August, DOC concentrations for all sites were 10–60 mg $l^{-1}$ higher than in the highest water level period (end of May–beginning of June). The highest seasonal fluctuation was found for the ridge (range 31–154 mg $l^{-1}$) and ryam (range 38–107 mg $l^{-1}$) sites. At the same time, the DOC concentration at the pool site did not change significantly during the whole season (range 27–53 mg $l^{-1}$).

### 14.3.3 Catchment Area, Water Flowing Direction, Discharge, and Carbon Loss

The bog flow drains into the stream, which is a 0.88 $km^2$ catchment (Fig. 14.1), with a span of 1.5 km from the west to the east and 0.4–1 km from the north to the south, characterized by ridge-hollow complex (30.5 and 49.3% of the catchment area, correspondingly), ryam (12.9% of the catchment area), and pools (7.3% of the catchment area). The eight main water flows form a beaded stream; an additional flow contributes to the main stream ~ 300 m to the river mouth. The water flow is directed south-east to the stream that is a confluence to the Mukhrina River.

The DOC loss is directly dependent on the water discharge rather than the DOC concentration. Even the smallest DOC concentration (28.95 mg $l^{-1}$) during the spring freshet causes high DOC loss because of the highest discharge, i.e., a massive flush of the DOC. Since end of May (May 27) till the end of June (June 28), while the snow melt water discharging, the amount of ~ 45% of DOC is flushed and 32.5% of seasonal discharge (94,978.7 $m^3$) occured. Totally, an amount of 9.8 tons of carbon has been lost from the catchment area for the spring–autumn season 2021, i.e., ~ 11.14 g C $m^{-2}$ $year^{-1}$.

## 14.4 Discussion

In this paper, we studied a seasonal dynamic of the DOC concentration for the different ombrotrophic peatland ecosystems and estimated a total seasonal loss of DOC with the surface runoff. The highest concentrations of DOC are associated with the driest ecosystems with the high tree coverage—ryam and ridge (the average seasonal water table is 41 and 30 cm below the surface, correspondingly) (Dyukarev et al. 2021a, b), where after spring freshet the thick aerobic zone (acrotelm) creates favorable environmental conditions and promotes the active organic matter decomposition and increased amount of DOC. There is evidence that peatland tree cover has been dynamic with greater cover during warmer periods in the Holocene (Velichko et al. 1998; Ratcliffe et al. 2017). Thus, expecting global warming may increase tree coverage at the omrotrophic bogs and, together with the same or increased amount of precipitation, trigger higher DOC export to the streams.

At the same time, there was not found any significant increase of DOC concentration during the whole season in the ecosystems collecting water from all catchments—pool and stream. Despite the similar proportion of the hollows, ridges, and ryams in the catchment area, the DOC concentration in the pools and stream did not exceed 55 mg $l^{-1}$ while the ridge and ryam showed concentrations over 100 mg $l^{-1}$. It looks like the DOC loss to a large extent is controlled by the water table rather than the amount of produced DOC—in the time of high water (beginning of spring and autumn), the DOC concentrations are low, causing the low DOC export; but in the time of low water level, warm temperatures, and corresponding high DOC concentration, the role of ryam and ridge ecosystems decreased. When the water table drops to the lowest level, active water flow stops, and water hold in the peat pores, these ecosystems do no contribute much to the increased amounts of DOC in the pools and stream.

The short-term blast DOC flush from the ridge and ryam ecosystems might be expected when the heavy rain event occurs after the protracted period without precipitation (Tiwari et al. 2022), but in this study, it has not been discovered due to the monthly sampling period. Therefore, to increase the precision of DOC monitoring, the high-frequency hydrochemical survey must be organized in the study area. It was shown that this monitoring type improves annual flux calculations, reveals seasonal relationships between discharge and DOC concentrations, and improves discharge measurements during low flow periods (Gandois et al. 2021).

The calculated annual specific DOC flux for the same area was 11.14 g C $m^{-2}$ $year^{-1}$ and comparable with our previous estimations of 7.7 g C $m^{-2}$ $year^{-1}$ (Bleuten et al. 2020). The difference in the results is caused by using the DOC concentration data for different years (2011–2016 in the last estimates vs 2021 in this paper) and the modeled discharge data in the (Bleuten et al. 2020) and the in-situ water discharge monitoring data in the current research. The DOC flux for the Mukhrino peatland is higher than this value for the Central Siberian Plateau (2.8–4.7 g C $m^{-2}$ $year^{-1}$) (Prokushkin et al. 2011). The differences are likely related to the peat chemical composition, but this analysis has not been carried out for our site.

The carbon accumulation rate for the Mukhrino peatland ranged from 5.66 to 189.52 g $m^2$ $year^{-1}$ for the last 9500 years (Tsyganov et al. 2021). Thus, the annual specific DOC flux is equal to 6–196% of the carbon accumulation rate, i.e., sometimes it exceeds the amount of carbon sequestrated in the peat and acts as a carbon source. The role of DOC loss with the surface runoff increases in case of warming and droughts where the low carbon accumulation (12 g $m^2$ $year^{-1}$) (Tsyganov et al. 2021) and the high DOC concentrations (till 153.5 mg $l^{-1}$) occur.

## 14.5 Conclusion

In this paper, we summarized the first-year results of systematic monitoring for the DOC concentrations and stream discharge at the Mukhrino peatland. We discovered the DOC seasonal variation for the different ecosystems at the ombrotrophic peatland, estimated the stream catchment area based on a digital elevation model and calculated the total amount of DOC loss with the surface runoff. The results discovered the seasonal pattern of DOC behavior: a massive DOC loss with the spring freshet, and further evenly distributed losing till October; the DOC concentrations at the different ecosystems reach their maximum in August with the highest values for the dry ecosystems and the lowest values for the wet ecosystems.

The future directions of research are: (1) the high-frequency DOC and discharge monitoring; (2) to elucidate control on temporal variability in DOC considering drought and storm flow events; (3) studying DOC concentration vertical distribution; (4) studying the organic isotopic compounds of the DOC to estimate the role of different ecosystems to the annual DOC loss.

**Acknowledgements** The work was performed within the framework of the state task of Yugra State university (1022090600010-3-1.5.1) according to the order of the Government of the Russian Federation No. 2515-r dated September 2, 2022, in order to implement the most important an innovative project of national importance aimed at creating a unified national system for monitoring climate-active substances. We appreciate Sergey Kopysov and Aleksandr Erofeev for their help with stream discharge monitoring. Special thanks to Wladimir Bleuten and Elena Lapshina for their ideas which laid as a basis of this paper.

## References

Armstrong A, Holden J, Luxton K, Quinton JN (2012) Multi-scale relationship between peatland vegetation type and dissolved organic carbon concentration. Ecol Eng 47:182–188. https://doi.org/10.1016/j.ecoleng.2012.06.027

Bleuten W, Zarov E, & Schmitz O (2020) A high-resolution transient 3-dimensional hydrological model of an extensive undisturbed bog complex in West Siberia. Mires Peat 26:1–25. https://doi.org/10.19189/MaP.2019.OMB.StA.1769

Bonnett SAF, Ostle N, Freeman C (2006) Seasonal variations in decomposition processes in a valley-bottom riparian peat-land. Sci Total Environ 370:561–573. https://doi.org/10.1016/j.scitotenv.2006.08.032

Clark JM, Ashley D, Wagner M., Chapman PJ, Lane SN, Evans CD, Heath-waite AL (2009) Increased temperature sensitivity of net DOC production from ombrotrophic peat due to water table draw-down. Global Change Biol 15:794–807. https://doi.org/10.1111/j.1365-2486.2008.01683.x

Dinsmore KJ, Billett MF, Skiba UM, Rees RM, Drewer J, Helfter C (2010) Role of the aquatic pathway in the carbon and greenhouse gas budgets of a peatland catchment. Global Change Biol 16:2750–2762. https://doi.org/10.1111/j.1365-2486.2009.02119.x

Dittmar T, Kattner G (2003) The biogeochemistry of the river and shelf ecosystem of the Arctic Ocean: a review. Mar Chem 83(3–4):103–120. https://doi.org/10.1016/S0304-4203(03)00105-1

Dyukarev EA, Godovnikov EA, Karpov DV, Kurakov SA, Lapshina ED, Filippov IV, Filippova NV, Zarov EA (2019) Net ecosystem exchange, gross primary production and ecosystem respiration in ridge-hollow complex at Mukhrino bog. Geography, Environment, Sustainability 12(2):227–244. https://doi.org/10.24057/2071-9388-2018-77

Dyukarev E, Filippova N, Karpov D, Shnyrev N, Zarov E, Filippov I, Filippov I, Voropay N, Avilov V, Artamonov A, Lapshina E (2021a) Hydrometeorological dataset of West Siberian boreal peatland: a 10-year record from the Mukhrino field station. Earth Syst Sci Data 13(6):2595–2605. https://doi.org/10.5194/essd-13-2595-2021

Dyukarev E, Zarov E, Alekseychik P, Nijp J, Filippova N, Mammarella I, Filippov I, Bleuten W, Khoroshavin V, Ganasevich G et al (2021b) The multiscale monitoring of peatland ecosystem carbon cycling in the middle Taiga zone of western Siberia : the Mukhrino Bog case study. Land 10:824. https://doi.org/10.3390/land10080824

Fenner N, Ostle N, McNamara N, Sparks T, Harmens H, Reynolds B, Freeman C (2007b) Elevated $CO_2$ effects on peatland plant community carbon dynamics and DOC production. Ecosystems 10:635–647. https://doi.org/10.1021/acs.analchem.5b04668

Fenner N, Freeman C, Lock MA, Harmens H, Reynolds B, Sparks T (2007a) Interactions between elevated $CO_2$ and warming could amplify DOC exports from peatland catchments. Environ Sci Technol 41:3146–3152

Field CB, Lobell DB, Peters HA, Chiariello NR (2007) Feedbacks of terrestrial ecosystems to climate change. Annu Rev Environ Resour 32:1–29. https://doi.org/10.1146/annurev.energy.32.053006.141119

Filippov IV, Lapshina ED (2008) Peatland unit types of lake-bog systems in the Middle Priob'ie (Western Siberia). In: Transactions of UNESCO Department of Yugorsky State University "dynamics of environment and global climate change", pp 115–124

Freeman C, Evans CD, Monteith DT, Reynolds B, Fenner N (2001) Export of organic carbon from peat soils. Nature 412:785. https://doi.org/10.1038/35090628

Freeman C, Fenner N, Ostle NJ, Kang H, Dowrick DJ, Reynolds B, Lock MA, Sleep D, Hughes S, Hudson J (2004) Export of dissolved organic carbon from peatlands under elevated carbon dioxide levels. Nature 430:195–198. https://doi.org/10.1038/nature02707

Frey KE, Smith LC (2005) Amplified carbon release from vast West Siberian peatlands by 2100. Geophys Res Lett 32(9):1–4. https://doi.org/10.1029/2004GL022025

Gandois L, Tananaev NI, Prokushkin A, Solnyshkin I, Teisserenc R (2021) Seasonality of DOC export from a Russian subarctic catchment underlain by discontinuous permafrost, highlighted by high-frequency monitoring. J Geophys Res Biogeosci 126:e2020JG006152. https://doi.org/10.1029/2020JG006152

Karssenberg D, Schmitz O, Salamon P, de Jong K, Bierkens MF (2010) A software framework for construction of process-based stochastic spatio-temporal models and data assimilation. Environ Model Softw 25(4):489–502. https://doi.org/10.1016/j.envsoft.2009.10.004

Kuhry P, Nicholson BJ, Gignac LD, Vitt DH, Bayley SE (1993) Development of Sphagnum-dominated peatlands in boreal continental Canada. Can J Bot 71:10–22. https://doi.org/10.1139/b93-002

McNamara NP, Plant T, Oakley S, Ward S, Wood C, Ostle N (2008) Gully hotspot contribution to landscape methane ($CH_4$) and carbon dioxide ($CO_2$) fluxes in a northern peatland. Sci Total Environ 404:354–360. https://doi.org/10.1016/j.scitotenv.2008.03.015

Moore TR (1987) Patterns of dissolved organic matter in subarctic peatlands. Earth surface processes and landforms 12(4):387–397. https://doi.org/10.1002/esp.3290120405

Nilsson M, Sagerfors J, Buffam I, Laudon H, Eriksson T, Grelle A, Klemedts-son L, Weslien P, Lindroth A (2008) Contemporary carbon accumulation in a boreal oligotrophic minerogenic mire; a significant sink after accounting for all C-fluxes. Global Change Biol 14:2317–2332. https://doi.org/10.1111/j.1365-2486.2008.01654.x

Olefeldt D, Roulet N, Giesler R, Persson A (2013) Total waterborne export and DOC composition from ten nested subarctic peatland catchments—importance of peatland cover, groundwater influence, and inter-annual variability of precipitation patterns. Hydrol Process 27(16):2280–2294. https://doi.org/10.1002/hyp.9358

Pokrovsky OS, Manasypov RM, Loiko S, Shirokova LS, Krickov IA, Pokrovsky BG, Kolesnichenko LG, Kopysov SG, Zamtsov VA, Kulizhsky SP, Vorobyev SN, Kirpotin SN (2015) Permafrost coverage, watershed area and season control of dissolved carbon and major elements in western Siberian rivers. Biogeosciences 12(21):6301–6320. https://doi.org/10.5194/bg-12-6301-2015

Prokushkin AS, Pokrovsky OS, Shirikova LS, Kortes MA, Viers J, Prokushkin SG, Amon RMW, Guggenberger G, McDowell WH (2011) Sources and the flux pattern of dissolved carbon in rivers of the Yenisey basin draining the Central Sbierian Plateau. Environ Res Lett 6(4):045212. https://doi.org/10.1088/1748-9326/6/4/045212

Ratcliffe JL, Creevy A, Andersen R, Zarov E, Gaffney PP, Taggart MA, Payne RJ (2017) Ecological and environmental transition across the forested-to-open bog ecotone in a west Siberian peatland. Sci Total Environ 607:816–828. https://doi.org/10.1016/j.scitotenv.2017.06.276

Roulet NT, Lafleur PM, Richard PJH, Moore TR, Humphreys ER, Bubier J (2007) Contemporary carbon balance and late Holocene carbon accumulation in a northern peatland. Global Change Biol 13:397–411. https://doi.org/10.1111/j.1365-2486.2006.01292.x

Sheng Y, Smith LC, MacDonald GM, Kremenetski KV, Frey KE, Velichko AA, Lee M, Beilman D, Dubinin P (2004) A high resolution GIS based inventory of the west Siberian peat carbon pool. Global Biogeochem Cycles 18(3):1–14. https://doi.org/10.1029/2003GB002190

Shirokova LS, Pokrovsky OS, Kirpotin SN (2013) Biogeochemistry of organic carbon, CO2, CH4, and trace elements in thermokarst water bodies in discontinuous permafrost zones of Western Siberia. Biogeochemistry 113:573–593. https://doi.org/10.1007/s10533-012-9790-4

Smith LC, MacDonald GM, Velichko AA, Beilman DW, Borisova OK, Frey KE, Kremenetski KV, Sheng Y (2004) Siberian peatlands a net carbon sink and global methane source since the early Holocene. Science 303(5656):353–356. https://doi.org/10.1126/science.1090553

Strack M, Zuback Y, McCarter C, Price J (2015) Changes in dissolved organic carbon quality in soils and discharge 10 years after peatland restoration. J Hydrol 527:345–354. https://doi.org/10.1016/j.jhydrol.2015.04.061

Tiwari T, Sponseller RA, Laudon H (2022) The emerging role of drought as a regulator of dissolved organic carbon in boreal landscapes. Nat Commun 13(1):1–11. https://doi.org/10.1038/s41467-022-32839-3

Tranvik LJ, Jansson M (2002) Terrestrial export of organic carbon. Nature 415:861–862. https://doi.org/10.1038/415861b

Tsyganov AN, Zarov EA, Mazei YuA, Kulkov MG, Babeshko KV, Yushkovets SY, Payne RJ, Ratcliffe JL, Fatyunina YA, Zazovskaya EP, Lapshina ED (2021) Key periods of peatland development and environmental changes in the middle taiga zone of Western Siberia during the Holocene. Ambio 50(11):1896–1909. https://doi.org/10.1007/s13280-021-01545-7

Velichko AA, Kremenetski KV, Borisova OK, Zelikson EM, Nechaev VP, Faure H (1998) Estimates of methane emission during the last 125,000 years in Northern Eurasia. Global Planet Change 16–17:159–180. https://doi.org/10.1016/S0921-8181(98)00032-0

Waddington JM, Roulet NT (1997) Groundwater flow and dissolved carbon movement in a boreal peatland. J Hydrol 191(1–4):122–138. https://doi.org/10.1016/S0022-1694(96)03075-2

Wickland KP, Neff JC, Aiken GR (2007) Dissolved organic carbon in Alaskan boreal forest: sources, chemical characteristics, and biodegradability. Ecosystems 10:1323–1340. https://doi.org/10.1007/s10021-007-9101-4

Zarov EA, Lapshina, ED, Kuhlmann I, Schulze E-D (2021) Dissolved organic carbon vertical movement and carbon accumulation in West Siberian peatlands. Biogeosci Discuss [preprint]. https://doi.org/10.5194/bg-2021-211

# Chapter 15
# Specifics of Accumulation and Profile Distribution of Organic Carbon in Soils of Park and Recreational Areas of Rostov Agglomeration

**P. N. Skripnikov, S. N. Gorbov, A. Y. Matetskaya, and V. A. Ivolgina**

**Abstract** Artificial tree plantations of different ages in the city limits of Rostov-on-Don, Aksai and Bataysk were studied. In total, 22 monitoring sites were studied on the territory of park and recreational zones of the Rostov agglomeration. Geobotanical and soil research was carried out for each monitoring site. The study shows a statistically significant increase in the organic carbon content in the 0–10 cm layer of calcic chernozems of city recreational zones. There was on average of $4.23 \pm 0.88\%$ regarding to the chernozems of reserved and native areas of the Rostov agglomeration, where the content of C org. was $2.86 \pm 0.63\%$. The comparison of the C org. content in the surface 10-cm layer for studied forest parks was concluded that, regardless of the location and floral composition features, the difference in the organic matter content is small: C org. $= 4.23 \pm 0.88\%$. It was possible to identify a pattern associated with the planted forests age and the accumulated organic carbon amount in the surface horizon. In forest phytocenoses aged 70–90 years, the content of C org. in the surface layer was significantly higher ($4.40 \pm 0.78\%$) than in younger ones ($3.42 \pm 0.76\%$), whose age was 30–50 years. The profile distribution of organic carbon under woody vegetation has specific features. Some monitoring sites demonstrate a sharp decrease in the content of C org. (by 2–2.5 times) during the transition from the A horizon to the underlying horizon. The rest are characterized by a smoother distribution of organic carbon throughout the humus-accumulative strata, which is associated with a larger total biomass of herbaceous plants forming under the canopy of trees at these monitoring sites.

**Keywords** Calcic chernozems · Urban forest parks · Organic matter · Humus · Natural soils

P. N. Skripnikov (✉) · S. N. Gorbov · A. Y. Matetskaya · V. A. Ivolgina
Academy of Biology and Biotechnology of D.I. Ivanovsky, Southern Federal University, Rostov-on-Don, Russian Federation
e-mail: skripnikov@sfedu.ru


M. Korneykova et al. (eds.), *Smart and Sustainable Urban Ecosystems: Challenges and Solutions*, Springer Geography, https://doi.org/10.1007/978-3-031-37216-2_15

## 15.1 Introduction

The woody vegetation planting in the steppe zone is one of the manifestations of anthropogenic interference in the functioning of natural ecosystems, correcting the steppe biocenoses circulation (Canedoli et al. 2020; Huot et al. 2017; Morel et al. 2015; Yang et al. 2014). Therefore, the research of soil cover of urban forest parks is an important object for of long-term anthropogenic impact studying on steppe ecosystems (Velts and Lupova 2013). Overall, the landscaping development in cities is one of the factors that ensure the soil organic matter accumulation (Kimpe and Morel 2000). This tendency is especially observed in native treeless areas and cities developing in unfavorable conditions, for example, in the arid zone (Vodyanitskii 2015). Woody phytocenoses affect the temperature and hydrological soil regimes, the atmospheric precipitation composition, the number and activity of various organic matter destructors, form a sub-morphological flora which is different from the steppe and provide additional dead vegetation to the soil (Demakov et al. 2017; Gorbov and Bezuglova 2014; Vysotskiy 1962). It leads to organic matter accumulation changes and also increases the intensity of carbon inorganic forms migration and part of the water-soluble organic matter in the soil profile. However, the impact of the same plant species on the soil manifests itself differently in different biogeocenoses and largely depends on the physico-geographical conditions (Karpachevskiy 1981). The purpose of this work is to study the influence of the change of plant formations on the features of the accumulation and profile distribution of organic carbon in the conditions of a large agglomeration of the steppe zone.

## 15.2 Objects and Methods

The objects of the study were the natural soils of the parks and recreational zones of the Rostov agglomeration, applying to one subtype-ordinary carbonate chernozems (calcic chernozems), experiencing minimal influence of the anthropogenic factor (the soil cover has been preserved since the period of agrogenic use, the absence of modern human economic activity). The internal spatial structure of recreational zones at the monitoring sites (with an area of 100 $m^2$) allocated considering the dominant tree species, and their age was studied; visual assessment of the condition of tree and shrub plantations was carried out, also floral and geobotanical descriptions were carried out. Monitoring sites were laid on watershed areas to minimize the impact of the relief factor on soil properties. In the selected recreational areas, 16 full soil profiles were made, and surface soil samples were additionally taken from a ten-centimeter layer in ten-fold repetition for each monitoring site. As a control site, the sites on the territories of the exposition of the Botanic Garden of the Southern Federal University "Priazovskaya Steppe" (sites 3.4–3.5) were selected. It represents a 60-year-old old reservoir that was artificially subjected to the repatriation of steppe and red book species, as well as a virgin plot on the territory of a natural monument

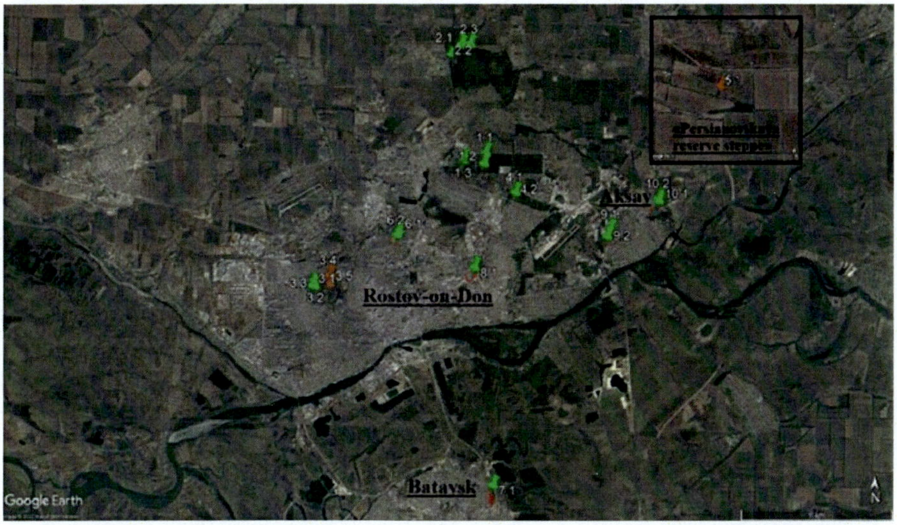


**Fig. 15.1** Map of the location of monitoring sites

with a reserved regime "Persianovskaya reserved steppe" (site 5.1) 10 km from the city of Novocherkassk (Yu and Myasnikova 2014). In total, 22 monitoring sites were investigated (Fig. 15.1 and Table 15.1).

Visual assessment of the state of tree and shrub plantations was carried out according to the method of Alekseyev (1989), which allows interpreting the habitus of the plant as an indicator of environmental conditions. With an indicator from 100 to 80%, the vital state of the tree stand is considered as healthy, at 79–50% is weakened, at 49–20% is severely weakened and at 19% and below is completely destroyed.

Geobotanical descriptions were carried out for each park to assess species biodiversity. The projective coverage of a single geobotanical description site was determined visually and expressed as a percentage. The taxation (forestry) indicators of the tree tier were determined by eye, and the diameter and height of each tree under study were determined using a tape measure. The abundance of herbaceous plant species is indicated on the Brown-Blank scale (Braun-Blanquet 1951):

The biomorphological structure of the flora was determined by the system of life forms of K. Raunkier and the ecological and morphological classification of biomorphs of I. G. Serebryakov. In the distribution of species by ecological-cenotic groups (ECG) and geographical element, the indications of the "Flora of the Lower Don" (Zozulina and Fedyayevoy 1984), the "Summary of the flora of the Exposition of the Botanic Garden of the SFU "Priazovskaya Steppe" (Shmarayeva et al. 2021) and the Internet version of the database "Flora of vascular Plants of Central Russia" were considered". The criterion for the distribution of the species of the studied flora by ecological groups is the ratio to the degree of moisture and salinity of soils. The nomenclature of vascular plant species is given according to the regional determinant of flora (Zozulina and Fedyayevoy 1984, 1985). The species diversity

**Table 15.1** Characteristic of studied monitoring sites

| Monitoring sites | Number of site | Coordinates | Characteristic |
|---|---|---|---|
| Temernitskaya grove | 1.1 | N 47.294859° E 39.762646° | Large artificial forest plantation on the north-eastern outskirts of the city |
| | 1.2 | N 47.289895° E 39.761479° | |
| | 1.3 | N 47.291334° E 39.747751° | |
| Shchepkinsky forest | 2.1 | N 47.343712° E 39.744084° | Large artificial forest plantation on the north-eastern outskirts of the city |
| | 2.2 | N 47.340411° E 39.737190° | |
| | 2.3 | N 47.344121° E 39.750227° | |
| Botanic Garden of the Southern Federal University | 3.1 | N 47.232851° E 39.645382° | Coniferous nurseries |
| | 3.2 | N 47.233300° E 39.648200° | |
| | 3.3 | N 47.236400° E 39.645800° | Forest belt of oak near arable land |
| | 3.4 | N 47.237514° E 39.656652° | Control fallow plot |
| | 3.5 | N 47.234234° E 39.657186° | Control fallow plot |
| Ordzhonikidze village | 4.1 | N 47.277120° E 39.783838° | Large artificial forest plantation on the north-eastern outskirts of the city |
| | 4.2 | N 47.277600° E 39.784600° | |
| «Persianovskaya reserve steppe» | 5.1 | N 47.504655° E 40.152729° | Control fallow plot |
| K. Chukovsky Park | 6.1 | N 47.258397° E 39.704121° | Old city park |
| | 6.2 | N 47.256614° E 39.702811° | |
| Forest plantation of Bataysk | 7.1 | N 47.256614° E 39.702811° | Forest belt near the cemetery |
| N. Ostrovsky Park | 8.1 | N 47.242500° E 39.756389° | Old city park |
| Forest plantations of Aksai | 9.1 | N 47.261178° E 39.850075° | Independently arisen forest area |
| | 9.2 | N 47.257851° E 39.848214° | |
| | 10.1 | N 47.274444° E 39.883333° | Forest belt on the south-western border of Aksay |

(continued)

**Table 15.1** (continued)

| Monitoring sites | Number of site | Coordinates | Characteristic |
|---|---|---|---|
| | 10.2 | N 47.275556°<br>E 39.883056° | |

of plant communities and their complexes were evaluated according to the indicators proposed in the works of Whittaker (1972), which have become classic in modern ecology: α-diversity – (internal diversity of habitat for a description representing a homogeneous community), β-diversity (diversity between different communities along the gradient of the environment).

The organic carbon content was determined by two methods:

(1) a method for determining humus in soil according to Tyurin in the Nikitin modification with a colorimetric ending according to Orlov-Grindel (Mineyev 2001). The method involves the oxidation of humus with an excessive amount of chromium mixture ($K_2Cr_2O_7 + H_2O + H_2SO_4$) in a drying oven heated to 150 °C, followed by considering the trivalent chromium formed during the reaction. Accounting is carried out at a wavelength of 590 nm, providing maximum absorption of $Cr^{3+}$ ions, in 10 mm cuvettes.
(2) high-temperature catalytic combustion on the carbon analyzer TOC-L CPN Shimadzu in the console for dry samples SSM-5000A. This method is based on high-temperature catalytic combustion of the sample and subsequent detection of the released carbon dioxide. The analysis of the sample realizes in two stages: Total carbon is determined by burning the sample at a temperature of 900 °C, inorganic—at 200 °C with the addition of orthophosphoric acid. Organic carbon is determined by subtracting from total—inorganic (Tagiverdiev et al. 2020). The advantage of this method is that the "side" result of the analysis is the determination of the amount of inorganic carbon in a dry sample.

The calculation of soil organic carbon reserves is carried out based on data obtained during laboratory analyses according to the following formula: reserve of carbon of C org. t/ha = $C * H * d$, where $C$ is the content of organic matter, %; $d$ is the thickness of the horizon (layer), cm; $H$ is the density of the soil, g/cm$^3$.

Statistical data processing was performed using the software Statistica for Windows 10.0, MS Excel.

## 15.3 Results and Discussions

In the study of vegetation, 94 species of woody and herbaceous plants from 39 families belonging to 78 genera and three classes were identified: coniferous Pinopsida, dicotyledonous Magnoliopsida and monocotyledonous Liliopsida.

Shchepkinsky Forest is characterized by a very uneven distribution of herbaceous vegetation, depending on which species of trees and shrubs were planted. In total, 17 species from 15 families of flowering plants were noted here. *Glechoma hederacea* L. is abundant in the herbaceous tier and *Geum urbanum* L. (Table 15.2).

The highest species diversity of plants was noted in the Temernitskaya grove: 50 species of flowering plants from 28 families. The tree and shrub layers are dominated by representatives of the Fabaceae and Sapindaceae families (*Robinia pseudoacacia* L., *G. triacanthos*, *Acer negundo* L., *A. tataricum* L., *A. fruticosa*) (Table 15.3). In the grove, the plots represent a stage of a succession series, transitional to wheatgrass (background species—*Elytrigia repens* Desv.). The herbaceous tier is dominated by representatives of weedy-shrubby, weedy-meadow and shrub-forest vegetation.

Exclusively *Picea abies* (L.) H. Karst and *Pinus pallasiana* D. Don. grow in the nursery of the Botanic Garden of the Southern Federal University. The floral composition of the remaining tiers is almost identical and includes 28 species of herbaceous plants, shrubs and lianas from 19 families. The herbage is dominated by a representative of shrub-forest vegetation *Viola odorata* L. In addition, representatives of weed

**Table 15.2** Dominant species of the tree tier of the studied monitoring sites

| Monitoring sites | Species | Abundance | Projective cover, % |
|---|---|---|---|
| «Shchepkinsky forest» | *Glechoma hederacea* | 3 | 30 |
| Botanic Garden SFU | *Viola odorata* | 5 | 75 |
| «Temernitskaya grove» | *Elytrigia repens* | 5 | 90 |
| K.Chukovsky Park | *Poa bulbosa, Stellaria media* | 5 | 85 |
| N. Ostrovsky Park | *Alliaria petiolata* | 5 | 85 |
| Ordzhonikidze park | *Poa bulbosa* | 4 | 55 |
| Forest plantations of Aksai | *Chaerophyllum temulum* | 5 | 90 |

**Table 15.3** Dominant species of the grassy tier of the studied monitoring sites

| Monitoring sites | Species | Average diameter (m) | Average height (m) |
|---|---|---|---|
| « Shchepkinsky forest» | *Quercus robur* | 0.18 | 12.1 |
| Botanic Garden SFU | *Picea abies, Pinus pallasiana* | 0.17 | 3.6 |
| « Temernitskaya grove» | *Robinia pseudoacacia* | 0.17 | 8.1 |
| K.Chukovsky Park | *Fraxinus excelsior* | 0.15 | 6.6 |
| N. Ostrovsky Park | *Morus alba* | 0.09 | 4.2 |
| Ordzhonikidze park | *Robinia pseudoacacia* | 0.25 | 10.6 |
| Forest plantations of Aksai | *Robinia pseudoacacia* | 0.19 | 7.6 |

communities are noted in the herbage: *Chelidonium majus* L., *Physalis alkekengi* L. (abundance from 1 to 2 b.). Weed-psammophilic, weed, synanthropic, sun-meadow, meadow-steppe, weed-fringe-meadow and characteristic of the communities that have developed in this area were found: *Bromus tectorum* L., *Fumaria schleicheri* Soy. -Will, *Cirsium setosum* (Willd.) Besser, *C. canum* (L.) All., *Clematis recta* L., *Cynoglossum officinale* L. and *Securigera varia* (L.) Lassen. On another site mainly plantings of such woody plants as *Quercus robur* L., *A. campestre* L. and *A. pseudoplatanus* L. are presented.

Monitoring sites in the parks of Chukovsky and Ordzhonikidze are floristically like each other. The herbaceous cover is represented by dominant species such as *Poa bulbosa* L., *Stellaria media* (L.) Vill. and *Taraxacum officinale* Webb. Mixed woody vegetation dominates on the sites of the N. Ostrovsky Park and forest plantations in Aksai. In all the above-mentioned artificial plantings, such species as *Fraxinus americana* L., *F. excelsior* L., *A. pseudoplatanus, A. tataricum, A. negundo, Sorbus hybrida* L., *Morus alba* L., *G. triacanthos, R. pseudoacacia, Prunus domestica* L., *Syringa vulgaris* L., *Ulmus pumila* L., *Sambucus nigra* L., *Juglans regia* L., *C. coggygria, Cornus sanguinea* L. and *Euonymus europaeus* L. were planted. In the park named after N. Ostrovsky, seedlings of *Celtis occidentalis* L. (abundance of 4 b.) and *Alliaria petiolata* (Bieb.) Cavara et Grande (abundance of 5 b., in bulk) are represented in the mass.

As a control site, a site was selected on the territory of the exposition of the Botanic Garden of the Southern Federal University "Priazovskaya Steppe", on which steppe vegetation grows naturally for the zone with the dominance of species of the Poaceae and Asteraceae families. Studies were also carried out on the territory of the protected area "Persianovskaya Steppe", which is a virgin area where such families as Asteraceae, Poaceae and Fabaceae occupy a leading place in the flora.

The results of the assessment of the vital condition of trees showed that the highest indicator of the vitality of trees was found at the sites of the Shchepkin Forest—95.27%. The lowest indicator of tree vitality was recorded in Ordzhonikidze Park. Here, tree stands grow near the highway, which can significantly affect their condition (Table 15.4).

The soils of forest parks, compared with soils under steppe vegetation, have increase the humus-accumulative stratum thickness. These soils were characterized by pronounced agronomically valuable structure formation of the soddy horizon and decrease the effervescence depth in the B and BC illuvial horizons. Chemical analysis of soil samples of the 0–10 cm layer showed the following results. There is a statistically significant (U-criterion, $p < 0.05$) increase in the content of organic carbon in soils under woody vegetation, and on average it is $4.23 \pm 0.88\%$ ($N = 200$) relative to the chernozems of reservoir and virgin areas of the Rostov agglomeration, where the content of C org. is $2.86 \pm 0.63\%$ ($N = 25$) (Fig. 15.2).

Comparing the monitoring sites for the content of C org. in the surface layer, it can be concluded, that, regardless of the spatial position and features of the floral composition, the content of organic carbon shows similar values and does not show statistically significant differences among themselves. The highest median values in organic

**Table 15.4** Vital condition of trees

| Monitoring sites | Number of trees, % | | | | Ln, %* | The vital state of the plantings |
|---|---|---|---|---|---|---|
| | Healthy | Weak | Severely weak | Completely destroyed | | |
| « Shchepkinsky forest» | 89.5 | 5.2 | 5.3 | 0 | 95.27 | Healthy |
| Botanic Garden SFU | 31.6 | 7.7 | 26.4 | 34.3 | 49.35 | Severely weak |
| « Temernitskaya grove» | 30 | 50 | 20 | 0 | 80 | Healthy |
| K.Chukovsky Park | 62.1 | 27.6 | 6.7 | 3.6 | 84.32 | Healthy |
| N. Ostrovsky Park | 38.5 | 46.1 | 7.7 | 7.7 | 74.23 | Weak |
| Ordzhonikidze park | 6.1 | 18.2 | 36.3 | 39.4 | 35.4 | Severely weak |
| Forest plantations of Aksai | 30.8 | 17.3 | 15.4 | 36.5 | 50.9 | Weak |

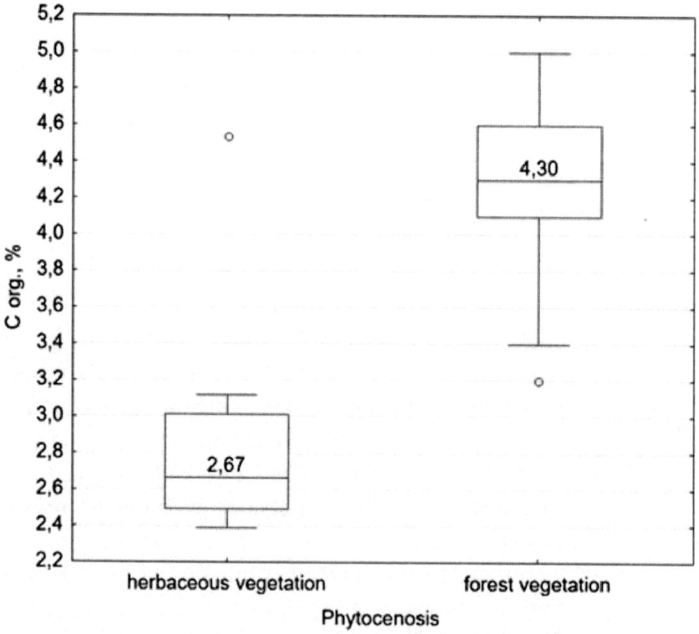


**Fig. 15.2** Characteristics of median and interquartile range for C org., % for the surface ten-centimeter layer of soils under herbaceous and woody phytocenoses

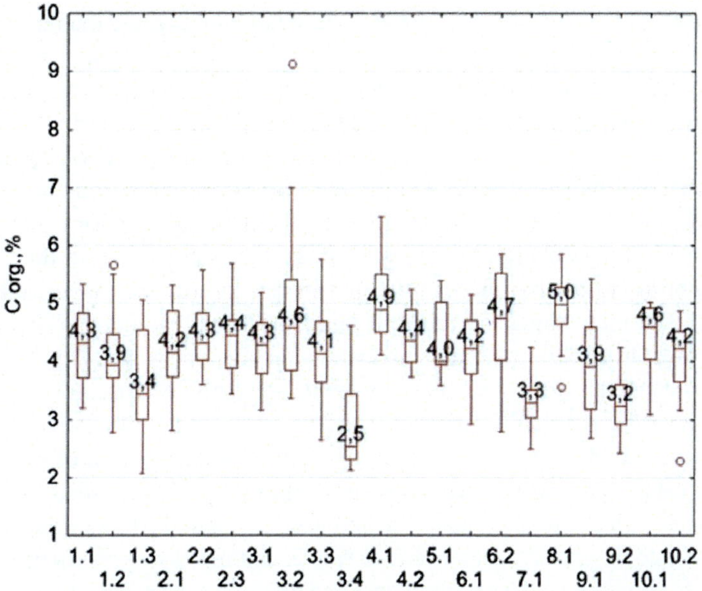


**Fig. 15.3** Characteristics of median and interquartile range for C org., % at the studied monitoring sites in the surface ten-centimeter layer (Temernitskaya grove: 1.1–1.3, Shchepkinskiy forest: 2.1–2.3, Botanic Garden: 3.1–3.2—softwood nurseries, 3.3—oak grove, 3.4—reservoir, Ordzhonikidze: 4.1–4.2, Persianovskaya steppe: 5.1, K. Chukovsky Park: 6.1–6.2, N. Ostrovsky Park: 8.1, Aksai: 9.1–10.2)

carbon content were found in Ordzhonikidze (4.1)—4.9%, K. Chukovsky (6.2)—4.7%, N. Ostrovsky (8.1)—5.0% and the nursery of the Botanic Garden (4.6)—4.6% (Fig. 15.3). Two of the above-mentioned monitoring sites, namely Ordzhonikidze 4.1 and Park N. Ostrovsky 8.1, have a statistically significantly higher content of C org. compared with all other studied sites. The content of organic carbon at the monitoring sites in Temernitskaya grove (1.3) and Aksai (9.1, 9.2), on the contrary, is significantly lower than in other forest parks of the Rostov agglomeration.

Such a low content of C org. can be explained by the later formation of woody vegetation here, as evidenced by the age of *Robínia pseudoacácia*, which is the dominant tree layer in these areas. Thus, at the first stage of development of a forest area, a small amount of dead ground and root mass of plants enters the soil. It is due to the youth of the forest phytocenosis, and part of the humus formed before forest reclamation measures is mineralized. It should also be considered that for a long time (until the mid-70s of the last century) the plot 1.3 of the Temernitskaya grove belonged to the state farm and was used in agricultural activities, which could lead to a decrease in humus reserves and depletion of the soil with organic carbon. The totality of data on the history of the formation of woody vegetation, as well as geobotanical surveys at sites 9.1 and 9.2 of Aksai allow us to determine the age of trees in this forest area in 20–30 years. Interestingly, this site was formed spontaneously, without carrying out forest reclamation measures. Thus, it can be concluded that the age of forest parks plays a decisive role in the accumulation of organic matter in the surface layer.

The profile distribution of total organic matter demonstrates two trends. On the one hand, peaks of humus accumulation in the surface turf horizon are possible, followed by a relatively sharp decrease in its content in the underlying thickness. It is demonstrated by soil profiles laid on the territory of coniferous nurseries of the Botanic Garden 3.1–3.2, Park N. Ostrovsky 8.1 and Aksai monitoring site 9.1. In the horizon AU content C org. decreases by 2–2.5 times compared to the turf horizon and further along the profile, and the distribution of organic matter of the studied sections has similar values. On the other hand, individual monitoring sites (Temernitskaya grove, Ordzhonikidze Park and Shchepkinskiy Forest sites, virgin and reservoir areas) are characterized by the absence of such a sharp decrease in C org. There are characterized by a smoother profile distribution of organic matter throughout the humus-accumulative strata. The graph of the profile distribution of organic carbon for the most representative sections is shown in Fig. 15.4. This distribution is associated with a higher total biomass of herbaceous plants forming under the canopy of trees, since the dying root system of weeds and meadow grasses provides direct entry of dead roots into the soil thickness and, therefore, a gradual decrease in humus content with depth. The different nature of the entry of dead plant into the soil leads to changes in the reserves of organic carbon. The largest reserves of organic matter in the meter thick were found in the parks of the Eastern Protective Ring: Shchepkin forest 2.2 = 284 t/ha, Temernitsky grove 1.1 = 280 t/ha and Ordzhonikidze = 270 t/ha. The lowest indicators were recorded in areas with young forest phytocenosis: Temernitskaya grove 1.3 = 178 t/ha Aksai 9.1 = 176 t/ha, as well as in the control reservoir chernozem of the Botanic Garden 3.4 = 175 t/ha (Fig. 15.5).

## 15.4 Conclusions

The data obtained now cannot entirely confirm the assumptions that the factor of different floral composition of forest parks is crucial in the process of accumulation of organic matter in the surface horizon. Thus, all the plant communities studied by us differ significantly in species diversity, the composition of the dominant species, both specially planted tree and shrub species and spontaneously appeared grasses and also in the degree of projective coverage and abundance. The presence of well-developed herbaceous vegetation does not always lead to an increase in this indicator in the turf horizon, which indicates the relevance of this study and the need for an integrated approach to assess the interrelationships of various components of recreational areas. Nevertheless, it was possible to identify a pattern associated with the age of planted forests and the content of organic carbon. In forest phytocenoses aged 70–90 years, the content of C org. in the surface layer is significantly higher ($4.40 \pm 0.78\%$) than in younger ones ($3.42 \pm 0.76\%$), whose age is 30–50 years. The difference between the soils of forest parks and steppe areas is also obvious in almost all the studied indicators. In forest phytocenoses, the content of organic carbon in the turf horizon

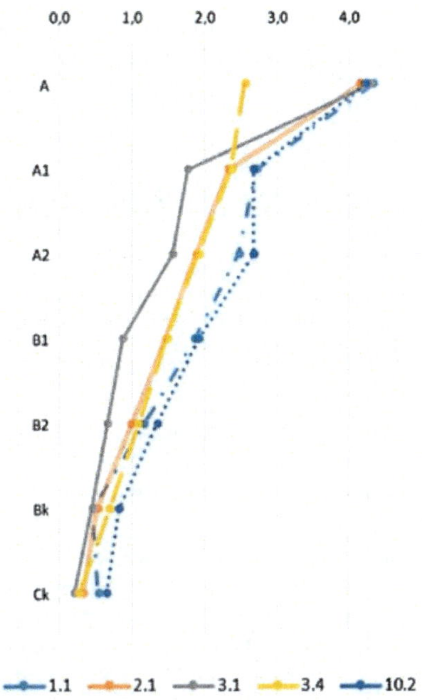


**Fig. 15.4** Profile distribution of C org., % in the studied soil profiles. Temernitskaya grove 1.1, Shchepkinsky forest: 2.1, nursery of coniferous trees of the Botanic Garden 3.1, reservoir plot 3.4, Aksai 10.2

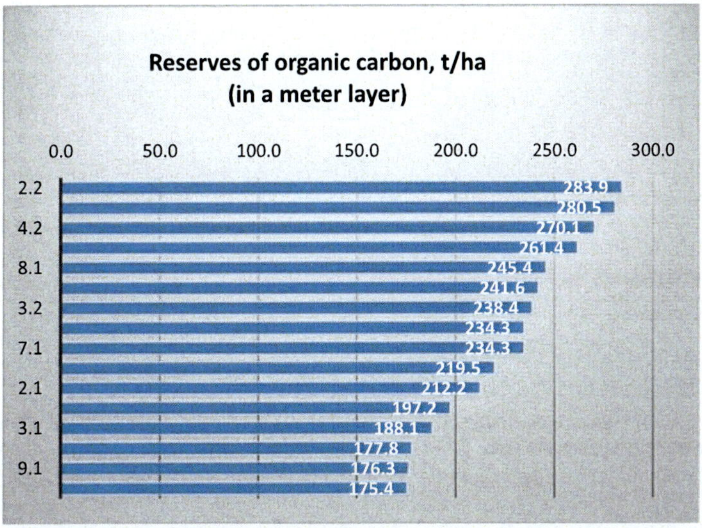


**Fig. 15.5** Reserves of organic carbon in the meter layer, t/ha. (Temernitskaya grove: 1.1–1.3, Shchepkinskiy forest: 2.1–2.3, Botanic Garden: 3.1–3.2—softwood nurseries, 3.3—oak grove, 3.4—reservoir, Ordzhonikidze: 4.1–4.2, Persianovskaya steppe: 5.1, K. Chukovsky Park: 6.1–6.2, N. Ostrovsky Park: 8.1, Aksai: 9.1–10.2)

significantly increases. The humus profile of chernozems acquires the features of dark gray forest soil, which is expressed in a sharp decrease in humus with depth compared to the upper horizons.

**Acknowledgements** The study was carried out at financial support by of the Russian Science Foundation (RSF) No. 23-27-00418, https://rscf.ru/project/23-27-00418/.

## References

Alekseyev VA (1989) Diagnostika zhiznennogo sostoyaniya derev'yev i drevostoyev. Lesovedeniye 4:38–53 (in Russian)

Braun-Blanquet J (1951) Pflanzensoziologie. Grundzüge der Vegetationskunde, 2nd edn. Springer, Wien, 631 p.

Canedoli C, Ferrè C, Abu El Khair D, Padoa-Schioppa E, Comolli R (2020) Soil organic carbon stock in different urban land uses: high stock evidence in urban parks. Urban Ecosyst 23:159–171. https://doi.org/10.1007/s11252-019-00901-6

De Kimpe CR, Morel JL (2000) Urban soil management: a growing concern. Soil Sci 165:31–40

Demakov, Yu.P., Nureyeva TV, Krasnov VG, Ryzhkov AA (2017) Ekologo-resursnyy potentsial lesnykh nasazhdeniy na priovrazhno-balochnykh zemlyakh srednego Povolzh'ya. Vestnik Povolzhskogo gosudarstvennogo tekhnologicheskogo universiteta. Seriya: Les. Ekologiya. Prirodopol'zovaniye 3(35):73–87. https://doi.org/10.15350/2306-2827.2017.3.73. (in Russian)

Gorbov SN, Bezuglova OS (2014) Specific features of organic matter in urban soils of Rostov-on-Don. Eurasian Soil Sci 47(8):792–800. https://doi.org/10.1134/S1064229314080043.EDNUEUYSD

Huot H, Joyner J, Córdoba A, Shaw RK, Wilson MA, Walker R, Muth TR, Cheng Z (2017) Characterizing urban soils in New York City: profile properties and bacterial communities. J Soils Sediments 17:393–407. https://doi.org/10.1007/s11368-016-1552-9

Karpachevskiy LO (1981) Les i lesnyye pochvy. Lesnaya promyshlennost', Moskva, 264 p. (in Russian)

Mineyev VG (2001) Praktikum po agrokhimii - 2-ye izd. Pod redaktsiyey akademika RASKHN Mineyeva V.G. Izdatel'stvo: Izd-vo MGU, Moscow, 689 p. (in Russian)

Morel JL, Chenu C, Lorenz K (2015) Ecosystems services provided by soils of urban, industrial, traffic and military areas (SUITMAs). J Soils Sediments 15:1659–1666. https://doi.org/10.1007/s11368-014-0926-0

Shmarayeva AN, Kuz'menko IP, Makarova LI (2021) Struktura sinantropnoy flory osobo okhranyayemoy prirodnoy territorii «Botanicheskiy sad Yuzhnogo federal'nogo universiteta». Stepi Severnoy Yevrazii: materialy IKH mezhdunarodnogo simpoziuma 9:866–872. https://doi.org/10.24412/cl-36359-2021-866-872. (in Russian)

Tagiverdiev SS, Gorbov SN, Bezuglova OS, Skripnikov PN (2020) The content and distribution of various forms of carbon in urban soils of southern Russia on the example of Rostov agglomeration. Geoderma Reg 21:00266. https://doi.org/10.1016/j.geodrs.2020.e00266

Velts NYu, Lupova IV (2013) Opyt monitoringa fitobioty Yuzhnogo Priural'ya. Vestnik OGU 10 (159):176–181. (in Russian)

Vodyanitskii YN (2015) Organic matter of urban soils: a review. Eurasian Soil Sci 48(8):802–811. https://doi.org/10.1134/S1064229315080116.EDNUFBZKX

Vysotskiy GN (1962) Izbrannyye sochineniya. T. 1–2. Moscow. (in Russian)

Whittaker RH (1972) Evolution and measurement of species diversity. Taxon 21(2–3):213–251

Yang L, Yuan L, Peng K, Wu S (2014) Nutrients and heavy metals in urban soils under different green space types in Anji, China. CATENA 115:39–46. https://doi.org/10.1016/j.catena.2013.11.008
Yu YO, Myasnikova MA (2014) Floristicheskiye osobennosti zalezhnykh uchastkov stepey Rostovskoy oblasti. APRIORI. Seriya: Yestestvennyye i Tekhnicheskiye Nauki 3:4 (in Russian)
Zozulina GM, Fedyayevoy VV (1984, 1985) Flora Nizhnego Dona (opredelitel'). Chast'1, chast' 2. /Pod red. Izd-vo RGU, Rostov-na-Donu, 280 s. 240 p. (in Russian)

# Chapter 16
# Microbial Activity of Technosols Based on Peat-Sand Mixtures in Different Climatic Zones

**Maria N. Vasileva, Maria V. Korneykova, Ekaterina V. Kozlova, Viacheslav I. Vasenev, and Yulia L. Sotnikova**

**Abstract** The microbiological parameters of technosols based on peat-sand mixtures, made in cities of different bioclimatic zones: Apatity (subarctic), Moscow (temperate humid), Rostov-on-Don (temperate dry), were determined for the first time. Such properties as microbial biomass, activity and specific respiration of technosols were evaluated by the technics of substrate-induced and basal respiration, as well as the physiological profile and functional diversity of soil microbial communities by the MicroResp technic. Fungal biomass was assessed using light luminescence microscopy. In technosols from different climatic zones, microbial biomass carbon and microbial activity (basal respiration) values were found to be lower compared to natural (background) reference soils. Fourteen months after the beginning of the experiment, the microbial biomass in Moscow and Rostov-on-Don decreased by around 1.5 and 2 times, respectively, compared to background soils, whereas in Apatity, on the opposite, it increased. Microbial respiration in the conditions of the north increased by 1.2 times, while in temperate climate it decreased by 1.2 (Moscow) and 3.4 times (Rostov-on-Don). The specific respiration (microbial metabolic quotient) $qCO_2$ was found to be sufficiently high (3.5–4.5 $CO_2$-C μg-1 Cmic $h^{-1}$) for the first two years of the study for all experimental stations. Alterations in the microbial community physiological profile of technosols in different bioclimatic zones were discovered, most noticeably demonstrated in Moscow. In terms of fungal biomass, the technosols of Rostov-on-Don are closest to the background soil, while in Moscow it is lower by 2 times and almost 3.5 times in Apatity. After 14 months in the technosols, created in temperate humid and temperate dry conditions, tended to change the functional diversity of microbial communities towards

M. N. Vasileva (✉) · M. V. Korneykova · E. V. Kozlova · Y. L. Sotnikova
RUDN University, Moscow, Russia
e-mail: vasilyeva_mn@pfur.ru
URL: http://www.rudn.ru

M. V. Korneykova
Kola Science Centre of Russian Academy of Science, Apatity, Russia

V. I. Vasenev
Wageningen University, Wageningen, Netherlands



M. Korneykova et al. (eds.), *Smart and Sustainable Urban Ecosystems: Challenges and Solutions*, Springer Geography, https://doi.org/10.1007/978-3-031-37216-2_16

natural reference. Under northern conditions, the physiological profile of the microbial community changed insignificantly. Alterations in the microbiological characteristics of technosols in various bioclimatic conditions revealed over a period of 14 months demonstrate a significant impact of the climatic factor on the formation of stable soil systems.

**Keywords** Human-made soil constructions · Bioclimatic gradient · Microbial specific respiration · Green infrastructure · Community level physiological profile

## 16.1 Introduction

Soil, as a main base of urban green infrastructure, strongly linked to the sustainable development goals, by providing the recreational and aesthetic function for comfort living (O'Riordan et al. 2021; United Nations 2019; Vasenev et al. 2016). The development of new construction technologies for urban soils and the adaptation of old ones can ensure the sustainability of technosols and at the same time support the functioning of the Urban Green Infrastructure.

With the increasing rate of urbanization, land use is changing and human impact on soils is growing (Chien and Krumins 2021; Demina et al. 2018; Ivashchenko et al. 2019). Urban soil remains crucial ecological asset and important component of the urban ecosystem. Due to the high dynamic and diversity, man-made soil constructions are an attractive object for assessing properties and functions for the purpose of further modelling (Hazelton et al. 2021). The analysis of the functioning of soil constructions in regions with different environmental conditions and the level of development of urban green infrastructure is especially relevant (Deeb et al. 2020; González-Méndez and Chávez-García 2020).

Soil is a source of ecosystem services—one of the carbon stores and a contributor to carbon emissions to the atmosphere through soil respiration (O'Riordan et al. 2021; Vasenev et al. 2017; Wang et al. 2011). The microbial component provides the main flux of $CO_2$ from the soil to the atmosphere (Saccá et al. 2017). The microbial community is highly sensitive to anthropogenic influences and is therefore an indicator of environmental change and is often used in assessing the health of terrestrial ecosystems (Ivashchenko et al. 2015; Sarzhanov et al. 2015; Wang et al. 2011).

Various materials are used for the construction of technosols, depending on the functions that they are required to perform and on the availability of materials (Fabbri et al. 2021; Smagin and Sadovnikova 2015). Natural materials with different organic matter content (in different proportions), such as sand, peat, displaced topsoil is used to create technosols. At the same time, technologies exist to create technosols based on mining waste to mitigate the effects of industrial pollution (Fabbri et al. 2021; Ivashchenko et al. 2021a; Slukovskaya et al. 2019; Uzarowicz et al. 2020). However, to fully evaluate the effectiveness of constructions, it is necessary to analyse materials comparable between regions that would be available and similar in properties.

Universal in design soil constructions could become a representative background for explaining the changes in the functioning of technosols in various bioclimatic zones.

The aim of the study is a comparative analysis of universal soil constructions for the creation of lawns with similar properties in different bioclimatic zones of Russia, based on microbiological indicators. The results of the study will identify the degree of influence of the factors determining the functioning of soils, and on this basis, adapt existing technologies for creating technosols to support sustainable urban development.

## 16.2 Research Area

The research area was carried out in cities located in different climatic zones: subarctic (Apatity), temperate continental with humid climate (Moscow) and temperate continental with dry climate (Rostov-on-Don) (Fig. 16.1).

In 2020, universal constructions (Unified Technosols) based on sand, peat and 'cut-off' (the surface fertile soil layer removed during construction) in a 1:1:1 ratio were created at the experimental sites in each city in three replicates. These types of substrates are used quite often for soil construction, especially for Moscow

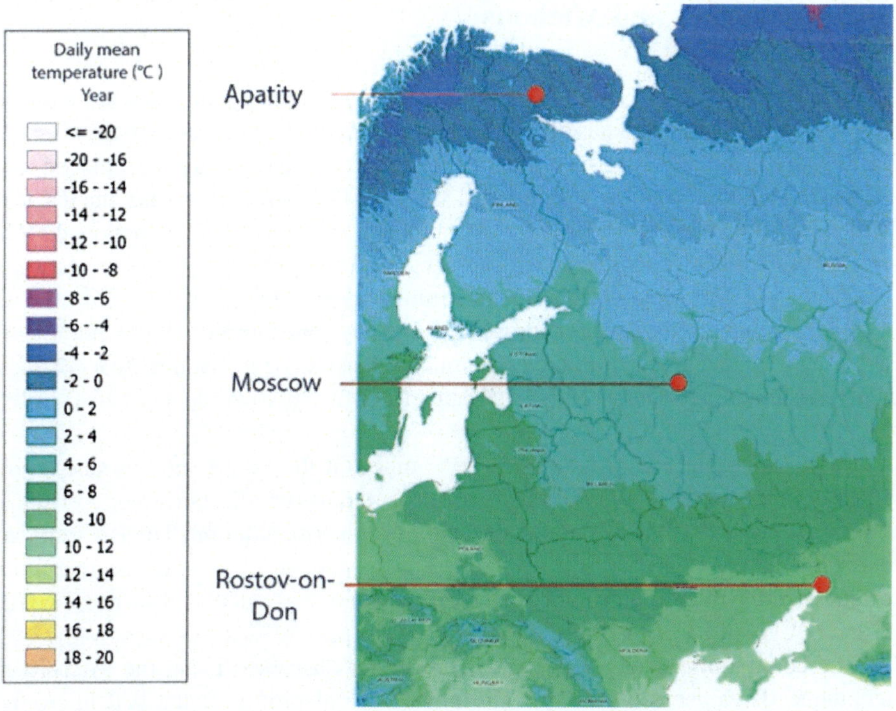


**Fig. 16.1** Research area

region, which is more studied regarding technosols design (Brianskaia et al. 2020; Ivashchenko et al. 2021a). The sizes of the experimental plots at the Apatity station were 1 by 2 m, in Moscow—3 by 3 m, in Rostov-on-Don—2 by 2 m. As a lawn covering, a grass mixture with a seeding rate of 60 g/m$^2$ was used. In Apatity, a grass mixture with a seeding rate of 160 g/m$^2$ and a technology for pre-growing seedlings was additionally used.

Before establishing the soil constructions, the following chemical characteristics were determined for the mixtures composing them. Total carbon at the northern station was 3.05, 2.25% at the Moscow station, and 1.75% at the southern station. The content of total nitrogen in the mixtures in Apatity was 0.05%, in Moscow—0.02%, in Rostov-on-Don—0.04%. $pH_{H_2O}$ was 6.24, 7.21 and 8.45, respectively.

A grass mixture of perennial ryegrass (*Lolium perenne*) 30%, red fescue (*Festuca rubra*) 35% and meadow bluegrass (*Poa pratensis*) 35% with a seeding rate of 60 g/m$^2$ was used to create lawn cover. Agrotechnical measures included the application of the mineral fertilizer Azophoska at a rate of 60 g/m$^2$ and active irrigation at all sites. In northern conditions, dolomite flour was applied in an amount of 300 g/m$^2$ to reduce soil acidity while creating soil constructions. Regular mowing was carried out at a frequency of once every 1–2 weeks between May and October, depending on the region.

## 16.3 Materials and Methods

The microbial biomass carbon in soil ($C_{mic}$) was measured by the substrate-induced respiration (SIR) method, based on capturing the respiratory response to the addition of a readily available substrate–glucose. Substrate samples (1 g) were placed in a 15 ml vial, glucose solution (0.1 ml/g soil, 0.5 mg/g) was added dropwise, the vial was sealed, and time was recorded. The vials were incubated (3–5 h of incubation, 22 °C), and the gas sample was collected with a syringe and injected into a gas chromatograph (KrystaL 2000 M, 'Meta-Chrom' manufacturer, Yoshkar-Ola, Russia) equipped with a thermal conductivity detector for measuring $CO_2$ concentration. Soil Cmic (μg C g-1 soil) was calculated from the equation SIR (μl $CO_2$ g $g^{-1}$ soil $h^{-1}$) × 40.04 + 0.37 (Anderson and Domsch 1978; Ananyeva et al. 2008; Anderson and Domsch 2010).

Basal (microbial) respiration was determined in the native (not enriched) soil (24 h of incubation, 22 °C). The BR was measured as for SIR, but water (0.1 mL/g soil) was added to the soil samples instead of a glucose solution. The BR rate was expressed in μg $CO_2$–C/g soil per hour. The specific respiration of the microbial biomass, or the microbial metabolic quotient ($qCO_2$), was calculated as the ratio BR/$C_{mic}$ = $qCO_2$ (μg $CO_2$–C/mg $C_{mic}$/h).

The community level physiological profile was assessed using the MicroResp technique (Creamer et al. 2016). The physiological profile of the soil microbial community is assessed using the MicroRespTM technique. Soil samples are placed in a well (945 μl volume) of a special 96-well plate and nutrient substrates belonging to the groups of amino acids, carbohydrates, carboxylic acids, and phenolic acids are added to each well. Each substrate is added to the wells of the plate in three replicates. In addition, a specially prepared indicator gel (agar, cresol red, potassium chloride and sodium hydrogen carbonate) was placed in the 450 μl wells of the other corresponding plate. The soil plate is covered with the indicator gel plate, separated by a membrane and incubated for 6 h at 25 °C. During incubation $CO_2$ is released from the soil, changing the pH and the colour of the gel (pink to yellow). The colour change is measured on a microplate reader ('Uniplan', 450λ nm filter). The absorbance units are converted into substrate-induced soil respiration, expressed in μg C $g^{-1}$ soil $h^{-1}$ (Chapman et al. 2007).

Fungal biomass was determined by light luminescence microscopy using white calcofluor. Accounting for spores and the length of the mycelium was carried out on a Biomed 5 PR LYUM luminescent microscope (Russia). Cells were desorbed from the soil using an MSV-3500 Vortex (Latvia) at a speed of 3500 rpm for 10 min. Fungal biomass (mg/g soil) was calculated assuming the spore density equals 0.837 g/$cm^3$, and the mycelium density equals 0.628 g/$cm^3$ (Polyanskaya and Zvyagintsev 2005). The content of fungal biomass per gram of dry soil was calculated considering its moisture content.

All studied parameters of soil constructions were measured twice, in 2020 and 2021 years. The results of samples analysis were compared with their natural analogues (forest ecosystem) in the respective cities and the results of the previous year.

## 16.4 Results and Discussion

During the first year of the experiment, the microbial biomass of technosols in the studied cities ranged from 150 to 230 μg C g-1 and was 2.8–5.3 times lower than its natural analogues (Fig. 16.2). After 14 months in Moscow and Rostov-on-Don, it decreased by more than 1.5 times, and in Apatity, the values showed a tendency to increase. However, in comparison with natural reference the microbial biomass values remained low. The trend of biomass values distribution in the climatic gradient differed for unified soil constructions and background soils. Thus, in technosols, the microbial biomass decreased in the 'north–south' direction, whereas in natural reference samples it decreased in the 'south–north' direction.

Basal respiration of natural references exceeded that of technosols by 1.8–2.3 times in different regions. After 14 months from the beginning of the experiment, it decreased by more than 3 times in Rostov-on-Don. In Moscow and Apatity, the values have not changed significantly. In general, in technosols, there was a tendency like natural ecosystems—decrease of basal respiration in climatic gradient Apatity–Moscow–Rostov-on-Don. Values of specific respiration of soil constructions in the first year of observations exceeded natural analogues by 1.5–3.2 times. The microbial metabolic coefficient $qCO_2$ remained high at 3.5–4.5 $CO_2$-C $mg^{-1}$ C mic $h^{-1}$ and did not differ significantly from the results of the previous year. The high values of

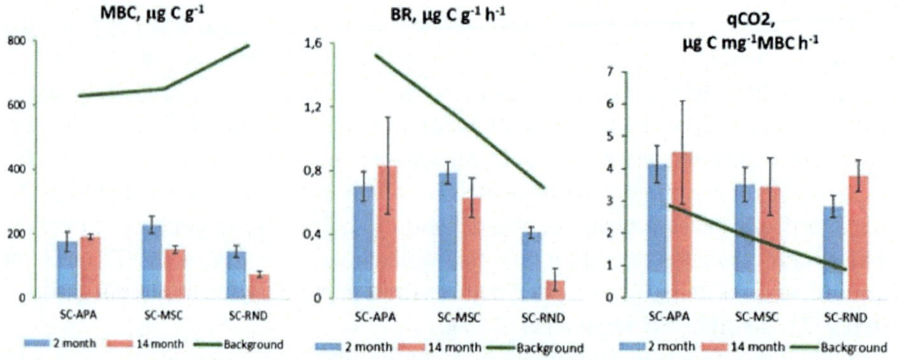


**Fig. 16.2** Microbial indicators of soil constructions: MBC—microbial biomass carbon, BR—basal respiration, $qCO_2$—specific respiration of microbial biomass (microbial metabolic coefficient); SC—soil constructions: SC-APA—Apatity, SC-MSC—Moscow, SC-RND—Rostov-on-Don; BG—background soil

this coefficient may indicate adverse conditions for the development of the microbial community in the soil (Ivashchenko et al. 2015).

The soil constructions of the studied towns showed a rather high level of carboxylic acid mineralization, from 32 to 69%, as in their natural analogues, where the contribution ranged from 47 to 62% (Fig. 16.3). The group of microorganisms that decompose carbohydrates in the technosols ranged from 15 to 35%.

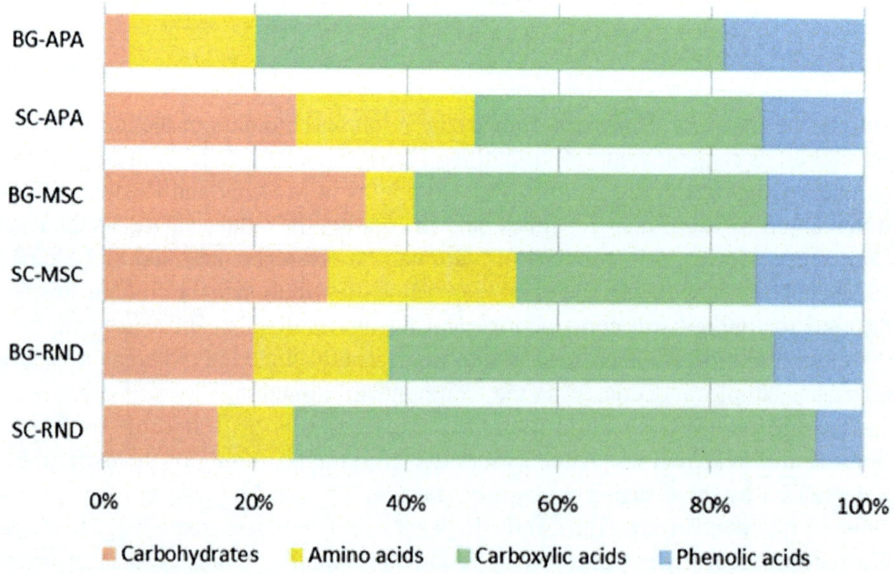


**Fig. 16.3** Community-level physiological profile of technosols in comparison with natural analogues in climatic gradient. SC—soil constructions: SC-APA—Apatity, SC-MSC—Moscow, SC-RND—Rostov-on-Don; BG—Background soil

In Rostov-on-Don and Moscow, their amount was only 1.2–1.3 times lower than in natural reference. For natural ecosystems of Apatity, the contribution of this group to the total respiration of microorganisms was only 3%, while for technosols, it was closer to that in Moscow and Rostov-on-Don and reached 26%.

After 14 months, the contribution to total respiration of the microorganisms decomposing carboxylic acids decreased, most notably in Moscow—by more than 2 times (Fig. 16.4), in Rostov-on-Don—by 12% and in Apatity—by 20%. There is a general trend for all stations to increase the proportion of microorganisms reacting to carbohydrate application—by 1.5–2.2 times. In Moscow, the response to the application of substrates of the amino acids group increased significantly, from 9 to 25%, as well as of difficult to decompose phenolic acids, from 1 to 14% of the total contribution to respiration.

According to the similarity of the functional structure of the microbial community of technosols, it can be noted that the technosols of Moscow were the closest to the background soils in 2020 year (they were in one large cluster). Soil constructions of Apatity were also quite close to their natural analogues (Fig. 16.5). In Rostov-on-Don, technosols differed comparatively sharply from soils of natural references.

The prevalence in the consumption of easily available substrates cannot be claimed with certainty to be a direct indication of the inefficiency of the microbial community. Despite some changes in the structure, the dominance of microorganisms decomposing carbohydrates and carboxylic acids remains, which is characteristic of both natural and urban ecosystems (Ivashchenko et al. 2021a, b).

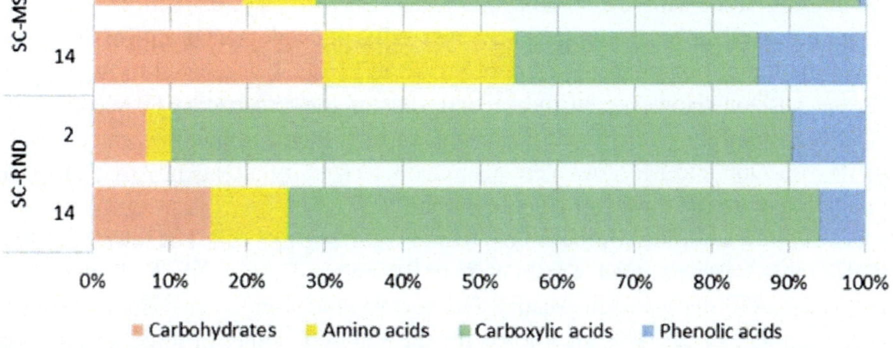


**Fig. 16.4** Community-level physiological profile in temporal dynamic. SC—soil constructions: SC-APA—Apatity, SC-MSC—Moscow, SC-RND—Rostov-on-Don; BG Background; 2–2 months after creation, 14–14 months after creation

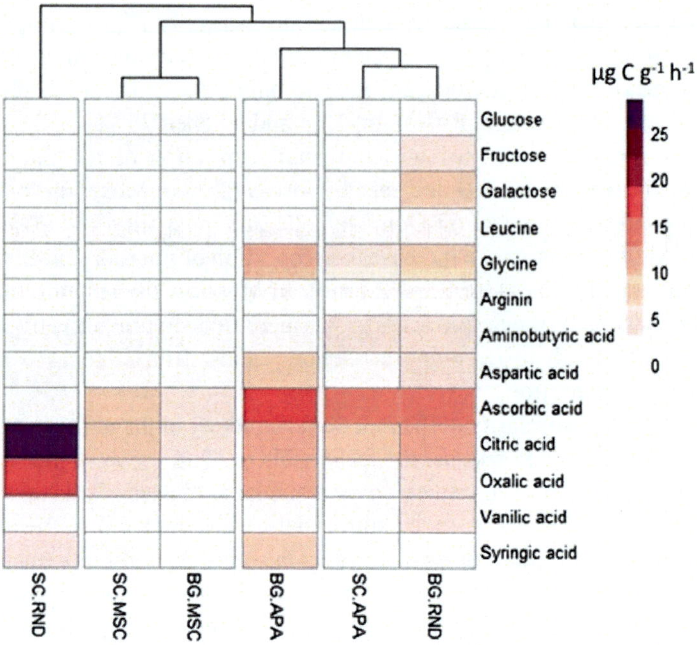


**Fig. 16.5** Similarities in the functional structure of the microbial community of technosols in 2020 year. SC-APA—Apatity, SC-MSC—Moscow, SC-RND—Rostov-on-Don; BG—background soil

In 2021, the similarity in the physiological profile of microbial communities of the soil constructions laid in Moscow and the background soils remained (Fig. 16.6). The physiological profile of technosols in Rostov-on-Don has become much closer to the structure of community of the background soils. The functional structure of the microbial community of the soil constructions in Apatity has deviated from its natural analogues but has become closer to the physiological profile of the microbial community of the Moscow stationary.

In summary, the most significant changes in the physiological profile of the microbial community of technosols were observed in Moscow compared to Apatity and Rostov-on-Don. However, in the second year of the experiment in Moscow and Apatity similarities in the functional profile of microbial communities can be noted. Rostov-on-Don and Moscow were also characterized by a tendency to change the functional structure towards natural communities.

The biomass of fungi varied from 0.073 to 0.790 mg/g of soil. The lowest biomass (0.078 mg/g soil) was noted in technosols of Rostov-on-Don, and the highest biomass reached 0.2473 mg/g soil in Apatity. For background soils, the maximum biomass of 0.790 mg/g of soil was also noted for the subarctic zone. In terms of biomass, the technosols of Rostov-on-Don were closest to the background, for the Moscow region, the biomass of fungi is 2 times lower, and in Apatity it is almost 3.5 times lower than in the background plot.

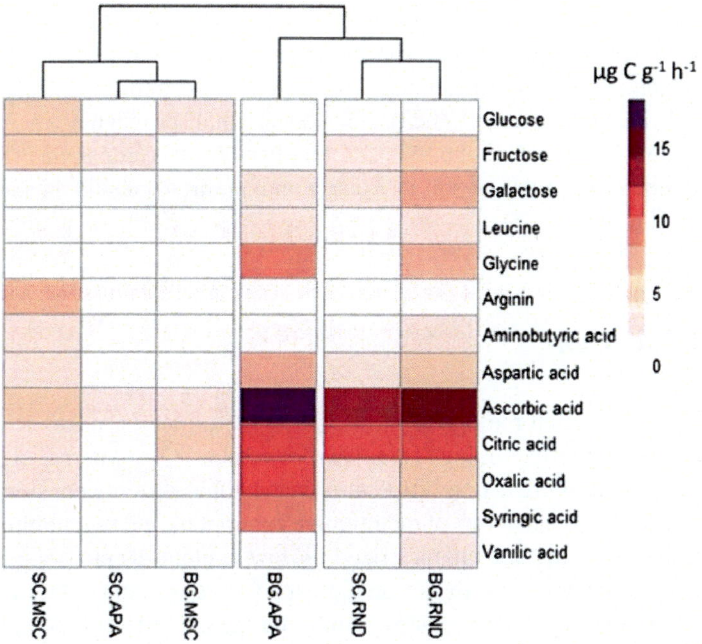


**Fig. 16.6** Similarities in the functional structure of the microbial community of technosols in 2021 year. SC-APA—Apatity, SC-MSC—Moscow, SC-RND—Rostov-on-Don; BG—Background

The smallest proportion of mycelium (43–45%) was noted for the soils of Rostov-on-Don and the largest (60–87%) for the soils of Apatity. The proportion of mycelium in the biomass of fungi for Moscow soil varied from 55 to 69%. The length of fungal mycelium varied from tens to hundreds of meters per gram of soil. A minimum of mycelium (25–27 m/g of soil) was also found in the soils of Rostov-on-Don and a maximum (118 and 544 m/g of soil) in the soils of Apatity.

The average length of the mycelium was equal to 41 and 105 m/g in Moscow soils. The dominance of mycelial forms in the north and its minimum amount in the south can be explained by the type of natural zone in which the objects of study are located (northern taiga and steppe, respectively), the prevailing vegetation and soil moisture. The proportion of thin (less than 3 μm in diameter) mycelium of fungi in the studied soils was relatively high (up to 41%) in the soils of Rostov-on-Don.

The proportion of thin fungal hyphae in the soils of Moscow and Apatity was significantly lower: 20 and 32%, respectively. For all studied samples, the number of unicellular fungal propagules (spores and yeasts) was $10^4$ cells/g of soil. However, in the technosols of Apatity, their number reached $10^5$ cells/g of soil. A high proportion of fine mycelium in soils may indicate a stressful state of the mycobiota community under extreme climatic conditions (Marfenina et al. 2016). The basidiomycete buckle mycelium of fungi was found rarely (about 7–10% of all hyphae), and only in the subarctic zone.

The majority of mycobiota propagules (from 79.6 to 100%) was represented by specimens of small size, 2–3 $\mu$m. The studied soils of Moscow and Rostov-on-Don had approximately the same proportion of small mycobiota propagules: 93.3–100% and 95.6–100%, respectively. The proportion of small propagules in Apatity was lower and amounted to 79.6–86.9%. The number of large propagules (5 and more $\mu$m) was about 103 cells/g of soil in Apatity and hundreds of cells in samples from Moscow and Rostov-on-Don.

## 16.5 Conclusion

The sensitivity of microbiological indicators to disturbances makes it possible to assess changes in the functioning of soil microorganisms. Over a period of 14 months, the soil constructions created in different climatic zones differed considerably from their natural analogues in terms of quantitative parameters and functioning of microbial communities. They are characterized by low content of microbial biomass as well as its high respiratory activity and high specific respiration value. This state of microbial communities of soil constructions indicates the instability of ecosystems at this stage of development. However, the tendency to change the properties of microbial communities towards their natural analogues is clearly visible. For example, the microbial biomass content and basal respiration have increased in soil constructions located in Apatity. Probably, microbial communities of northern ecosystems, having a shorter life cycle, adapt faster to changing environmental conditions and as a result, over the period under study are more similar in their properties to their natural analogues. In the technosols of Moscow and Rostov-on-Don, despite the decrease in the values of the indicators mentioned above, the microbial metabolic coefficient is more stable over time. These cities are also characterized by a trend towards a change in the physiological profile of microbial communities towards their natural analogues, with similarities in the structure of microbial communities noted for Apatity and Moscow. The quantitative indicators of soil fungal communities showed an ambiguous result, currently demonstrating only trends in the formation of a microbial community in technosols: the soils of the subarctic zone are characterized by the highest biomass and length of the fungal mycelium, while the soils of the southern region are characterized by the minimum values.

**Acknowledgements** Soil microbial analysis and assessment of Technosols' functions were funded by RSF grant No 17-77-20046. Soil chemical analysis was supported by the state task 122022400109-7 (FMEZ-2022-0011).

## References

Ananyeva ND, Susyan EA, Chernova OV, Wirth S (2008) Microbial respiration activities of soils from different climatic regions of European Russia. Euro J Soil Biol 44(2):147–157. https://doi.org/10.1016/j.ejsobi.2007.05.002

Anderson JPE, Domsch KH (1978) A physiological method for the quantitative measurement of microbial biomass in soils. Soil Biol Biochem 10:215–221. https://doi.org/10.1016/0038-0717(78)90099-8

Anderson TH, Domsch KH (2010) Soil microbial biomass: the eco-physiological approach. Soil Biol Biochem 42(12):2039–2043. https://doi.org/10.1016/j.soilbio.2010.06.026

Brianskaia IP, Vasenev VI, Brykova RA, Markelova VN, Ushakova NV, Gosse DD, Gavrilenko EV, Blagodatskaya EV (2020) Analysis of volume and properties of imported soils for prediction of carbon stocks in soil constructions in the Moscow metropolis. Eurasian Soil Sci 53(12):1809–1817. (in Russian)

Chapman SJ, Campbell CD, Artz RRE (2007) Assessing CLPPs using MicroRespTM—a comparison with biology and multi-SIR. J Soils Sediments 7(6):406–410. https://doi.org/10.1065/jss2007.10.259

Chien SCH, Krumins JA (2021) Natural versus urban global soil organic carbon stocks: a meta-analysis. Sci Total Environ 807:150999. https://doi.org/10.1016/j.scitotenv.2021.150999

Creamer RE, Stone D, Berry P, Kuiper I (2016) Measuring respiration profiles of soil microbial communities across Europe using MicroRespTM method. Appl Soil Ecol 97:36–43. https://doi.org/10.1016/j.apsoil.2015.08.004

Deeb M, Groffman PM, Blouin M, Egendorf SP, Vergnes A, Vasenev V, Cao D, Walsh D, Morin T, Séré G (2020) Using constructed soils for green infrastructure-challenges and limitations. Soil 6(2):413–434. https://doi.org/10.5194/soil-6-413-2020

Demina S, Vasenev V, Ivashchenko K, Ananyeva N, Plyushchikov V, Hajiaghayeva R, Dovletyarova E (2018) Microbial properties of urban soils with different land-use history in New Moscow. Soil Sci 183(4):132–40. https://doi.org/10.1097/SS.0000000000000240

Fabbri D, Pizzol R, Calza P, Malandrino M, Gaggero E, Padoan E, Ajmone-Marsan F (2021) Constructed Technosols: a strategy toward a circular economy. Appl Sci 11(8):3432. https://doi.org/10.3390/app11083432

González-Méndez B, Chávez-García E (2020) Re-thinking the Technosol design for greenery systems: challenges for the provision of ecosystem services in semiarid and arid cities. J Arid Environ 179:104191. https://doi.org/10.1016/j.jaridenv.2020.104191

Hazelton P, Murphy B (2021) Understanding soils in urban environments. CSIRO Publishing, 186 p

Ivashchenko K, Ananyeva N, Vasenev V, Sushko S, Seleznyova A, Kudeyarov V (2019) Microbial C-availability and organic matter decomposition in urban soils of megapolis depend on functional zoning. Soil Environ 38(1):31–41. https://doi.org/10.25252/SE/19/61524

Ivashchenko K, Ananyeva N, Vasenev V, Ryzhkov O, Kudeyarov V, Valentini R (2015) Soil microbial biomass and gas-production activity ($CO_2$) in chernozems of different land use. Soil-Water J: 43–50. https://doi.org/10.21657/tsd.47684

Ivashchenko K, Lepore E, Vasenev V, Ananyeva N, Demina S, Khabibullina F, Vaseneva I , Selezneva A, Dolgikh A, Sushko S, Marinari S, Dovletyarova E (2021a) Assessing soil-like materials for ecosystem services provided by constructed Technosols. Land 10(11):1185. https://doi.org/10.3390/land10111185

Ivashchenko K, Sushko S, Selezneva A, Ananyeva N, Zhuravleva A, Kudeyarov V, Makarov M, Blagodatsky S (2021b) Soil microbial activity along an altitudinal gradient: vegetation as a main driver beyond topographic and edaphic factors. Appl Soil Ecol 168(5):104197. https://doi.org/10.1016/j.apsoil.2021.104197

Marfenina OE, Nikitin DA, Ivanova AE (2016) The structure of fungal biomass and diversity of cultivated micromycetes in Antarctic soils (Progress and Russkaya Stations). Eurasian Soil Sci 49(8):934–941

O'Riordan R, Davies J, Stevens C, Quinton JN, Boyko C (2021) The ecosystem services of urban soils: a review. Geoderma 395(1):115076. https://doi.org/10.1016/j.geoderma.2021.115076

Polyanskaya LM, Zvyagintsev DG (2005) The content and composition of microbial biomass as an index of the ecological status of soil. Eurasian Soil Sci 38:625–633

Saccá MLA, Caracciolo AB, di Lenola M, Grenni P (2017) Ecosystem services provided by soil microorganisms. In: Soil biological communities and ecosystem resilience. Springer International Publishing, 9–24.https://doi.org/10.1007/978-3-319-63336-7_2

Sarzhanov DA, Vasenev VI, Sotnikova YL, Tembo A, Vasenev II, Valentini R (2015) Short-term dynamics and spatial heterogeneity of $CO_2$ emission from the soils of natural and urban ecosystems in the central chernozemic region. Eurasian Soil Sci 48(4):416–424. (in Russian)

Slukovskaya MV, Vasenev VI, Ivashchenko KV, Morev DV, Drogobuzhskaya SV, Ivanova LA, & Kremenetskaya IP (2019) Technosols on mining wastes in the subarctic: efficiency of remediation under Cu-Ni atmospheric pollution. Int Soil Wat Conserv Res 7(3):297–307. https://doi.org/10.1016/j.iswcr.2019.04.002

Smagin AV, Sadovnikova NB (2015) Creation of soil-like constructions. Eurasian Soil Sci 48(9):981–990. (in Russian)

Uzarowicz HŁ, Wolińska A, Błońska E, Szafranek-Nakonieczna A, Kuźniar A, Słodczyk Z, Kwasowski W (2020) Technogenic soils (Technosols) developed from mine spoils containing Fe Sulphides: microbiological activity as an indicator of soil development following land reclamation. Appl Soil Ecol 156. https://doi.org/10.1016/j.apsoil.2020.103699

Vasenev VI, Smagin AV, Ananyeva ND, Ivashchenko KV, Gavrilenko EG, Prokofeva TV, Paltseva A, Stoorvogel JJ, Gosse DD, Valentini R (2017) Urban soil's functions: monitoring, assessment and management. In: Adaptive soil management: from theory to practices. Springer, Singapore, 359–409.https://doi.org/10.1007/978-981-10-3638-5_18

Vasenev VI, Stoorvogel JJ, Plyushchikov VG, Ananyeva ND, Ivashchenko KV, Romzaykina ON (2016) Basal respiration as a proxy to understand spatial trends in $CO_2$ emissions in the Moscow region. RUDN J Agron Animal Industr 4:94–107. https://doi.org/10.22363/23122797XX2016644944107

Wang M, Markert B, Shen W, Chen W, Peng C, Ouyang Z (2011) Microbial biomass carbon and enzyme activities of urban soils in Beijing. Environ Sci Pollut Res 18(6):958–967. https://doi.org/10.1007/s11356-011-0445-0

# Chapter 17
# Strategies for Phytoremediation of Polluted Sites in the Subarctic

**M. A. Pistsova, A. V. Gromova, M. V. Slukovskaya, Yu. A. Dvornikov, and A. L. Gurinov**

**Abstract** Pollution by potentially toxic elements (PTE) leads to the formation of industrial barrens. Remediation of polluted sites located in the Subarctic is hampered by unfavorable climatic conditions. Soil erosion results in considerable loss of topsoil and is also an important pathway for lateral migration of sediment-associated PTEs. Here, phytoremediation is considered as an appropriate management to reduce these risks. The strategy is tested to an area impacted by a Cu/Ni smelter in the Kola Peninsula. Soil properties and, in particular, susceptibility to water erosion are determined using the RUSLE equation. Annual soil losses in the territory are 4–10 mm; the values are related to the conformation of the terrain. From the calculated soil loss values, three strategies (for hilltops, slopes, and depressions) are proposed to protect the territory from degradation by limiting the mobilization, transport, and accumulation of metals.

**Keywords** Industrial barren · Phytoremediation · Soil erosion assessment · Environmental pollution

M. A. Pistsova (✉)
Lomonosov Moscow State University, GSP-1, Leninskie Gory, Moscow 119991, Russian Federation
e-mail: pistsovama@my.msu.ru

A. V. Gromova · Yu. A. Dvornikov · A. L. Gurinov
Peoples' Friendship University of Russia (RUDN University), 6 Miklukho-Maklaya Street, Moscow 117198, Russian Federation

M. V. Slukovskaya (✉)
Laboratory of Nature-Inspired Technologies and Environmental Safety of the Arctic, Kola Science Centre, Russian Academy of Sciences, 184209 Apatity, Russia
e-mail: m.slukovskaya@ksc.ru

I.V. Tananaev Institute of Chemistry and Technology of Rare Elements and Mineral Raw Materials, Kola Science Centre, Russian Academy of Sciences, 184209 Apatity, Russia

M. A. Pistsova · A. L. Gurinov
Institute of Geography, Russian Academy of Sciences, 29, Staromonetny Lane, Moscow 119017, Russia



M. Korneykova et al. (eds.), *Smart and Sustainable Urban Ecosystems: Challenges and Solutions*, Springer Geography, https://doi.org/10.1007/978-3-031-37216-2_17

## 17.1 Introduction

Environmental protection (and mainly pollution control) is always an important problem for industrialized regions, and it is especially acute in the Russian Subarctic, where industrial pressure affects vulnerable ecosystems. Long-term pollution of soils by potentially toxic elements (PTE) not only increases the potential toxicity of soils, but also the contamination levels of waterbodies. In polar regions, long-term intensive pollution of soils by PTE has resulted in the formation of industrial barrens (IB)—open, desolate landscapes (projective vegetation cover less than 10%) that have developed around point sources of industrial pollution (Kozlov and Zvereva 2007). In the current paper, we analyzed a case of an IB in the vicinity of the town of Monchegorsk (Murmansk region).

The IB formed by the impact of the Cu/Ni smelter (Kola Mining and Metallurgical Company (MMC), formerly «Severonickel») in Monchegorsk (Kola Peninsula, NW Russia) is one of the most extensive. The MMC plant remains active nowadays, and its annual emissions are estimated to be up to 40,000 tons which can be deposited over vast areas up to 300 km away from the source (Moore 1978; Simms et al. 2003). The PTE content in the topsoil can reach extremely high values which are comparable to the processed Cu/Ni ore (Ni—0.24–4.2%, Cu—0.36–5.8%) (Kashulina 2017) and exceed the threshold limit value by 1–3 orders of magnitude (Kashulina 2018).

Despite an ecological policy aimed at reducing the environmental effect, revegetation in the Subarctic takes decades and centuries. Migration and accumulation of PTEs are associated with soil erosion and are controlled by landscape characteristics (e.g., slope steepness and vegetation). Soil erosion and PTE transit predominate on elevated surfaces and slopes, while topographic depressions act as reservoirs of polluted sediments.

We propose a specific phytoremediation strategy for each type of landscape. These strategies are based on the qualitative and quantitative assessment of the landscape (inc. soil erosion rates) as well as the specific ecological properties of plant species: attitude to light, humidity (flooding and draught), soil chemical composition, and wind resistance (based on the root system).

## 17.2 Study Area

The study area (Fig. 17.1) is located on the Kola Peninsula beyond the Arctic Circle. The climate is characterized by cold with no dry season and cold summers (Dfc by Köppen-Geiger classification) (Beck et al. 2018). The indigenous vegetation cover is formed mainly by plants of northern taiga species: *Picea abies* (L.) H. Karst and *Pinus sylvestris* L. (Manninen et al. 2015), *Salix sp.*, and *Betula pubescens* Ehrh. (Koptsik et al. 2016). Leptic Albic/Entic Podzol is the dominating soil type in the area (Dvornikov et al. 2022). Wetlands are also common—they occupy the low

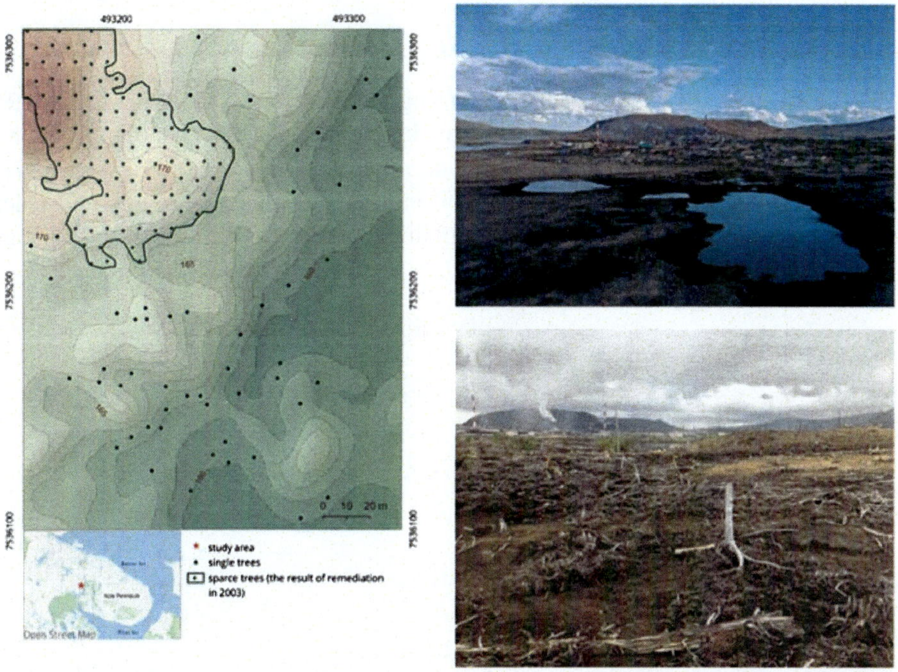


**Fig. 17.1** Monchegorsk industrial barren: topography (DTM was obtained as a result of photogrammetric UAV data processing (Dvornikov et al. 2022))

hypsometric levels and act as a place of peat accumulation; Histosol is the most typical soil type here.

Quaternary sediments are represented by moraine sandy loams with pebbles and boulders, as well as anthropogenic sediments (Smirnova et al. 2017a). The study area is in the suburban area and is characterized by an irregular semi-natural bumpy topography of glacial origin: small, partly isolated depressions, and hollows alternating with moraine crests and hummocks. The elevation varies from 159 to 177 m above sea level.

Soil type and topography explain up to 80% of the variance in the Cu/Ni content distribution (Fernandez et al. 2003). But it should be noted that the entire territory is extremely polluted: high acidity or alkalinity, poor vegetation, lack of the nutrients (Ca, Mg, Mn K, and P), and high concentrations of Ni and Cu (Slukovskaya et al. 2017).

The northwestern part of the site was managed with phytoremediation in 2003. On the part of the site, the degraded Podzol subsoils were covered by peat (Koptsik et al. 2016).

## 17.3 Materials and Methods

Geochemical studies were carried out during the field work. A field-portable XRF analyzer was used to assess the content of nickel, copper, and other chemical elements in soils (Dvornikov et al. 2022). PTE concentrations (Zn, Cd, Pb, Hg, Cu, Co, Ni, As, and Mn) were then estimated by soil indices: Average single pollution index (PIavg) (Inengite et al. 2015), Nemerow pollution index (PI Nemerow), and the total pollution index (Zc) (Romzaykina et al. 2020) to define more and less polluted areas.

$$\text{PIAvg} = \frac{1}{m}\sum_{i=1}^{m}\text{PI}i \tag{17.1}$$

$$\text{PI} = \frac{Ci}{B} \tag{17.2}$$

where

PI$i$ single pollution index of a particular PTE,
$n$ number of studied PTE,
$B$ background content of the PTE (Rossii 1997).

Contamination level is considered as the following: $\text{PI}_{\text{avg}} \leq 1$—low, $1 \leq \text{PI}_{\text{avg}} \leq 3$—moderate, $\text{PI}_{\text{avg}} \geq 3$—heavy.

$$\text{PI Nemerow} = \sqrt{\frac{(\frac{1}{n}\sum_{i-1}^{n}\text{PI})^2 + \text{PI}_{\text{max}}^2}{n}} \tag{17.3}$$

where

PI single pollution index of a particular heavy metal,
$\text{PI}_{\text{max}}$ maximal value of the single pollution index of all PTEs,
$n$ number of studied PTE.

Contamination level is considered as the following: PI *Nemerow* $\leq 0.7$—clean, $0.7 \leq$ PI *Nemerow* $\leq 1$—warning limit, $1 \leq$ PI *Nemerow* $\leq 2$—slight pollution, $2 \leq$ PI *Nemerow* $\leq 3$—moderate pollution, $3 \leq$ PI *Nemerow*—heavy pollution.

$$\text{Zc} = \sum K_{ci} - (n-1) \tag{17.4}$$

$$K_{ci} = \frac{Ci}{Cb} \tag{17.5}$$

where

$K_{ci}$ concentration coefficient of the *i*-th PTE,
$n$ number of PTEs,
$Ci$ an actual content of the *i*-th PTE in soils, mg kg$^{-1}$.

Contamination level is considered based on Methodology guidelines "Hygienic evaluation of soil in residential areas" from 1999-04-05:

> Zc < 16-permissible, 16 ≤ Zc ≤ 32—moderately dangerous, 32 ≤ Zc ≤ 128 dangerous, Zc > 128—extremely dangerous (Hygienic evaluation of soil in residential areas).

We used the well-known and commonly used (mostly for arable lands) empirical model Revised Universal Soil Loss Equation (RUSLE) to calculate soil erosion rates [Revised Universal Soil Loss Equation (2001), 53]. The Revised Universal Soil Loss Equation (RUSLE) estimates soil loss from erosion (A) caused by rainfall and associated surface runoff. RUSLE considers six parameters: climatic erosivity (*R*), soil erodibility (*K*), slope length and steepness (LS), cover management (*C*), and support practices (*P*) to calculate soil loss (Renard et al. 1997; Revised Universal Soil Loss Equation 2001).

$$A = R * K * \mathrm{LS} * C * P \tag{17.6}$$

R-factor is a measure of the total annual erosive rainfall for a specific location, as well as the distribution of erosive rainfall throughout the year. According to global rainfall erosivity dataset (Nikolova 2016), R-factor for the research area is 191 (MJ mm)/(ha h yr).

K-factor is the soil erodibility factor—the susceptibility of soil particles to detachment and transport by rainfall and runoff. Soil texture is the principal factor affecting K, but structure, organic matter content, and permeability are also important. Williams and Renard (1983) as cited in Chen et al. (2011) proposed the following equation for K-factor calculating (Benavidez et al. 2018):

$$\begin{aligned} K = 0.2 + 0.3\exp\left(0.0256 * \mathrm{Sa} * \left(1 - \frac{\mathrm{Si}}{100}\right)\right) * \left(\frac{\mathrm{Si}}{\mathrm{Cl} + \mathrm{Si}}\right)^{0.3} \\ * \left(1.0 - \frac{0.25 * \mathrm{C}}{\mathrm{C} + \exp(3.72 - 2.95\mathrm{C})}\right) \\ * \left(1.0 - \frac{0.7 * \mathrm{SN}}{\mathrm{SN} + \exp(-5.51 + 22.9\mathrm{SN})}\right) \end{aligned} \tag{17.7}$$

The correction factor converting *K*-value into SI units of metric ton hours per MJ per mm equals to 0.1317. We used averaged granulometric composition to calculate the factor according to Eq. 17.2. We used *K*-factor values that characterize typical soils in the study area: 0.013 for Podzol and 0.1 for Histosol. The *K*-factor value for Histosol was chosen according to Moore (1978) based on: (i) the similarity of natural environment with the Northwest Territories in Canada and (ii) the idea that *K*-factor value for the European Histosols should exceed 0.03 (Nikolova 2016).

The LS (slope length-gradient) factor represents a ratio of soil loss under given conditions to that at a site with the "standard" slope steepness of 9% and slope length of 22.13 m. The steeper and more prolonged the slope is, the higher are the erosion risks. The LS-factor was calculated based on DTM derivatives (slopes and float accumulation). The high-resolution (1.5 m) DTM was obtained as a result of photogrammetric UAV data processing (Dvornikov et al. 2022; Zhang et al. 2017).

C-factor is the crop or land cover management factor and is used to determine the relative effectiveness of soil and crop management systems in terms of preventing soil loss. RUSLE was developed for agricultural areas; however, it can be adapted for non-agricultural conditions. The study area is almost devoid of vegetation. Forest stand density is ensured in the northwestern part of the site only, where remediation was carried out in 2003. According to May and Place (2005), these sites were assigned a value of 0.13, which is typical for transitional woodland and shrubs. Outside this area, no protection of soil from erosion is ensured. Therefore, the value of the C-factor here is 1.0 (despite the presence of individual trees).

P is the support or land management practice factor, e.g., contouring, terracing, and strip cropping. For areas where there is no support practice (as in our area) the P factor is set to 1.0 (Simms et al. 2003).

For the selection of plant species, we have created the database with attitude of plant species to environmental controls (attitude to light, hardiness zone, moisture resistance, and draught tolerance) considering the accumulation capacity of Ni and Cu. For the selection of plant species, specific criteria for each strategy are defined (environmental controls and the erodibility factors), and the list of plants for each strategy reflects compliance with the chosen criteria.

## 17.4 Results

### *17.4.1 Soil Loss Estimated by RUSLE*

The value of the LS-factor varies from hundredths of a unit to nearly 8 (Fig. 17.2). The maximum values are confined to the steepest slopes, and the minimum values characterize sub-horizontal and gently sloping surfaces. The minimum LS-values act as a lowering factor in the equation (values less than 1). Maximum values indicate areas of maximum soil loss and exceed minimum values by 1–2 orders of magnitude. By the coincidence of the visual pattern of the LS and A images, it is also clear that geomorphological factors determine the spatial variability of erosion rates in the study area.

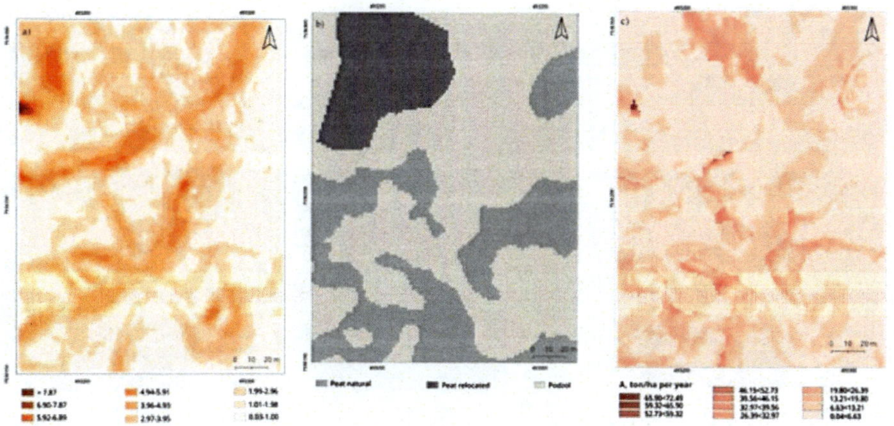

**Fig. 17.2** LS-factor (**a**), soil types (**b**), and soil loss values (**c**)

As a result, the maximum values of A (up to 72 tons/ha per year) characterize the steepest and longest slopes where soils are classified as Podzols, and the vegetation cover is reduced.

When converting A (t/ha per year) to erosion rates (mm per year), it was found that 4–10 mm of soil layer is washed away from the area per year. Due to the lack of vegetation cover, the soil has neither the time nor the opportunity to condense, being constantly devoid of organic matter source and exposed to external aggressions.

### *17.4.2 Soil Pollution by PTE*

Based on the calculations of the soil pollution indices (average single pollution index (PIavg), Nemerow pollution index (PI Nemerow), and the total pollution index (Zc), the following results were obtained (Fig. 17.3).

There are lower concentrations of heavy metals on hilltops compared to the topographic depressions represented by all three indices. In some cases, the concentrations of heavy metals on hilltops are five times lower than concentrations of the topographic depressions considering the Average single pollution index and the Nemerow pollution index. Regarding the total pollution index, the difference in heavy metal concentration reach up to 95 times and can be explained by the migration of heavy metals from the upper surfaces to depressions due to the factors of erodibility and the higher capability of Histosols to accumulate heavy metals compared to Podzols (Paltseva et al. 2022).

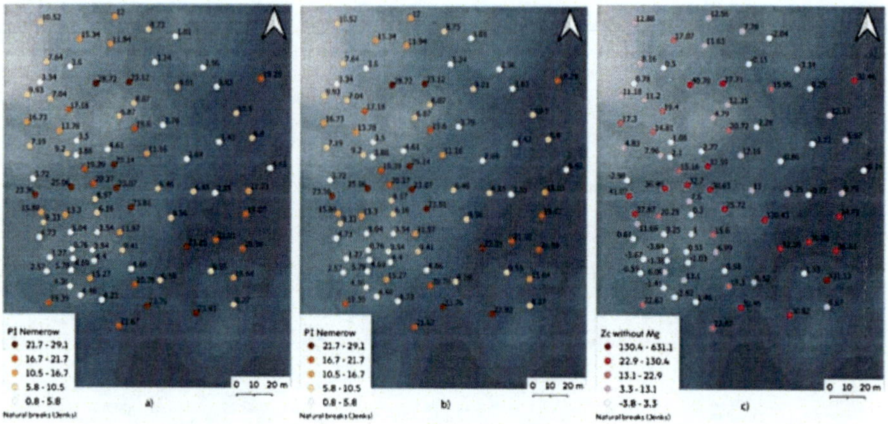


**Fig. 17.3** Spatial representation of the soil pollution of the IB estimated by three soil pollution indices (Average single pollution index (PIavg) **a**, Nemerow pollution index (PI Nemerow) **b**, the total pollution index (Zc) **c**) considering the heights of the relief

## *17.4.3 Phytoremediation Strategies*

While considering the issue of remediation of the industrial barren on the Kola Peninsula, it is important to consider (i) the high content of PTE (Cu and Ni), (ii) soil erosion, and (iii) climate conditions. Based on the data obtained on soil loss values, three phytoremediation strategies were developed: for hilltop areas, for slopes, and for depressions. Due to the harsh climate conditions on the Kola Peninsula (Dvornikov et al. 2022), only local plant species can be used, metal-hyperaccumulating plant species have not been implemented for phytoremediation strategies, as no suitable plant species was identified so far in the area. Phytoremediation strategies are chosen based on factors which are used in RUSLE: K-factor, LS-factor, C-factor, and P-factor. Were also considered hardiness zone, sun exposure, moisture resistance, and draught tolerance.

Local species able to cope with soil pollution are used for different remediation strategies on the Monchegorsk IB. In order to create more sustainable ecosystems for phytoremediation, it is more beneficial to use different forms of plants. However, the use of trees only or perennials is also possible depending on the area ($m^2$) and steepness. The database containing plant species that are tolerant to pollution and arctic climate conditions was proposed (Table 17.1).

*Phytoremediation of Hilltop Areas.* Hilltop areas are characterized by flat to gently undulating terrain (up to 5–7° steep) with an elevation ranging from 165 to 177 m a.s.l. Rain and snowmelt water transfers soil mass from the upper surfaces to the bottoms of depressions. Therefore, the values of the PTE content increase from top to bottom. Soil moisture in hilltop areas is generally lower than in the rest of the territory. However, the substrate (soil type) is an important factor in this case.

The A-value is higher near the edges which require plant species with strong root systems. For the unvegetated plots with a higher LS-factor (higher than 4.94)

**Table 17.1** Plant species for the industrial barren remediation strategies. *Legend* Attitude to light: shade-tolerant (s-t), heliophyte (h); attitude to flooding and draught: resistant (r), non-resistant (n-r), hardly withstands (hw), short flooding (sf), short draught (sd); root system strength: relatively strong (rs), strong (s); root system type: surface (s), deep (d)

| Plant species name | Accumulated elements | Height (max) (m) | Crown diameter (max) (m) | Attitude to light | Attitude to flood | Attitude to draught | Root system strength | Root system type (WUR collection) | References |
|---|---|---|---|---|---|---|---|---|---|
| *Picea abies (L.) Karst* | Ni, Co, Zn, Cu | 5 | 0.6 | s-t | r | r | rs | s | Orlova et al. 2013) |
| *Picea obovata Ledeb* | Ni, Co, Zn, Cu | 5 | 0.6 | s-t | r (sf) | r | rs | s | Chernen'kova et al. 2014) |
| *Pinus sylvestris* | Ni, Cu, Pb, Cd | 5 | 0.6 | h | r | r | s | d | Golubeva and Glukhova 2015) |
| *Betula nana* | Ni, Cu, Pb, Cd | 0.5 | 0.5 | h | r | r (sd) | rs | s | Orlova et al. 2013) |
| *Betula pubescens* | Cu, Pb, Zn | 4 | 1 | h | r | hw | rs | s | Koptsik et al. 2016) |
| *Betula pendula Roth* | Cu, Pb, Zn | 5 | 0.5 | h | r | r | rs | s | Pfadenhauer and Klötzli, 2020) |
| *Populus tremula* | Pb, Cd, Co, Zn, V, Sr | 6 | 0.5 | h | r (sf) | r | rs | s | Kozlov and Zverev 2022) |
| *Salix arctica* | Cd, Zn | 0.3 | 0.5 | s-t | r | hw | rs | s | Orlova et al. 2013) |
| *Salix schwerinii* | Cd, Zn | 5 | 5 | s-t | r | hw | rs | s | Terebova et al. 2014) |
| *Larix sibirica* | Ni, Cu | 5 | 1 | h | r | hw | rs | s | Renard et al. 1997) |

(continued)

**Table 17.1** (continued)

| Plant species name | Accumulated elements | Height (max) (m) | Crown diameter (max) (m) | Attitude to light | Attitude to flood | Attitude to draught | Root system strength | Root system type (WUR collection) | References |
|---|---|---|---|---|---|---|---|---|---|
| *Alnus incana* | Ni, Cu, Pb | 5 | 2 | h | r | hw | rs | s | Smirnova et al. 2017a) |
| *Salix polaris* | Ni, Cu, Cd, Zn | 0.1 | 3–6 | s-t | r | hw | rs | s | Pfadenhauer and Klötzli, 2020) |
| *Salix phylicifolia L* | Ni, Cu, Cd, Zn | 0.5 | 0.5 | s-t | r | hw | rs | s | Koroleva and Kopeina 1937) |
| *Salix glauca L* | Ni, Cu, Cd, Zn | 0.5 | 0.5 | s-t | r | hw | rs | s | Koroleva and Kopeina 1937) |
| *Salix caprea* | Cd, Co, Cr, Ni, Cu, Al, Fe, As, S, Ca | 3 | 3 | s-t | r | hw | rs | s | Koroleva and Kopeina 1937) |
| *Salix cinerea L* | Ni, Cu, Cd, Zn | 2 | 2 | s-t | r | hw | rs | s | Koroleva and Kopeina 1937) |
| *Festuca rubra L* | Zn, Cu, Pb, Ni, Fe | 0.2 | – | h | r | hw | s | d | Slukovskaya et al. 2017) |
| *Festuca pratensis Huds* | Zn, Cu, Pb, Ni, Fe | 0.2 | – | h | r | hw | s | d | Koroleva and Kopeina 1937) |
| *Bromopsis inermis (Leyss.) Holab* | F, Cu, Ni | 0.7 | – | h | r | r | s | d | Koroleva and Kopeina 1937) |
| *Dactylis glomerata (L.)* | Cu, Ni | 0.6 | – | h | hw | r | s | d | Koroleva and Kopeina 1937) |

(continued)

**Table 17.1** (continued)

| Plant species name | Accumulated elements | Height (max) (m) | Crown diameter (max) (m) | Attitude to light | Attitude to flood | Attitude to draught | Root system strength | Root system type (WUR collection) | References |
|---|---|---|---|---|---|---|---|---|---|
| *Elytrigia repens (L.)* | Cu, Ni | 0.5 | – | h | r | r | s | s | Koroleva and Kopeina 1937) |
| *Calamagrostis lapponica* | Cu, Ni | 0.5 | – | s-t | r | r | s | d | Koroleva and Kopeina 1937) |
| *Calamagrostics phragmitoides Hartm* | Cu, Ni | 0.5 | – | s-t | r | r | s | d | Kozhin and Sennikov 2016) |
| *Deschampsia caespitosa* | Cu, Ni | 0.3 | – | h | r | r | s | d | Koroleva and Kopeina 1937) |
| *Poa arctica* | Cd, Pb, Cu, Zn | 0.1 | – | s-t | n-r | r | rs | s | Pfadenhauer and Klötzli, 2020) |
| *Lolium perenne* | Cd, Pb, Cu, Zn | 0.2 | – | h | r | r | s | d | Tarasova et al. 2020) |
| *Calamagrostis epigeios (L.) Roth* | Ni, Pb, Cu | 0.8 | – | h | r | r | s | d | Smirnova et al. 2017a) |
| *Chamaenerion angustifolium (L.) Scop* | Pb, Zn, Cd, Cu, Ni | 0.8 | – | h | hw | r | rs | s | Koroleva and Kopeina 1937) |

denser perennial plantings are needed combined with trees and bushes. The plants-heliophytes resistant to winds and draught with a strong root system, that are able to grow on Histosols, can be used for phytoremediation. The tree species that are more tolerant to specific conditions on such landforms are *Pinus sylvestris* and *Populus tremula. Populus tremula* is highly recommended for the polluted sites as it gives shoots even without any improvement of the soil quality (Kozlov and Zverev 2022). The perennials for the hilltop surfaces are *Bromopsis inermis* (Leyss.) Holab, *Dactylis glomerata* L., *Elytrigria repens* L., *Deschampsia caespistosa* L., *Lolium perenne* L., *Calamagrostics epigeios* (L.) Roth., and *Chamaerion angustifolium* (L.) Scop.

*Phytoremediation of Slopes.* The slopes are the most dynamic areas on the territory and diversified both in steepness and exposure. However, most of the slopes are about 15–20° steep. Therefore, it is important to consider when choosing plants their root systems which must be able not to gain a foothold and develop on the slope only, but also to maintain their stability in the future. By stabilizing slopes, root systems decrease erosion, PTE mobility, and availability (Koptsik et al. 2016).

The slope remediation strategy may include plants of different life forms. For example, grass vegetation reduces the A-value by about 20 times (Morel et al. 2015), based on CP-factors in RUSLE (with a constant $P = 1$). However, the grass cover is characterized by seasonality, so erosion become active during spring snowmelt.

*Exposure to the sun influences the moisture of the slope. However, the elevation changes do not provide much shadow from one hill to another one. Therefore, sun exposure is not the main factor considering strategies.* Slopes are exposed to higher winds and erosion rate than depressions. Plant species for slopes are *Picea abies (L.) Karst*, *Picea obovata* Ledeb., *Pinus sylvestris, Betula pendula Roth, Betula nana, Betula pubescens, Populus tremula, Salix caprea, Salix cinerea L., Salix schwerinii, Larix sibirica, Elytrigia repens* (L.), *Poa arctica, Dactylis glomerata L., Lolium Perenne, Chamaenerion angustifolium* (L.) Scop., *Calamagrostis epigeios (L.) Roth.*, and *Festuca rubra. Festuca rubra* prevents the migration of Cu and Ni if it is grown on the substrate whose toxicity module equals 0.01 (Slukovskaya et al. 2017). Hence, it is a beneficial plant for slope stabilization; however, the initial soil should be covered by another layer of less toxic substrate (it is relevant to each phytoremediation strategy).

*Phytoremediation of Depressions.* Moraine depressions and hollows occupy elevations below 160 m above sea level and are the wettest areas. Accumulation predominates here, and it leads to a hyperconcentration of PTE transferred by lateral migration. The plants used here must be resistant not only to high PTE concentrations, but also to almost constant waterlogging. Limiting factors will play a decisive role in the selection of plants. The ability of root systems to limit erosion does not matter here. Thus, in depressions a wetland ecosystem can be obtained with plants capable of absorbing PTE. Such plants include: *Betula pendula* Roth., *Betula pubescens, Salix arctica, Salix shwerinii, Salix polaris, Salix phylicifolia* L., *Salix glauca* L., *Salix cinerea* L., and *Salix caprea. Betula pubescens* and *Salix caprea* are able to maintain their chemical composition even under the extreme conditions of an IB (Koptsik et al. 2016). Perennials meeting the depression criteria are *Calamagrostics lapponica, Calamagrostics phragmitoides, Calamagrostics epigeios* (L.) Roth., *Deschampsia caespitosa, Poa arctica, Elytrigia repens (L.), and Lolium perenne* (Fig. 17.4).

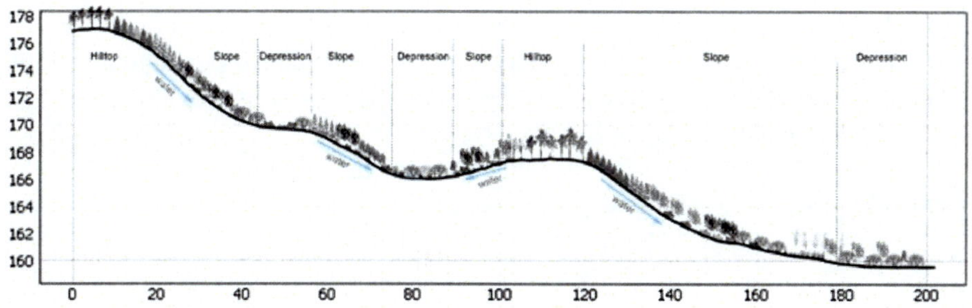


**Fig. 17.4** Phytoremediation strategies visualization

## 17.5 Discussion

The K-factor showed little variability because the averaged data of the composition of the Podzols was used. In addition, a problem was detected in determining the K-factor for soils saturated with organic content (Revised Universal Soil Loss Eq. 2001; Wall et al. 2002). The values of K-factor offered for different regions and types of land-use are sometimes several times different (Fernandez et al. 2003; May and Place 2005).

Soil loss values corresponded to the average values for European Part of Russia or slightly exceeded them. The exceedances of the average values were obtained due to the absence of vegetation cover in most of the territory, as well as relatively low values of soil density (we have 0.4–1.0 g/cm$^3$ for Histosols and 1.0–1.6 g/cm$^3$ for Podzols).

In fact, there is no natural vegetation on the territory, and the result of remediation is a sparse forest stand. The C-factor used (May and Place 2005) successfully characterizes the territory. However, there are other estimates of the C-factor (e.g., (Fernandez et al. 2003)), which are used for natural vegetation and urban areas. During the existence of a modern copper-nickel smelter (25 years), erosional truncation in different parts of the study area ranged from 10 to 25 cm. The average values of soil losses (A-value) are similar to the results of other researchers (Li et al. 2018; Tsymbarovich et al. 2020)—the order of magnitude, as well as variability. Despite the opinion (Gorbacheva 2011) that the RUSLE gives an overestimate, our results reflect the relationship between soil erosion and pollution rates.

Plant species for remediation strategies are chosen based on their natural areal, tolerance to soil pollution and criteria for the specific conditions of each type of landforms. However, there are still other factors that influence plants liveability, i.e., nutrient supply, air quality, soil quality, interrelationship between different plants of a particular plot, biotic stress, etc. (Schulze et al. 2019).

Probably, there is not only lateral migration down the slope, but also deep into the soil. Hence, the PTE concentrations do not only depend on the location of erosion and accumulation sites.

The presence of IB in the suburbs of Monchegorsk strongly affects the quality of life of residents. Remediation of the site and long-term effect of management should consider the provision of ecosystem services (Morel et al. 2015). Such services include the benefits for the biodiversity and for the health of population resulting from the restoration and protection of the landscape (foremost vegetation cover and soils). Currently, the quality of these ecosystem services is very low, but well-chosen phytoremediation strategies in line with environmental monitoring are very much in demand.

## 17.6 Conclusions

The maximum rates of soil erosion within the territory (on steep slopes devoid of vegetation cover) exceed 10–12 times the rates of erosion on very gentle slopes and subhorizontal surfaces and 3–4 times—on slopes of medium steepness. Vegetation further reduces the rate of erosion, sometimes even by 1–2 orders of magnitude (on the highest flat-topped hills). Intensive soil erosion is associated with the absence of vegetation cover. In addition, slow revegetation rate in subarctic conditions and under industrial pollution leads to continued soils degradation.

The geomorphological characteristics of the territory are decisive for the erosion process, due to the high variability of the LS-factor. Therefore, when choosing a remediation strategy, it is necessary to take into account the features of the topography of the territory. The full RUSLE equation or the LS-factor (for an approximate estimate) can be used for small areas with low variability of soil types and other factors. The use of local plant species or plant species that are able to survive in arctic climate conditions allows them to spread throughout the territory with low additional plants care measures. The suggested species could be recommended for phytoremediation strategies in the Kola Peninsula IB territory. Providing suitable phytoremediation strategies will decrease the erosion processes and stabilize the slopes, as well as slow down PTE migration gradually to acceptable PTE concentrations. In addition to pollution mitigation, which is the priority of these sites, phytoremediation strategies should improve the ecosystem services provided by the area, including biodiversity, carbon storage, and landscape. However, these services must be evaluated to attest to their effectiveness.

**Acknowledgements** This research was supported by RUDN University Strategic Academic Leadership Program (Dvornikov Yu.A., Gurinov A.L.) and by the state assignment № FMGE-2019-0005 of the IG RAS. Field work was supported by the Russian Science Foundation Project #21-77-10111.

## References

Beck HE, Zimmermann NE, McVicar TR, Vergopolan N, Berg A, Wood EF (2018) Present and future Köppen-Geiger climate classification maps at 1-km resolution. Sci Data 5:180214. https://doi.org/10.1038/sdata.2018.214

Benavidez R, Jackson B, Maxwell D, Norton K (2018) A review of the (Revised) Universal Soil Loss Equation (R/USLE): with a view to increasing its global applicability and improving soil loss estimates. Hydrol Earth Syst Sci Discuss 1–34. https://doi.org/10.5194/hess-2018-68

Capilleri PP, Motta E, Raciti E (2016) Experimental study on native plant root tensile strength for slope stabilization. Procedia Eng 158:116–121. https://doi.org/10.1016/j.proeng.2016.08.415

Chernen'kova TV, Bochkarev YN, Friedrich M, Boettger T (2014) The impact of natural and anthropogenic factors on radial tree growth on the northern Kola Peninsula. Contemp Probl Ecol 7(7):759–769. https://doi.org/10.1134/S199542551407004X

Dvornikov Y, Slukovskaya M, Yaroslavtsev A, Meshalkina J, Ryazanov A, Sarzhanov D, Vasenev V (2022) High-resolution mapping of soil pollution by Cu and Ni at a polar industrial barren area using proximal and remote sensing. Land Degrad Dev

Fernandez C, Wu JQ, Mccool D, Stockle C (2003) Estimating water erosion and sediment yield with GIS, RUSLE, and SEDD. J Soil Water Conserv 58

Golubeva EI, Glukhova EV (2015) Phytomelioration in recovery of pine forests on sandy land-scapes on Tersky coast of the white sea. Lomonosov Moscow State University

Gorbacheva EN (2011) Rusle model implementation for estimation intensivity of water erosion proceses on Belarussian soil cover. Soil Sci Agrochem (2):42–51 (In Russ.)

Inengite AK, Abasi CY, Walter C (2015) Application of pollution indices for the assessment of heavy metal pollution in flood impacted soil. Int Res J Pure Appl Chem 8:175–189

Jenson SK, Domingue JO (1988) Extracting topographic structure from digital elevation data for geographic information system analysis. Photogramm Eng Rem Sens 54(11):1593–1600

Kashulina GM (2017) Extreme pollution of soils by emissions of the copper–nickel industrial complex in the Kola Peninsula. Eurasian Soil Sci 50:837–849. https://doi.org/10.7868/s0032180x17070036

Kashulina GM (2018) Monitoring of soil contamination by heavy metals in the impact zone of copper-nickel smelter on the Kola Peninsula. Eurasian Soil Sci 51:467–478. https://doi.org/10.7868/s0032180x1804010x

Koptsik GN, Koptsik SV, Smirnova IE (2016) Alternative technologies for remediation of technogenic barrens in the Kola Subarctic. Eurasian Soil Sci 49(11):1294–1309. ISSN 1064-2293. © Pleiades Publishing, Ltd.

Koptsik GN, Koptsik SV, Smirnova IE, Sinichkina MA (2021) Effect of soil degradation and remediation in technogenic barrens on the uptake of nutrients and heavy metals by plants in the Kola Subarctic. Eurasian Soil Sci 54(8):1252–1264. ISSN 1064-2293. © Pleiades Publishing, Ltd.

Koroleva NE, Kopeina EI (2018) The floodplain meadows of class Molinio-Arrhenatheretea TX. 1937 in the Varzuga RI VER valley (Murmansk region). Rastitel'nost' Rossii 2018(34):85–100. https://doi.org/10.31111/vegrus/2018.34.85

Kozhin MN, Sennikov AN (2016) The Russian larch (*Larix archangelica*, Pinaceae) in the Kola Peninsula. Memoranda Soc. Fauna Flora Fennica 92:79–91

Kozlov MV, Zverev V (2022) Suitability of European Aspen (Populus tremula) for rehabilitation of severely polluted areas. Russ J Ecol 53(3):181–190. ISSN 1067-4136. © Pleiades Publishing, Ltd.

Kozlov MV, Zvereva EL (2007) Industrial barrens: extreme habitats created by non-ferrous metallurgy. Rev Environ Sci Biotechnol 6:231–259. https://doi.org/10.1007/s11157-006-9117-9

Li C, Grayson R, Holden J, Li P (2018) Erosion in peatlands: recent research progress and future directions. Earth-Sci Rev 185. https://doi.org/10.1016/j.earscirev.2018.08.005

Manninen S, Zverev V, Bergman I, Kozlov M (2015) Consequences of long-term severe industrial pollution for above ground carbon and nitrogen pools in northern taiga forests at local and regional scales. Sci Total Environ 536:616–624. https://doi.org/10.1016/j.scitotenv.2015.07.097

May L, Place C (2005) A GIS-based model of soil erosion and transport. Freshwater Forum 23

Meijer G, Bengough G, Knappett J, Loades K, Nicoll B (2019) Measuring the strength of root-reinforced soil on steep natural slopes using the cork-screw extraction method. Forests 10(12). https://doi.org/10.3390/F10121135

Moore TR (1978) Soil formation in Northwestern Canada. Ann Assoc Am Geogr 68(4):518–534; 48-61. https://doi.org/10.1111/j.1467-8306.1978.tb01214.x

Morel JL, Chenu C, Lorenz K (2015) Ecosystem services provided by soils of urban, industrial, traffic, mining, and military areas (SUITMAs). J Soils Sediments 15:1659–1666

Nikanov AN, Dorofeev VM, Megorskiy VV, Zhirov VK (2020) Environmental aspects of the accumulation of mineral elements in the organism of the population living in areas of intensive industrial activities in the European part of the arctic zone of Russia. J Chem Inform Model 53(9). FRC KSC RAS

Nikolova E (2016) Soil erosion modelling using RUSLE and GIS in the Republic of Macedonia. Tesi di laurea Magistrale. ING I - Scuola di Ingegneria Civile, Ambientale e Territoriale. https://www.politesi.polimi.it/handle/10589/132018

Orlova MA, Lukina NV, Tutubalina OV, Smirnov VE, Isaeva LG, Hofgaard A (2013) Soil nutrient's spatial variability in forest-tundra ecotones on the Kola Peninsula, Russia. Biogeochemistry 113(1–3):283–305. https://doi.org/10.1007/s10533-012-9756-6

Paltseva A, Slukovskaya M, Romzaykina O, Sarzhanov D, Drogobuzhskaya S, Dvornikov Y, Vasenev V (2022) Proximal sensing of soil pollution by heavy metals using a portable X-ray fluorescence analyzer in subarctic industrial barren: limitations and perspectives. In: Rakshit A, Ghosh S, Vasenev V, Pathak H, Rajput VD (eds) Soils in urban ecosystem. Springer, Singapore. https://doi.org/10.1007/978-981-16-8914-7_11

Panagos P, Borrelli P, Meusburger K et al (2017) Global rainfall erosivity assessment based on high-temporal resolution rainfall records. Sci Rep 7:4175. https://doi.org/10.1038/s41598-017-04282-8

Pfadenhauer JS, Klötzli FA (2020) Vegetation of the polar zone. In: Global vegetation. Springer, Cham. https://doi.org/10.1007/978-3-030-49860-3_14

Popova MB, Manakhov DV, Kizeev AN, Ushamova SF, Lipatov DN, Chirkov AY, Orlov PS, Mamikhin SV (2020) Contents and distribution of 137Cs in Podzols in the area of the Kola nuclear power plant. Eurasian Soil Sci 53(7):986–994. https://doi.org/10.1134/S10642293200 7011X

Renard K, Foster G, Weesies G, McCool D, Yoder D (1997) Predicting soil erosion by water: a guide to conservation planning with the revised universal soil loss equation (RUSLE), USDA Agriculture Handbook No. 703. United States Government Printing Office, Washington, DC, USA

Revised Universal Soil Loss Equation. Version 2. (RUSLE2). HANDBOOK. Prepared by the USDA RUSLE Development Team: March (2001)

Romzaykina ON, Vasenev VI, Paltseva A, Kuzyakov YV, Neaman A, Dovletyarova EA (2020) Assessing and mapping urban soils as geochemical barriers for contamination by heavy metal(loid)s in Moscow megapolis. J Environ Qual. https://doi.org/10.1002/jeq2.20142

Wageningen University and Research Image collections, Root systems drawings. https://images.wur.nl/digital/collection/coll13/search. Last accessed 2022/07/20

Rossii G (1997) Engineering and environmental surveys for construction (SR 11-102-97). Gosstroy Rossii, Moscow

Saltan NV, Sviatkovskaya EA (2021) Tree health of *Larix sibirica* Ledeb. in the railway impact zone on Kola Peninsula. In: Advanced technologies for sustainable development of urban green infrastructure. SSC. Springer geography. Springer, Cham. https://doi.org/10.1007/978-3-030-75285-9_1

Schulze E-D, Beck E, Buchmann N, Clemens S, Müller-Hohenstein K, Scherer-Lorenzen M (eds) (2019) Plant ecology. Springer, Berlin/Heidelberg, Germany, pp 257–299

Simms AD, Woodroffe CD, Jones BG (2003) Application of RUSLE for erosion management in a coastal catchment, southern NSW 2003. https://ro.uow.edu.au/scipapers/34

Slukovskaya MV, Vasenev VI, Ivashchenko KV, Dolgikh AV, Novikov AI, Kremenetskaya IP, Ivanova LA, Gubin SV (2020) Organic matter accumulation by alkaline-constructed soils in heavily metal-polluted area of subarctic zone. J Soils Sediments. https://doi.org/10.1007/s11368-020-02666-4

Slukovskaya MV, Kremenetskaya IP, Ivanova LA, Vasilieva TN (2017) Remediation in conditions of an operating copper-nickel plant: results of perennial experiment. Non-Ferrous Metals 43(2):20–26. https://doi.org/10.17580/nfm.2017.02.04

Slukovskaya MV, Novichonok EV, Kremenetskaya IP, Mosendz IA, Drogobuzhskaya SV, Markovskaya EF (2017) Application of festuca rubra l. in phytoremediation: complex evaluation of the technogenic soil effect on the plant. Uchenye zapiski of Petrozavodsk State University 4(165):70–80

Smirnova OV, Bobrovsky M, Khanina L (2017a) Boreal forests. In: Smirnova O, Bobrovsky M, Khanina L (eds) European Russian forests. Plant and vegetation, vol 15. Springer, Dordrecht. https://doi.org/10.1007/978-94-024-1172-0_3

Smirnova O, Bobrovsky M, Khanina L (2017b) Natural conditions and general descriptions of forest vegetation and forest soils. In: Smirnova O, Bobrovsky M, Khanina L (eds) European Russian forests. Plant and vegetation, vol 15. Springer, Dordrecht. https://doi.org/10.1007/978-94-024-1172-0_1

State Geological Map of the Russian Federation. Scale 1: 1,000,000 (third generation). The Baltic series. Sheet Q-(35), 36—Apatites. Explanatory note. Astafyev BYu, Bogdanov YuB, Voinova OA, Warriors AS, etc. (eds). VSEGEI Cartographic Factory, St. Petersburg, p 436. + 5 incl.

Research Institute of human ecology and environmental hygiene named after A.N. Synin (Rusakov NV, Tonkopiy NI, Velikanov NL), E.I. Martsinovsky Institute of Medical Parasitology and Tropical Medicine (Romanenko NA, Novosiltsev GI, Gannushkina LA, Dremova VP, Chromenkova EP, Grimaylo LV, Kozyreva TG, Evdokimov VI, Zemlianskiy OA, Evdokimov VV, Volishev AN, Gorokhov VV), "RADON" LLC (Simonov VD), All-Russian Research Institute of Environmental Protection (Matveev YN). Methodology guidelines 2.1.7.730-99 "Hygienic evaluation of soil in residential areas" from 1999-04-05

Tarasova E, Drogobuzhskaya S, Tapia-Pizarro F, Morev DV, Brykov VA, Dovletyarova EA, Slukovskaya M, Navarro-Villarroel C, Paltseva AA, Neaman A (2020) Vermiculite-Lizardite industrial wastes promote plant growth in a peat soil affected by a Cu/Ni smelter: a case study at the Kola Peninsula, Russia. J Soil Sci Plant Nutr 20(3), 1013–1018. https://doi.org/10.1007/s42729-020-00188-z

Terebova EN, Androsova VI, Markovskaya EF, Kaipiainen E (2014) The estimation of Salix schwerinii capacity for phytoremediation of polluted areas. https://doi.org/10.13140/2.1.4810.4000

Tsymbarovich P et al (2020) Soil erosion: an important indicator for the assessment of land degradation neutrality in Russia. Int Soil Water Conserv Res 8(4):418–429. https://doi.org/10.1016/j.iswcr.2020.06.002

Wall GJ, Coote DR, Pringle EA, Shelton IJ (eds) RUSLEFAC—revised universal soil loss equation for application in Canada: a handbook for estimating soil loss from water erosion in Canada. Research Branch, Agriculture and Agri-Food Canada, Ottawa. Contribution No. AAFC/AAC2244E, p 117

Wischmeier WH, Smith DD (1978) Predicting rainfall erosion losses—a guide to conservation planning. U.S. Department of Agriculture, Agriculture Handbook No. 537

Zhang H, Wei J, Yang Q, Baartman J, Gai L, Yang X, Yu J, Ritsema C, Geissen V (2017) An improved method for calculating slope length (λ) and the LS parameters of the Revised Universal Soil Loss Equation for large watersheds. Geoderma 308:36–45

# Chapter 18
# Remediation of Urban Soils with Nitrogen and Microbiological Fertilizers when Contaminated by Oil, Gasoline and Fuel Oil

**Tatiana Minnikova, Anna Ruseva, and Sergey Kolesnikov**

**Abstract** Rostov-on-Don is a major agro-industrial center in the South of Russia and one of the key points on the way to the Black Sea coast of Russia. At the same time, the share of motor transport (passenger cars) is 88%, trucks (small and medium-sized)—7% of the total transport, respectively. Commercial oil and its derivatives are regularly transported by road and rail through the territory of the city as a transport hub. Considering that diesel fuel and gasoline are used as fuel for cars and trucks, and the transportation of commercial oil often passes through the territory of Rostov-on-Don, remediation of oil-polluted urban soils is necessary. The work objective was to assess the impact of nitrogen and microbiological fertilizers on the ecological state of urban soils in Rostov-on-Don. To assess the ecological state of soils, an experiment was modeled under controlled laboratory conditions. For the experiment, the level of oil pollution of 5% of the soil mass was studied; the incubation period was 30 days. The ecological state was assessed by indicators of phytotoxicity, enzymatic activity, and the total number of soil bacteria. It was found that oil, fuel oil and gasoline pollution caused inhibition of biological parameters by 23–49%. When applying complex mineral fertilizer as nitrogen fertilizer and "Baikal EM-1" as microbiological fertilizer, a decrease in phytotoxicity of contaminated soils with fuel oil and gasoline was observed by 7–75% compared to no treatment soil. In the soil with gasoline, the introduction of complex mineral fertilizer and "Baikal EM-1" caused stimulation of catalase activity, dehydrogenases and the number of soil bacteria by 9–55% compared to contamination with gasoline alone. As a result of the study, it was found that the introduction of complex mineral fertilizer and "Baikal EM-1" into haplic chernozem contributed to the restoration of the soil ecological state when contaminated with light fraction petroleum products such as gasoline, and for fuel oil and oil, it is necessary to adjust the exposure period and doses of ameliorants.

---

T. Minnikova (✉) · A. Ruseva · S. Kolesnikov
Academy of Biology and Biotechnology Named After D.I. Ivanovsky, South Federal University, Rostov-on-Don, Russian Federation
e-mail: loko261008@yandex.ru



M. Korneykova et al. (eds.), *Smart and Sustainable Urban Ecosystems: Challenges and Solutions*, Springer Geography, https://doi.org/10.1007/978-3-031-37216-2_18

**Keywords** Haplic chernozem · Rostov-on-Don · Pollution · Complex mineral fertilizer · "Baikal EM-1" · Enzymatic activity · Phytotoxicity · Total number of bacteria

## 18.1 Introduction

Rostov-on-Don is a major agro-industrial center in the South of Russia and one of the key points on the road to the Caucasus and the Black Sea coast of Russia. More than 267.3 million people are projected to be transported through Rostov-on-Don in 2022, with the length of the network of dedicated lanes for route vehicles in 2021–2022—93.0 km (The Program... 2021). At the same time, the share of motor transport (passenger cars) is 88%, and trucks (small and medium-sized)—7% of the total number of vehicles. Commercial oil and its derivatives are regularly transported through the territory of the city as a transport hub (Al'bekov 2011; Sablina 2020). Diesel and gasoline are used as fuel for cars and trucks; fuel oil is used as boiler fuel for sea and river transport, in heat generators in agriculture, for various heating systems, furnaces, technological installations and steam heating systems. Pollution with oil and petroleum products occurs in any city or settlement (Gospodarek et al. 2021). In the case of an oil leak in an abnormal mode and in an emergency, contamination of surface waters and soil occurs, and the sanitation or remediation of oil-polluted soils is necessary. Petroleum products such as diesel fuel, gasoline and motor oil have a negative impact on soil biochemical processes and increase phytotoxicity (Caravaca and Rodán 2013; Gürtler et al. 2022; Kolesnikov et al. 2010; Rusin et al. 2017, 2018; Swapna et al. 2021; Wyszkowski et al. 2020, 2022). Restoration of soils with this type of pollution is complicated by a number of parameters: physico-chemical properties of oil and petroleum products (viscosity, density, octane number, impurity content, etc.), quantitative characteristics of oil and petroleum products (volume and concentration), soil buffering capacity and is also directly related to the biological properties of the soil (Ahmad et al. 2020). Therefore, bioremediation is the most effective method of restoring oil-contaminated soils. Within the framework of bioremediation measures, several main directions are used: stimulation of native hydrocarbon-oxidizing biota by applying organic and mineral fertilizers, bioaugmentation when other hydrocarbon-oxidizing microorganisms are added to soil native biota, phytoremediation when plants are planted in contaminated soil (Alori et al. 2022; Sales da Silva et al. 2020). Humic and nitrogen fertilizers were previously used effectively when the soil was polluted with oil of various concentrations (Minnikova et al. 2019). These fertilizers acted as stimulators of the native biota: nitrogen fertilizers equalized the ratio of C:N, and humic substances stimulated the native biota. In addition, mineral sorbents (glauconite) and organic sorbents (biochar) affect the soil condition during oil pollution along with bacterial preparations (Hazim and Al-Ani et al. 2019; Koshlaf and Ball 2017; Minnikova et al. 2021, 2022a, b, c; Ruseva et al. 2022).

In this regard, it is relevant to study the ability of oil-polluted urban soils (for example, in Rostov-on-Don) to remediate after the use of nitrogen fertilizers and microbiological preparations combining strains of various microorganisms (bacteria, fungi and actinomycetes).

The aim is to assess the ecological state of urban soils after bioremediation with nitrogen and microbiological fertilizers when contaminated with oil, gasoline and fuel oil.

## 18.2 Materials and Methods

*Soil.* To assess the ecological state of soils during remediation of contamination of haplic chernozem by oil, fuel oil and gasoline, an experiment was modeled under controlled laboratory conditions. Soil samples were taken from the top layer of 0–5 cm of arable land located in experimental plots of the Botanical Garden of the Southern Federal University (Fig. 18.1).

The soil was dried to an air-dry basis; organic and inorganic residues (roots, stones and other debris) were removed. Then, the soil was sifted through a sieve with a cell size of 3 mm.

*Experiment simulation.* For the experiment, 300 g of pre-prepared air-dry soil was moistened, and oil 5% of the soil weight and ameliorants were added. Complex mineral fertilizer and microbiological preparation—"Baikal EM-1"—were used in

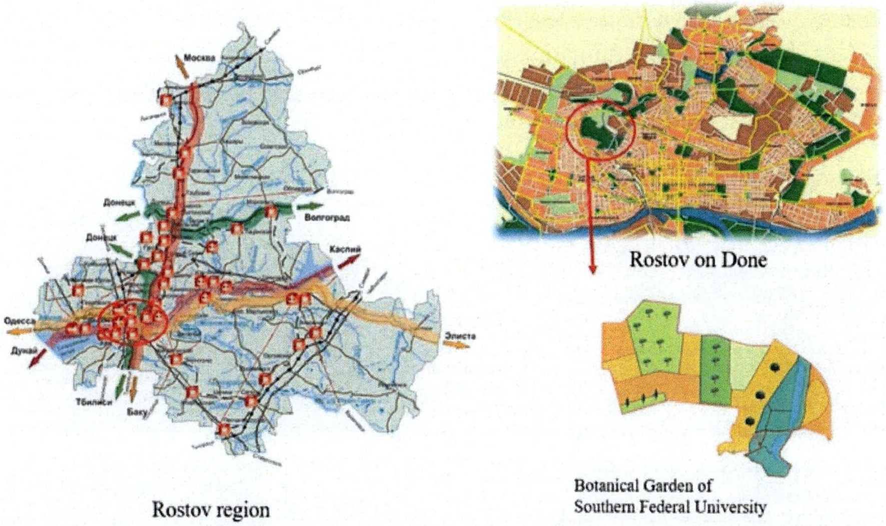


**Fig. 18.1** Map of the location of the soil sampling site (Botanical Garden of the Southern Federal University (Rostov-on-Don)

the study. Complex mineral fertilizer was applied to the soil after wetting and oil contamination, and "Baikal EM-1" was applied to dry soil, and then, oil was applied.

Complex mineral fertilizer (CMF) contains N, P, and K in approximately equal proportions (15:15:15). The fertilizer manufacturer is the "Nov-Agro" manufacturing and trading company (Novaya Melnitsa, Novgorod Region). Complex mineral fertilizer was introduced into the soil in an amount equivalent to compensate 0.75 g for the lack of nitrogen in the oil-contaminated soil and equalize the ratio between carbon and nitrogen.

The composition of «Baikal EM-1» includes bacterial strains (*Paenibacillus pabuli, Azotobacter vinelandii, Lactobacillus casei, Clostridium limosum, Cronobacter sakazakii, Rhodotorulla mucilaginosa and Cryptococcu*), hybrid yeast (*Saccharomyces, Candida lipolitica, Candida norvegensis and Candida guilliermondii*) and fungi (*Aspergillus, Penicillium and Actinomycetales*). "Baikal EM-1" produced by LLC "Scientific and Production Association EM-CENTER" was used in this study (Ulan-Ude, Republic of Buryatia). One-percent solution "Baikal EM-1" with a volume of 100 ml was introduced into the soil. When adding the solution, we achieved soil moisture content of 25–30%.

The incubation period of soils contaminated with oil, fuel oil and gasoline after the introduction of nitroammophos and "Baikal EM-1" at a temperature of 24–25°C, and a humidity of 25–30% was 30 days long. The experiment scheme and the stages of work are shown in Fig. 18.2.

*Measurement.* The ecological state was assessed by indicators of phytotoxicity, enzymatic activity, and the total number of soil bacteria (Table 18.1) (Kazeev et al. 2016). Measurements were taken after 30 days of soil incubation. When assessing the phytotoxic properties of the soil, garden radish (*Raphanus sativus* L. var. *sativus*) of the "Zhara" variety was used as a test object.

As a result of the determination of biological indicators, the integral indicator of the biological state (IIBS) was evaluated as an average to show the reflective response of all biological indicators to the state of the soil.

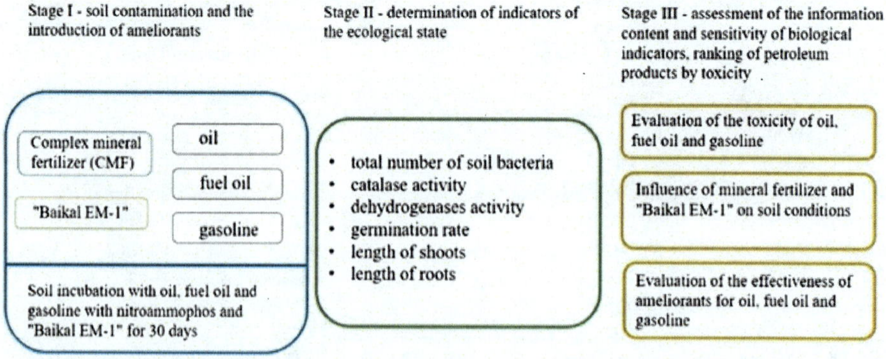


**Fig. 18.2** Scheme and stages of the experiment

**Table 18.1** Methods for assessing the ecological state of oil-contaminated soils

| No. | Indicator | Units of measurement | Method of measurement | References |
|---|---|---|---|---|
| | Total number of bacteria | $10^9$ in gram of soil dry weight | Luminescent microscopy with solution of acridine orange, 40X | (Zvyagintsev 1991) |
| | Catalase activity ($H_2O_2$: $H_2O_2$—oxidoreductase, EC 1.11.1.6) | ml $O_2$ per gram of soil dry weight in 1 min | By the rate of decomposition of 3% hydrogen peroxide, after contact with the soil (temperature, 20–22 °C) | (Galstyan 1978) |
| | Dehydrogenases activity (substrate: NAD (P)—oxidoreductase, EC 1.1.1) | mg of triphenylformazane (TPF) per gram of soil dry weight for hour | According to the rate of conversion of triphenyltetrazolium chloride (TPC) to TPF<br>PE 5800VI spectrophotometer at a wavelength of 540 nm | (Khaziev 2005) |
| | The germination rate of radish seeds (*Raphanus sativus L.* var. sativus) | % of germination seeds of control | Germination of radish (*Raphanus sativus* L.) after 7 days of the experiment | (Zvyagintsev et al. 2005) |
| | The length of the radish shoots (*Raphanus sativus* L. var. sativus) | $10^{-3}$ m | Of length of the shoots in radish (*Raphanus sativus* L.) after 7 days of the experiment | (Zvyagintsev et al. 2005) |
| | The length of the radish roots (*Raphanus sativus L.* var. sativus) | $10^{-3}$ m | Of length of the roots in radish (*Raphanus sativus* L.) after 7 days of the experiment | (Zvyagintsev et al. 2005) |

*Statistical Processing.* Significance was tested at the $p < 0.05$, ** $p < 0.001$, *** $p < 0.0001$ level (using the software package STATISTICA 12.0). Statistics (mean values, dispersion) were determined, and the reliability of different samples was established by using dispersion analysis (Student's *t*-test).

## 18.3 Results

It was found that oil, fuel oil and gasoline pollution caused inhibition of biological parameters by 34–92%. The change in the total number of bacteria in the soil contaminated with oil, fuel oil and gasoline is shown in Fig. 18.3. In clean soil without pollutants, complex mineral fertilizer and "Baikal EM-1" did not have a significant effect on the number of bacteria.

The introduction of oil, fuel oil and gasoline caused inhibition of the bacteria number by 64, 77 and 62%, respectively. Complex mineral fertilizer and "Baikal EM-1" did not have a significant effect on the number of bacteria in oil-contaminated soil. In the soil with fuel oil, complex mineral fertilizer and "Baikal EM-1" caused stimulation by 14 and 27%. In the soil with gasoline, the introduction of complex mineral fertilizer and "Baikal EM-1" caused a stimulation of the number of soil bacteria by 17 and 12%.

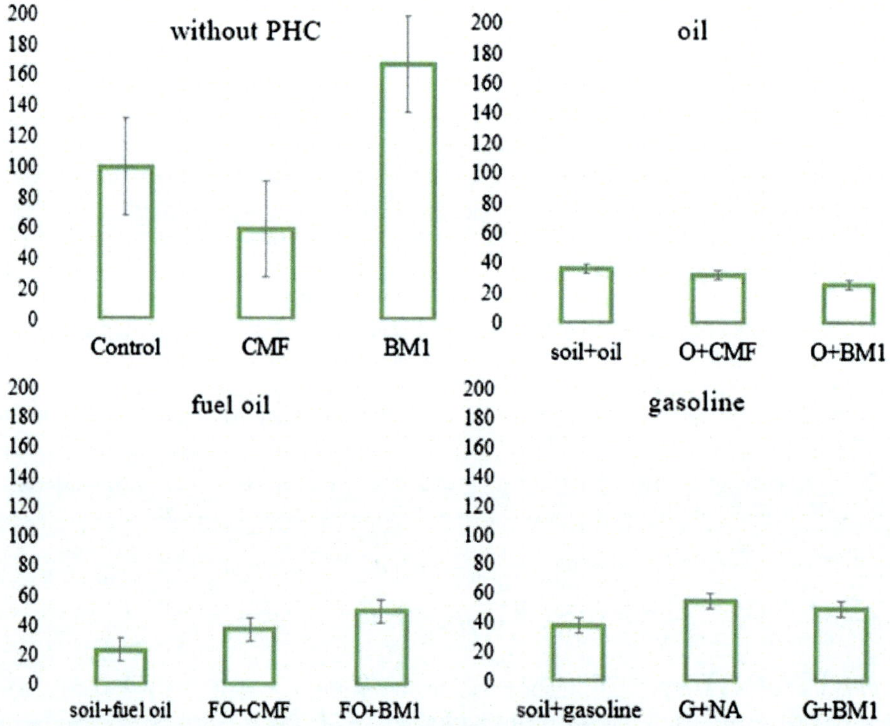


**Fig. 18.3** Total number of bacteria in haplic chernozem when contaminated with oil, fuel oil and gasoline after the introduction of complex mineral fertilizer and "Baikal EM-1", % of control. PHC—petroleum hydrocarbons; O—oil; FO—fuel oil; G—gasoline; CMF—complex mineral fertilizer; BM1—"Baikal EM-1"

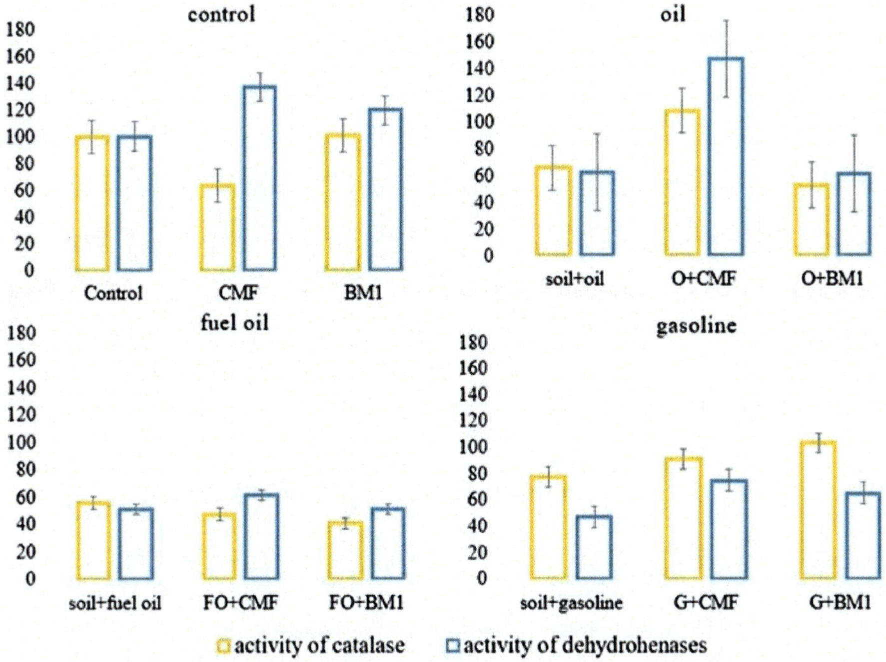


**Fig. 18.4** Changes in the activity of catalase and dehydrogenases of haplic chernozem when polluted with oil, fuel oil and gasoline after the introduction of complex mineral fertilizer and "Baikal EM-1", % of control. PHC—petroleum hydrocarbons; O—oil; FO—fuel oil; G—gasoline; CMF—complex mineral fertilizer; BM1—"Baikal EM-1"

The activity of catalase and dehydrogenases during the introduction of complex mineral fertilizer and "Baikal EM-1" varied unequally (Fig. 18.4). If the catalase activity during the introduction of complex mineral fertilizer was inhibited by 37%, then the dehydrogenases activity was stimulated by 37% in control. In addition, stimulation of dehydrogenase activity was detected when introducing "Baikal EM-1".

Oil, fuel oil and gasoline inhibited enzyme activity by 23–49%. When complex mineral fertilizer was introduced into oil-contaminated soil, catalase and dehydrogenase activity was stimulated by 43 and 85%. "Baikal EM-1" had no effect on the activity of enzymes in oil-contaminated soil. In the case of contamination with fuel oil, the stimulation of dehydrogenases activity was established with the introduction of complex mineral fertilizer (10%). With gasoline contamination and the introduction of complex mineral fertilizer and "Baikal EM-1", catalase activity was stimulated by 12 and 26%. The activity of dehydrogenases was also stimulated by 28 and 18% with the introduction of complex mineral fertilizer and "Baikal EM-1".

Changes in phytotoxic parameters such as germination, length of shoots and roots are shown in Fig. 18.5. Complex mineral fertilizer in pure soil inhibited germination and slowed down the growth of the radish root system by 66 and 64% relative to the control. At the same time, "Baikal EM-1" did not have a significant impact. Oil

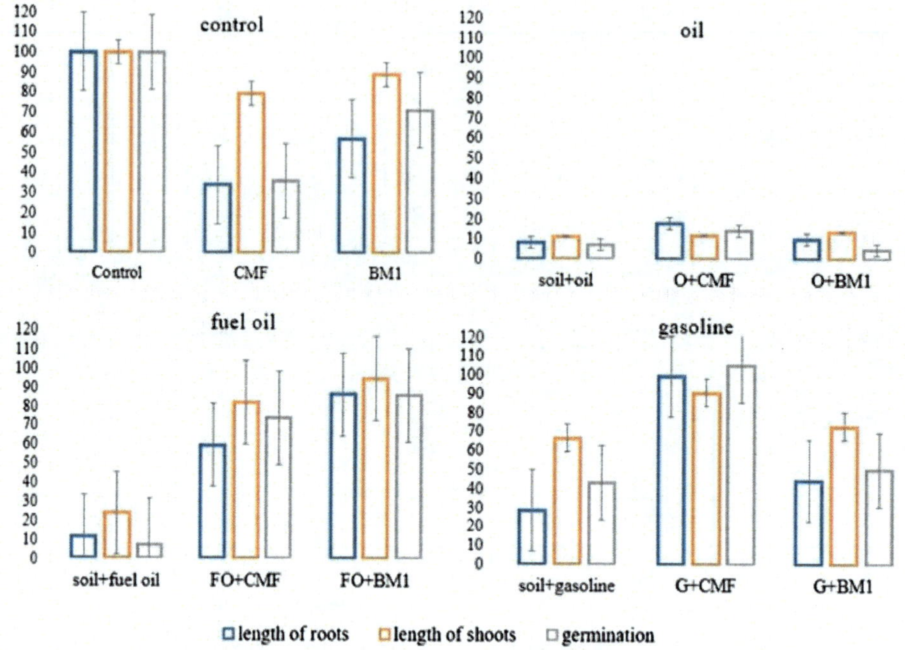


**Fig. 18.5** Changes in phytotoxic indicators of haplic chernozem when contaminated with oil, fuel oil and gasoline after the introduction of complex mineral fertilizer and "Baikal EM-1", % of control. PHC—petroleum hydrocarbons; O—oil; FO—fuel oil; G—gasoline; CMF—complex mineral fertilizer; BM1—"Baikal EM-1"

causes the greatest inhibitory effect on germination rate, and the length of shoots and roots of radish were decreased by 89–92%. At the same time, fuel oil also reduced germination by 93%, as well as oil, but the length of shoots and roots was longer. Gasoline reduced germination of radish by 57%, length of shoot—by 34%, and length of root—by 72%.

The introduction of complex mineral fertilizer stimulated germination, shoot and root length in fuel oil-polluted soil by 67, 58 and 48%. "Baikal EM-1" with the same type of pollution stimulated germination, the length of shoots and roots by 79, 71 and 75%. In the soil with gasoline, the introduction of complex mineral fertilizer made it possible to reach the level of uncontaminated soil (control group), causing stimulation of germination, length of shoots and radish roots by 62, 64 and 71%. The introduction of "Baikal EM-1" stimulated the germination of radish on soil contaminated with gasoline by 16%.

The integral indicator of biological state (IIBS) was calculated based on six indicators of soil biological activity (Fig. 18.6). It was established that complex mineral fertilizer in uncontaminated soil at a dose applicable to oil, fuel oil and gasoline pollution had a toxic effect on biota. This is due to the fact that with the optimal functioning of the soil, the balance of C:N is in the ratio of 1:9–1:12 (Kurganova

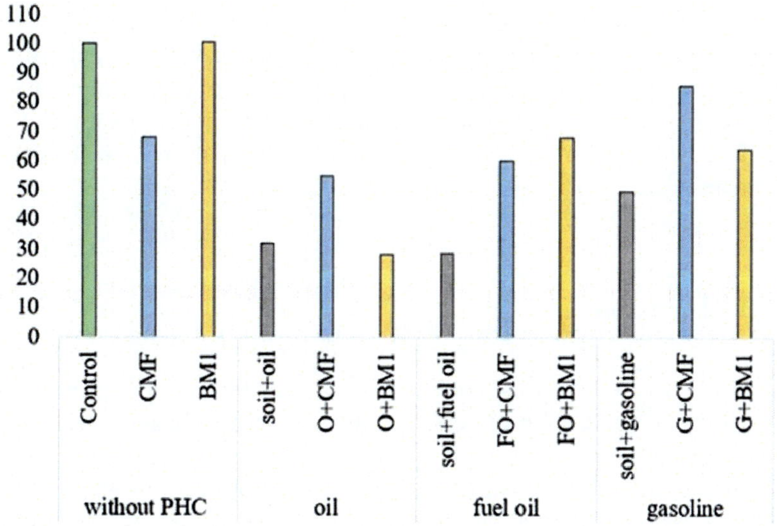


**Fig. 18.6** Integral indicator of the biological state of haplic chernozem when contaminated with oil, fuel oil and gasoline after the introduction of complex mineral fertilizer and "Baikal EM-1". Note: PHC—petroleum hydrocarbons; O—oil; CO—crude oil; G—gasoline; CMF—complex mineral fertilizer; BM1—"Baikal EM-1"

et al. 2021; Lukin 2016; Terpelets and Plitin 2015), and with the addition of petroleum hydrocarbons, an increase in C with respect to N is observed (Degtyareva et al. 2020; Slyusarevsky et al. 2018; Timergazina and Perekhodova 2012).

Complex mineral fertilizer should compensate for the lack of nitrogen in the soil and optimize the microbiological and enzymatic activity of the soil. In oil-contaminated soil, the application of complex mineral fertilizer was 27% more effective than when applying "Baikal EM-1". The greater efficiency of applying complex mineral fertilizers to oil-contaminated soil compared to bacterial preparations is because in soil with a disturbed C:N ratio, even the introduced microorganisms cannot grow and multiply (Minnikova et al. 2019; Tereshchenko et al. 2004).

In the soil with fuel oil, the difference between the efficiency of using "Baikal EM-1" and complex mineral fertilizer was only 8%. The use of Baikal EM-1 for soil with fuel oil and gasoline is possible, but when the soil was contaminated with oil, the biological indicators of the soil were not restored to the control level and were more inhibited. Thus, several petroleum products according to the effectiveness of complex mineral fertilizer use (% of control): gasoline (86) > fuel oil (60) > oil (55). Several petroleum products according to the effectiveness of the "Baikal EM-1" use (% of control): fuel oil (68) > gasoline (64) > oil (28).

Remediation of fuel oil (heavy fraction of oil) is also associated with several difficulties due to its high viscosity and density, which directly affects the growth and development of plants and the functioning of soil biota (Odukoya et al. 2019). Compared with oil and fuel oil, gasoline (light fraction of oil) is a mixture of hydrocarbons with different evaporability and less toxic effects on soil biota (Vasilyeva et al. 2013, 2020). In this regard, even if the C:N ratio is normalized (after complex mineral fertilizer use) and the native biota is stimulated with the help of "Baikal EM-1", it is not possible to restore the ecological state for every oil product. Thus, the effectiveness of the use of complex mineral fertilizer and "Baikal EM-1" in soil contaminated with oil and petroleum products depends on the chemical nature of the contaminant.

## 18.4 Conclusion

As a result of the study, it was found that when the soil was contaminated with oil and gasoline, the application of complex mineral fertilizer was 27 and 22% more effective than when applying "Baikal EM-1". When contaminated with fuel oil, it is advisable to use both "Baikal EM-1" and complex mineral fertilizer. Petroleum hydrocarbon products form the following series according to the effectiveness of complex mineral fertilizer application (% of control): gasoline (86) > fuel oil (60) > oil (55); petroleum hydrocarbon products form the following series according to the effectiveness of "Baikal EM-1" application (% of control): fuel oil (68) > gasoline (64) > oil. "Baikal EM-1" into haplic chernozem helps to restore the soil when contaminated with gasoline and fuel oil, and when soil is contaminated with oil, a longer incubation period is required, and it is also necessary to adjust the doses of ameliorants.

**Acknowledgements** The study was financially supported by the grant of the President (MK-175.2022.5) and the project of the Ministry of Science and Higher Education of the Russian Federation to support the "Laboratory of Young Scientists" within the framework of the development program of the Interregional Scientific and Educational Center of the South of Russia (LabNOC-21-01AB, FENW-2021-0014), with the financial support of the project of the Strategic Academic Leadership Program of the Southern Federal University ("Priority 2030") for the creation of a youth laboratory of ecobiotechnologies for diagnosing and protecting soil health (No. SP-12-23-01).

## References

Ahmad AA, Muhammad I, Shah T, Kalwar Q, Zhang J, Liang Z, Mei D, Juanshan Z, Yan P, Zhi DX, Jun LR (2020) Remediation methods of crude oil contaminated soil. World J Agric Soil Sci 4:3. https://doi.org/10.33552/WJASS.2020.04.000595

Al'bekov AU (2011) Terminal and logistics infrastructure of the transport complex of the Rostov region. National Interests: Priorities Secur 36:10–15

Alori ET, Gabasawa AI, Elenwo CE, Agbeyegbe OO (2022) Bioremediation techniques as affected by limiting factors in soil environment. Front Soil Sci. https://doi.org/10.3389/fsoil.2022.937186

Caravaca F, Rodán A (2013) Assessing changes in physical and biological properties in soil contaminated by oil sludges under semiarid Mediterranean conditions. Geoderma 117:53–61. https://doi.org/10.1016/S0016-7061(03)00118-6

Degtyareva IA, Motina TY, Babynin EV, Ezhkova AM, Davletshina AY (2020) Ecological and toxicological assessment of the process of bioremediation of oil-contaminated soil. Theor Appl Ecol 3:196–202. https://doi.org/10.25750/1995-4301-2020-3-196-202
Galstyan AS (1978) Unification of methods for studying the activity of soil enzymes. Eurasian Soil Sci 2:107–114
Gospodarek J, Rusin M, Barczyk G, Nadgórska-Socha A (2021) The effect of petroleum-derived substances and their bioremediation on soil enzymatic activity and soil invertebrates. Agronomy 11(1):80. https://doi.org/10.3390/agronomy11010080
Gürtler S, Filho CR, Sanches ID, Magalhães LA, Alves MN, Oliveira WJ (2022)
Hazim RN, AL-Ani MA (2019) Effect of petroleum hydrocarbons contamination on soil microorganisms and biodegradation. Rafidain J Sci. https://doi.org/10.33899/rjs.2019.159391
https://www.python.org
Kazeev KS, Kolesnikov SI, Akimenko YV, Dadenko EV (2016) Methods for biodiagnostics of terrestrial ecosystems. SFU Publishing House, Rostov-on-Don, p 356
Khaziev FK (2005) Methods of soil enzymology. Institute of Biology of the UFA Scientific Center, Nauka, Moscow, Russia, pp 252
Kolesnikov SI, Gaivoronskii VG, Rotina EN, Kazeev KS, Val'kov VF (2010) Assessment of soil tolerance toward contamination with black oil in the South of Russia on the basis of soil biological indices: a model experiment. Eurasian Soil Sci 43(8):929–934
Koshlaf E, Ball AS (2017) Soil bioremediation approaches for petroleum hydrocarbon polluted environments 3:25–49. https://doi.org/10.3934/microbiol.2017.1.25
Kurganova IN, Gerenyu VOL, Smolentseva EN, Semenova MP, Lichko VI, Smolentsev BA (2021) Effect of land use type on the physical properties of chernozems in the forest-steppe zone of Western Siberia. Soil Sci 55(9):1061–1075. https://doi.org/10.31857/S0032180X21090045
Lukin SV (2016) Dynamics of the main indicators of fertility and productivity of arable soils in the Belgorod region. Agriculture 3:20–22
Minnikova TV, Kolesnikov SI, Denisova TV (2019) Effect of nitrogen and humic fertilizers on the biochemical state of oil-contaminated chernozem. South Russia: Ecol Dev 14(2):189–201. https://doi.org/10.18470/1992-1098-2019-2-189-201
Minnikova T, Kolesnikov S, Minkina T, Mandzhieva S (2021) Assessment of ecological condition of haplic chernozem calcic contaminated with petroleum hydrocarbons during application of bioremediation agents of various natures. Land 10:169. https://doi.org/10.3390/land10020169
Minnikova T, Kolesnikov S, Ruseva A, Kazeev K, Minkina T, Mandzhieva S, Sushkova S (2022a) Influence of the biochar on petroleum hydrocarbon degradation intensity and ecological condition of haplic chernozem. Eurasian J Soil Sci 11(2):157–166. https://doi.org/10.18393/ejss.1037798
Minnikova T, Ruseva A, Kolesnikov S (2022b) Assessment of ecological state of soils in southern Russia by petroleum hydrocarbons pollution after bioremediation. Environ Proc 9:49. https://doi.org/10.1007/s40710-022-00604-9
Minnikova TV, Ruseva AS, Kolesnikov SI, Revina SY, Gaivoronsky VG (2022c) Influence of biochar on the ecological state of ordinary chernozem when polluted with oil, gasoline and fuel oil. Agrochemistry 9:84–93. https://doi.org/10.31857/S0002188122090095
Odukoya J, Lambert R, Sakrabani R (2019) Understanding the impacts of crude oil and its induced abiotic stresses on agrifood production: a review. Horticulturae 5(2):47. https://doi.org/10.3390/horticulturae5020047
Ruseva AS, Minnikova TV, Kolesnikov SI, Revina SY, Evstegneeva NA, Trufanov DA (2022) Influence of "Baikal EM-1" on the ecological state of ordinary chernozem when polluted with oil, fuel oil and gasoline. Russian Agric Sci 4:43–50. https://doi.org/10.31857/S2500262722040081

Rusin M, Gospodarek J, Nadgórska–Socha A, Barczyk G (2017) Effect of petroleum–derived substances on life history traits of black bean aphid (*Aphis fabae* Scop.) and on the growth and chemical composition of broad bean. Ecotoxicology 26:308–319. https://doi.org/10.1007/s10646-017-1764-9

Rusin M, Gospodarek J, Nadgórska–Socha A, Barczyk G, Boligłowa E, Dabioch M (2018) Effect of petroleum–derived substances on life history traits of bird cherry–Oat aphid (*Rhopalosiphum padi* L.) and on the growth and chemical composition of winter wheat. Environ Sci Pollut Res 25:27000–27012. https://doi.org/10.1007/s11356-018-2723-6

Sablina E (2022) River transport went for export. Argus oil transport. Argus media 2020. 69 p. Available on-line 9 Sept 2022. https://www.argusmedia.com/-/media/Files/sample-reports/argus-neftetransport-russian-language-version-2020-08-14.ashx?la=en&hash=8DC1705DC37AA1B32BD6878B98C4167E38D92358

Sales da Silva IG, Gomes de Almeida FC, Padilha da Rocha e Silva NM, Casazza AA, Converti A, Asfora Sarubbo L (2020) Soil bioremediation: overview of technologies and trends. Energies. https://doi.org/10.3390/en13184664

Slyusarevsky AV, Zinnatshina LV, Vasilyeva GK (2018) Comparative ecological and economic analysis of oil-contaminated soil reclamation methods by in situ bioremediation and mechanical soil replacement. Ecol Ind Russia 22(11):40–45. https://doi.org/10.18412/1816-0395-2018-11-40-45

Swapna AA, Vijayammal R, Radha D (2021) Effects of spent engine oil on soil characteristics and selected phytochemicals in Amaranthus Hybridus. https://doi.org/10.21203/rs.3.rs-939651/v1

Tereshchenko NN, Lushnikov SV, Pyshieva EV (2004) Biological nitrogen fixation as a factor of acceleration of oil hydrocarbons degradation in soil and methods of stimulation the process. Biotechnol Russia 5:95–109

Terpelets VI, Plitin YS (2015) Humus state of leached chernozem in the agrocenoses of the Azov-Kuban lowland: monograph. KubGAU, Krasnodar, pp 127

The Program of integrated development of the transport infrastructure of the municipality "city of Rostov-on-Don" until 2035 416 p. Available on-line 9 Sept 2022. https://rostov-gorod.ru/upload/medialibrary/0cc/ПКРТИ.pdf

Timergazina IF, Perekhodova LS (2012) On the problem of biological oxidation of oil and oil products by hydrocarbon-oxidizing microorganisms. Oil and gas geology. Theory Pract 7(1):15

Vasilyeva GK, Strizhakova ER, Bocharnikova EA, Semenyuk NN, Yatsenko VS, Slyusarevsky AV, Baryshnikova EA (2013) Oil and oil products as soil pollutants. Technology of combined physico-biological treatment of polluted soils. Russian Chem J 57(1):79–104

Vasilyeva G, Kondrashina V, Strijakova E, Ortega-Calvo JJ (2020) Adsorptive bioremediation of soil highly contaminated with crude oil. Sci Total Environ 706:135739. https://doi.org/10.1016/j.scitotenv.2019.135739

Wyszkowski M, Wyszkowska J, Borowik A, Kordala N (2020) Contamination of soil with diesel oil, application of sewage sludge and content of macroelements in oats. Water, Air, Soil Pollut 231. https://doi.org/10.1007/s11270-020-04914-2

Wyszkowski M, Wyszkowska J, Kordala N, Borowik A (2022) Effects of coal and sewage sludge ashes on macronutrient content in maize (Zea mays L.) grown on soil contaminated with eco-diesel oil. Materials. 15. https://doi.org/10.3390/ma15020525

Zvyagintsev DG (1991) Methods of soil microbiology and biochemistry. D.G. Zvyagintsev. M.: Moscow state university, pp 304

Zvyagintsev DG, Babieva IP, Zenova GM (2005) Biology of soils, ed. 3rd, rev. and additional - M.: Moscow University Publishing House, 448 p

# Chapter 19
# Tree Health of *Betula Pubescens* L. for Sustainable Urban Ecosystems (in the Railway Impact Zone on Kola Peninsula)

**Natalya V. Saltan and Ekaterina A. Sviatkovskaya**

**Abstract** The environmental conditions on Kola Peninsula are affected by industry and transport with substantial consequences for tree health. The research aimed to assess the resistance of *Betula pubescens* in the railway impact zone in five settlements: Murmansk, Olenegorsk, Apatity, Polyarnye Zori, and Kandalaksha. The methodology included determining the state of plants, content of photosynthetic pigments in leaves, fluctuating asymmetry, and morphological parameters of the leaf lamina. Tree health assessment of *B. pubescens* showed state deterioration in 2021, which is associated with sanitary care for plantings. On forecourt areas revealed increase of photosynthetic pigments in the leaves birch not associated by railway transport activity. An assessment of changes in the stability of plant development using FA index revealed that the largest deviations from the norm were observed in Olenegorsk in 2018–2019 (FA = 0.058–0.062) and Kandalaksha in 2021 (FA = 0.060). Leaf size did not depend on the level of railway load. The correlation analysis showed a general trend for most cities is the negative dependence FA index on the size of morphological features. Based on the research outcomes, *B. pubescens* can be recommended for introduction to the protective plantings along the railway track and used for greening railway stations due to high resistance to pollution and long lifetime.

**Keywords** The rail station areas · Murmansk region · *B. pubescens* · Ecological features · Greening

N. V. Saltan (✉) · E. A. Sviatkovskaya
Polar-Alpine Botanical Garden-Institute of Kola Science Centre of the Russian Academy of Sciences, Apatity, Russia
e-mail: saltan.natalya@mail.ru

E. A. Sviatkovskaya
URL: http://pabgi.ru

 
M. Korneykova et al. (eds.), *Smart and Sustainable Urban Ecosystems: Challenges and Solutions*, Springer Geography, https://doi.org/10.1007/978-3-031-37216-2_19

## 19.1 Introduction

The urban environment is formed under the influence of various factors, both natural and anthropogenic. Anthropogenic factors (increased dustiness, gas pollution, changes in temperature, water, light regimes, pollution, etc.) determine the specificity of the ecological environment of the city, which differs sharply from the natural one, where the biological and ecological features of plants used for urban greening were formed (Kulagin and Shagieva 2005). The toxic effects of pollution on the tree health can be expressed through plant injury, morphological changes, decrease of vegetative and reproductive growths, early senescence, and even mortality (Dorsey et al. 2015; Viehweger 2014).

The vulnerability of urban ecosystems of the subarctic region of Russia (Murmansk region) is due to harsh soil and climatic conditions and an intensively developed industrial complex including heavy metals pollution. Mining and industrial enterprises are local sources. Most of the extracted raw materials and products are transported by rail. In addition to transporting chemicals, the maintenance and operation of rolling stock use hazardous materials that, if leaked, lead to environmental pollution (Borda-de-Agua et al. 2017; Kazantsev 2015). The problem of improving the urban environment is very relevant for the Kola Arctic. Green spaces that perform health-improving, environmental, and aesthetic functions optimize the urban environment (Gontar et al. 2010; Supuka 1997). A properly selected assortment of resilient plants allows you to decorate the urban landscape while maintaining sanitation and environmental efficiency. Railway transport's influence on plants growing in the right-of-way remains poorly understood. The study of plants used for landscaping railway stations located near cities is relevant to identifying resistant species. In urban plantings, a typical representative is a native species—*B. pubescens*.

The purpose of this work was to assess the resistance of *B. pubescens* to railway transport's impact using numerous indicators (e.g., tree state, the content of photosynthetic pigments in leaves, fluctuating asymmetry, morphological parameters of the leaf lamina) from different locations of the Kola Peninsula.

## 19.2 Objects and Method

### *19.2.1 Location*

The research was carried out in 2018–2021 at observation plots located in areas adjacent to railway stations, in the immediate vicinity of the tracks in Murmansk, Olenegorsk, Apatity, Polyarnye Zori, and Kandalaksha (Fig. 19.1). The tree nursery of the Polar-Alpine Botanical Garden-Institute (PABGI) (1 km from the city of Apatity) was selected as a natural reference (control).

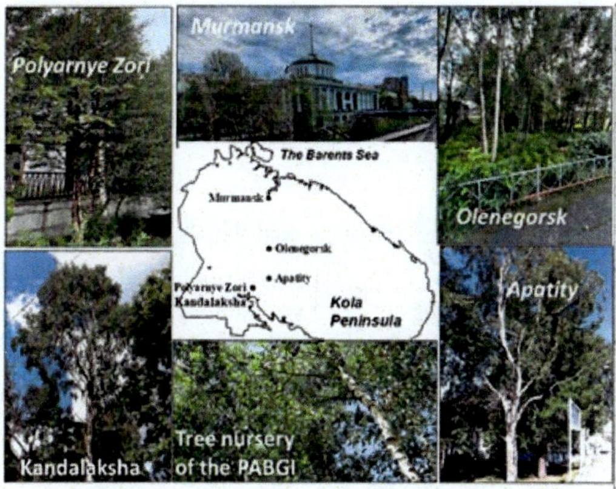


| Location | Coordinates |
|---|---|
| Murmansk | 68°58′N<br>33°04′E |
| Olenegorsk | 68°08′N<br>33°18′E |
| PABGI | 67°34′N<br>33°18′E |
| Apatity | 67°33′N<br>33°22′E |
| Polyarnye Zori | 67°22′N<br>32°29′E |
| Kandalaksha | 67°09′N<br>32°25E |

**Fig. 19.1** Location and view of observation plots

### *19.2.2 Characteristics of Site*

The Oktyabrskaya railway has been operating in the Murmansk region for more than 100 years (Transport 2008). The Murmansk region of service of the Oktyabrskaya railway includes 5 main railway stations (Murmansk, Olenegorsk, Apatity, Polyarnye Zori, Kandalaksha). The largest nodes are located in Murmansk and Apatity; Olenegorsk and Kandalaksha are junction stations with an average traffic load; Polyarnye Zori—with a minimum.

### *19.2.3 Experimental Species*

Middle-aged (40–50 years) plantations of *B. pubescens* growing on the territory of railway stations in 5–10 m from the railway tracks.

### *19.2.4 Data Collection and Data Analyzes*

In August 2018, 2021, the state of *B. pubescens* was evaluated at each site according to the study's methodology, Table 19.1 (Nikolaevskiy and Yakubov 2008).

To determine the pigment fund at the end of the growing season (August) 2018, 2019, 2021 birch leaves were sampled. The content of leaf plastid pigments was determined in alcohol extracts (96%) by optical density at the absorption maxima of chlorophylls *a* and *b*, carotenoids using a UV-1800 spectrophotometer from Shimadzu, Japan at wavelengths $\lambda = 665$, 649, and 470, respectively; all calculations were made for raw samples (Lichtenthaler and Wellburn 1983).

**Table 19.1** Description of classes of the tree state

| Classes of the tree state | Main signs | Additional signs |
|---|---|---|
| Without signs of weakening | The needles are green, dense canopy, the foliage is 100% | |
| Weakened | The needles are green, up to 25% of dry branches | Local damage to branches, trunk |
| Middle-weakened | Crown thinning, dry branches 25–50%, the needles are smaller | Local damage to branches, trunk, presence of stem pests |
| Highly weakened | The crown is very thinned, dry branches 50–75%, the needles are smaller than the previous class | Local damage to branches, trunk, presence of stem pests, water shoots on the trunk and branches |
| Shrinking | In crown more than 75% dry branches, the needles are small, light yellow, falls off prematurely | The trunk and branches are inhabited by pests and diseases, a partially withered tree |
| Deadwood | No needles, 100% dry branches | The trunk and branches are affected by pests and fungi |

Phytoindication methods are widely used to assess technogenic pollution (Mandra et al. 2019; Neverova 2009). One of these methods is the determination of fluctuating asymmetry (FA) to assess the state of ecosystems (Zakharov 2000). Fluctuating asymmetry is a deviation from ideal bilateral symmetry caused by developmental instability and genetic problems during development under the influence of environmental stress, primarily of anthropogenic origin (Kozlov et al. 1996). The strength of the impact of the stress factor contributes to the deviation of the FA index from the norm (Zorin and Korosov 2007), as a result of which the FA index at the macroscopic level is used to assess the stability of the development of the organism (Zakharov 2001). To assess fluctuating asymmetry, vegetative organs (f. e., leaf lamina) of woody plants are often used.

Birch leaf samples were collected at the end of the growing season in July 2018, 2019, and 2021. For the study, 100 leaves were selected from each site. Measurements of the main leaf parameters (length, width, and area) were performed. To estimate fluctuating asymmetry of leaves set of five biometric traits, which characterize that the stability of leaf formation in ontogenesis was applied (1—the width of the half of the sheet (the measurement was carried out in the middle of the leaf blade; 2—length of the second vein from the base of the leaf of the second order; 3—distance between the bases of the first and second veins of the second order; 4—distance between the ends of these veins; 5—the angle between the main vein and the second vein from the base of the leaf of the second order) (Guidelines for the implementation of quality assessment environment as living beings (stability assessment development of living organisms on the level of asymmetry of morphological structures) 2003). All measurements were carried out manually. The calculation integral index of fluctuating asymmetry of the morphological traits of leaf lamina has been produced using normalized difference algorithm (Zakharov 2000):

**Table 19.2** Point scale for assessing the quality of the environment by indicators of a violation of the stability of development

| FA index | Environment state | Point |
|---|---|---|
| < 0.04 | Conditional norm | I |
| 0.04–0.044 | Initial, minor deviations from the norm | II |
| 0.045–0.049 | Average level of deviations from the norm | III |
| 0.050–0.054 | Significant deviations from the norm | IV |
| > 0.055 | Critical state | V |

$$\text{FA} = \frac{1}{m * n} \sum_{i=1}^{m} \sum_{j=1}^{n} \frac{|(Lij - Rij)|}{|(Lij + Rij)|}$$

where *Lij* and *Rij* are values of the *j*-th character in the *i*-th leaf to the left and right side of the plane symmetry; *m*—number of leaves; $n = 5$ (morphological traits).

A 5-point system was developed to assess the degree of disturbance of plant development stability, Table 19.2 (Zakharov 2000). The FA values corresponding to the first point are observed, as a rule, in a sample of plants from favourable growing conditions. The fifth point characterizes the critical state of the environment when the plants are in a very depressed state.

Mathematical processing of the results was done using standard software packages for statistical calculations (Microsoft Office Excel 2016). The correlation coefficient ($r$) was calculated by the square method (Pearson's method) for the significance level of 0.05.

## 19.3 Results and Discussion

Over a three-year observation period tree, health assessment of *B. pubescens* showed its state deterioration in 2021 (Fig. 19.2). In 2018, healthy plants without signs of weakening were found in all cities except Murmansk. Weakened plants prevailed in Apatity (50%) and Olenegorsk (48%). Middle-weakened plants prevailed in Murmansk (57%). The maximum number of deadwoods was found in Kandalaksha (14%).

In 2021, there was a lack of healthy plants in Polyarnye Zori, and vice versa, its appearance in Murmansk (29%). In other cities, this indicator has not changed much. In Apatity, due to the state deterioration of birch, weakened plants were absent, the proportion of highly weakened plants (from 16 to 33%) increased, shrinking plants (17%), and dead wood (17%) appeared. In Kandalaksha, the proportion of middle-weakened plants increased; dead wood was not observed due to its removal. In Olenegorsk, the indicator of middle-weakened plants increased, in the absence of highly weakened ones. Shrinking plants were transformed into dead wood. In Murmansk, despite the appearance of plants without signs of weakening, middle-weakened plants turned into highly weakened ones.

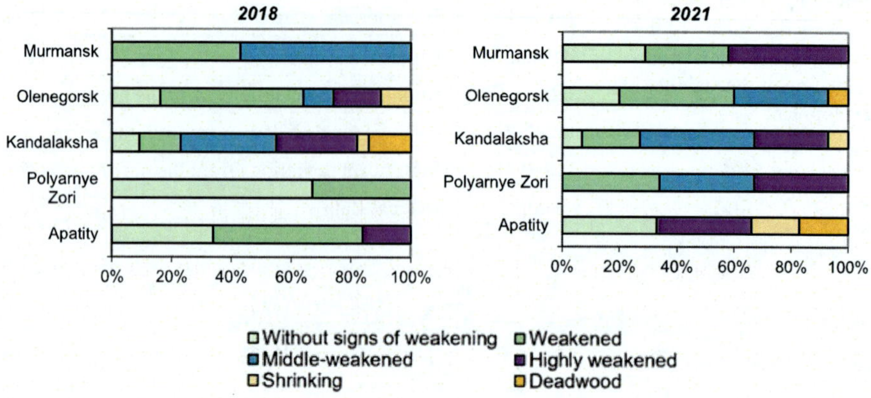


**Fig. 19.2** State of *Betula pubescens* in the cities

A high level of chronic air pollution by technogenic emissions has a negative impact on the whole complex of physiological processes in plants, including photosynthesis (Biswal et al. 2011; Vass and Aro 2007). There are several opinions about the effect of anthropogenic pollution on the photosynthetic activity of plants. Some authors point to its suppression under conditions of pollution (Balandaykin 2014; Bukharina et al. 2013). Others, on the contrary, show an increase in the content of green pigments in plant leaves in response to pollution, or the absence of changes (Voskresenskaya et al. 2014; Yashin and Zaitsev 2015). This is due to different types and levels of pollution, with the heterogeneity of environmental conditions of growth and the plant species specificity.

Our research has shown that the content of photosynthetic pigments in birch leaves varied significantly depending on the location and year of the study (Fig. 19.3). In the control plot, the content of chlorophyll *a* was higher than in some cities only in 2018; in all other years, the content was lower. The dynamics of pigment distribution over the years in the forecourt areas differed. In Apatity and Kandalaksha, the content of chlorophyll *a* increased yearly to the maximum values in 2021 was observed. In Polyarnye Zori, this indicator did not change significantly. In Murmansk and Olenegorsk, same dynamics was revealed, while in Murmansk in 2018–2019, content of chlorophyll a was higher than in other locations.

The distribution of the amount of chlorophyll *b* differed from that of chlorophyll *a* (Fig. 19.3). In 2018, the excess chlorophyll *b* content compared to the control was noted in Apatity and Kandalaksha. In other years, this parameter (as well as chlorophyll *a*) in the control was lower than in the railway station areas. In 2021, in all cities, an increase in chlorophyll *b* was detected to the maximum values over a three-year observation period. The amount of chlorophyll *b* was stable in Polyarnye Zori.

Carotenoids perform a protective function, shielding chlorophyll from photooxidation. Therefore, an increase in the content serves as a signal of an unfavourable state of plants. The content of carotenoids significantly correlated with chlorophyll

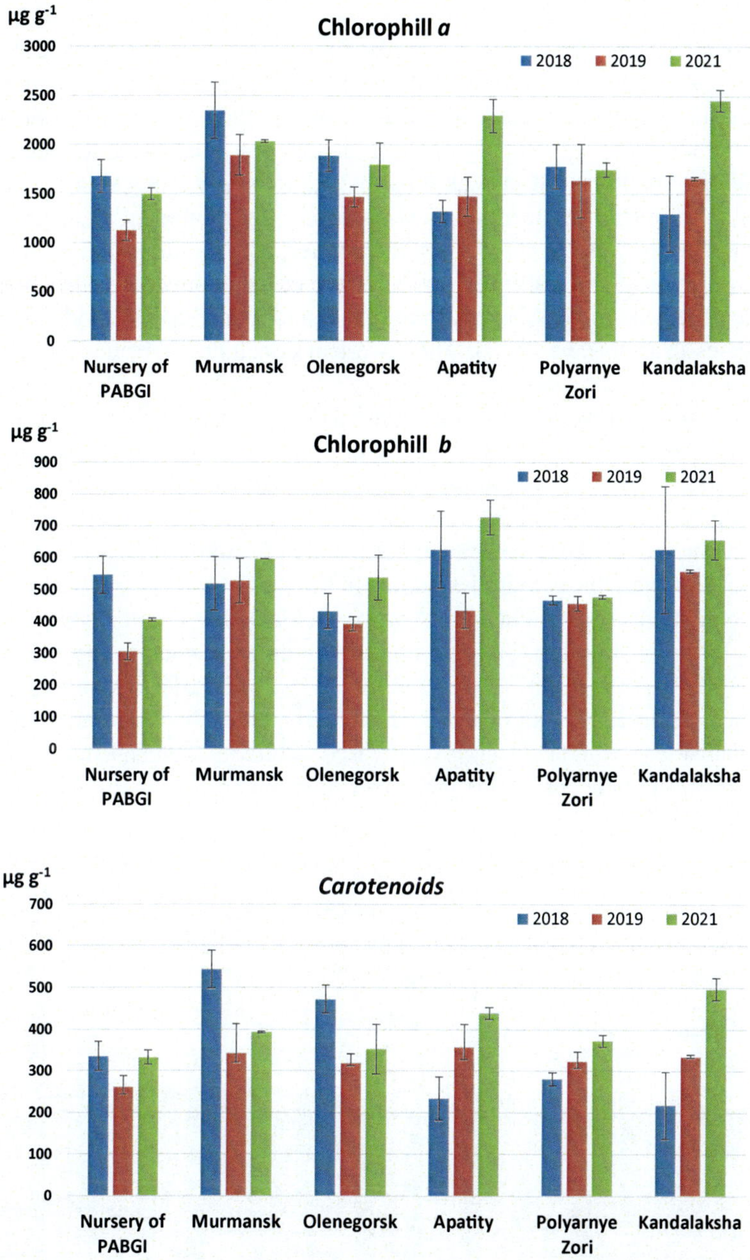


**Fig. 19.3** Mean content ($\mu gg^{-1}$ fresh weight) of chlorophyll *a*, *b*, carotenoids in leaves of *B. pubescens*

*a* ($r$ = 0.81–0.99, except for Polyarnye Zori) and did not correlate with chlorophyll *b* ($r$ = −0.38–0.53 ($p < 0.05$), except for control). The content of carotenoids was higher—compared to Apatity, Polyarnye Zori, and Kandalaksha—in the control only in 2018. In 2019, the content of carotenoids in the territories near the railway station did not differ significantly; in 2021, an increase in carotenoids was detected, especially in Apatity and Kandalaksha (what could be a warning sign).

Thus, in the territories near the railway stations, a predominant increase in photosynthetic activity was observed. In some studies, two trends in the change in the content of green pigments under the influence of an urbanized environment have been identified: an increase in resistant plants and a decrease in moderately resistant species (Sarsatskaya 2017). Thus, we can conclude about the stability of *B. pubescens*.

An assessment of the sustainability of birch development based on the FA index in the studied areas with varying degrees of impact of railway transport showed significant differences (Fig. 19.4). During the study period, the most stable FA was found for trees growing in the city of Polyarnye Zori (0.041–0.044), corresponding to a slight deviation from the development norm and indicating low-level environmental pollution. Low variability was also shown in the three-year dynamics of the photosynthetic pigments' content here. In Apatity, the FA index varied slightly compared to other locations, but due to growth in 2021, it corresponded to the average level of deviation from the norm. At the control site in 2018–2019, FA approached critical values, but in 2021, it significantly decreased and reached the conditional norm. In Murmansk and Kandalaksha, FA increased, while in Kandalaksha, it reached a critical level. In Olenegorsk, this parameter dropped sharply in 2021, despite high values in 2018–2019.

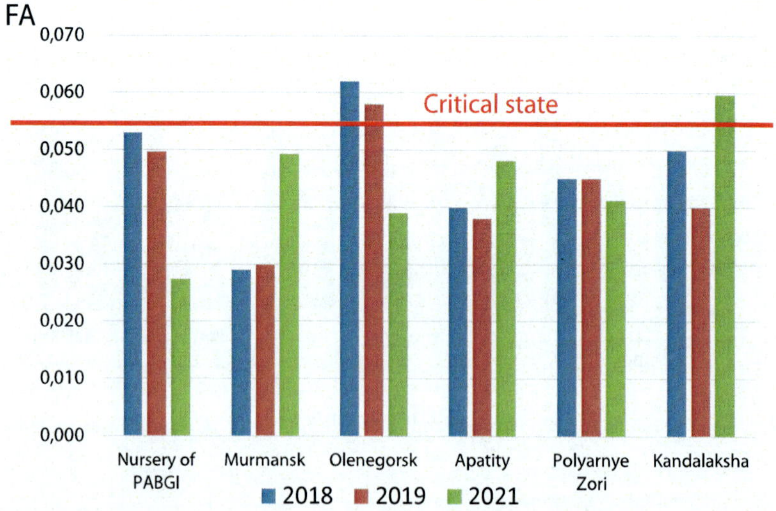


**Fig. 19.4** Integral index of fluctuating asymmetry of leaves of *B. pubescens*

Some researchers found that leaf fluctuating asymmetry does not serve as sensitive indicator of environmental stress, because it is influenced by many factors, including weather conditions (Kozlov et al. 2022; Sandner and Matthies 2017).

An assessment of changes in the morphological parameters of the leaf lamina showed that in the period 2018–2019, the minimum values of length, width, and area are typical for birches growing in Olenegorsk and Polyarnye Zori (Table 19.3). In 2021, they grew noticeably, approached the indicators in other cities and even exceeded them in leaf length. As a result, in 2021, the differences in morphological parameters between the plots were insignificant.

The level of changes in the anatomical structure of the assimilation organs of plants depends on the concentration and toxicity of pollutants, on the duration of exposure and also on the sensitivity of the species. Some researchers have shown that in the process of adaptation to negative environmental conditions in plants, leaf parameters can change, both upwards (Sokolova and Tingaeva 2008) and downwards (Bukharina and Dvoeglazova 2010). As our studies have shown, leaf size did not depend on the level of anthropogenic load, but rather associated with the characteristics of the place of growth (weather conditions, soil conditions, etc.)

The correlation analysis of all the studied parameters of *B. pubescens* leaves demonstrated the heterogeneity of the obtained dependences. Trees growing in

**Table 19.3** Dynamics of leaf morphological parameters of *B. pubescens* in 2018–2021

| Place of selection | Year | Leaf length, cm | Leaf width, cm | Leaf area, $cm^2$ |
|---|---|---|---|---|
| Murmansk | 2018 | 6.06 ± 0.12 | 4.52 ± 0.08 | 17.39 ± 0.61 |
| | 2019 | 6.20 ± 0.11 | 5.04 ± 0.13 | 18.51 ± 0.86 |
| | 2021 | 5.53 ± 0.10 | 4.54 ± 0.09 | 17.33 ± 0.48 |
| Olenegorsk | 2018 | **4.45 ± 0.15** | **3.69 ± 0.09** | **11.05 ± 0.63** |
| | 2019 | **4.41 ± 0.09** | **3.29 ± 0.06** | **8.18 ± 0.29** |
| | 2021 | 6.09 ± 0.17 | 4.23 ± 0.14 | 16.65 ± 0.36 |
| Apatity | 2018 | 5.91 ± 0.13 | 5.08 ± 0.14 | 21.59 ± 0.98 |
| | 2019 | 5.47 ± 0.19 | 5.06 ± 0.16 | 18.49 ± 1.10 |
| | 2021 | 5.23 ± 0.14 | 4.46 ± 0.14 | 16.03 ± 0.86 |
| Polyarnye Zori | 2018 | **4.53 ± 0.09** | **3.46 ± 0.08** | **10.71 ± 0.48** |
| | 2019 | **4.18 ± 0.08** | **3.44 ± 0.09** | **9.19 ± 0.38** |
| | 2021 | 5.49 ± 0.01 | 4.42 ± 0.09 | 14.32 ± 0.61 |
| Kandalaksha | 2018 | 5.90 ± 0.13 | 4.97 ± 0.06 | 20.52 ± 0.46 |
| | 2019 | 6.10 ± 0.10 | 4.63 ± 0.10 | 16.60 ± 0.55 |
| | 2021 | 5.54 ± 0.10 | 4.13 ± 0.07 | 14.88 ± 0.49 |
| The tree nursery PABGI | 2018 | 6.03 ± 0.25 | 5.14 ± 0.26 | 20.86 ± 1.66 |
| | 2019 | 5.52 ± 0.17 | 4.21 ± 0.13 | 14.04 ± 0.76 |
| | 2021 | 6.01 ± 0.23 | 4.59 ± 0.17 | 16.44 ± 0.89 |

Low values are in bold

Olenegorsk and Polyarnye Zori have similar functional relationships: an FA index with morphological parameters and chlorophyll *b* content (high significant negative correlation coefficients), and accordingly, high significant positive coefficients between the length, width, and area of the leaf plate and the amount of chlorophyll *b*. In the control plot, significant positive relationships were obtained between the content of photosynthetic pigments (to a lesser extent, carotenoids) and morphometric features; that is, the content of pigments increased in proportion to the leaf size.

In Apatity and Kandalaksha, the distribution of correlation coefficients was similar to the previous locations in terms of the ratio of FA to metric indicators. The difference was a high negative correlation between the content of chlorophyll *a* and leaf parameters. A positive dependence of the FA index on chlorophyll *b* was obtained, and in Apatity, there was a positive dependence also on chlorophyll *a*. Thus, for these areas, the synthesis of the main chlorophyll *a* increased with a decrease in the leaf lamina. No significant patterns were found in Murmansk, except for single high correlation coefficients.

Thus, the pollution level caused by rail transport did not significantly affect the studied parameters. It can be stated that in most cities, the FA index negatively correlates with the size of morphological features leaves (length, width, and area) (Table 19.4).

Thus, the conducted studies have shown that the native species (*B. pubescens)* was resistant to the impact of railway transport. The dynamics of changes in the parameters of the birch was not related to the activities of the railway transport. Thus, in birches growing in Murmansk (the highest traffic load), no deviations in development were revealed based on the FA index, leaf size varied slightly over the study period. According to the state of plants, two different qualitative transitions were noted: the appearance of healthy plants and highly weakened ones. The content of photosynthetic pigments showed no significant dynamics, except for carotenoids, which decreased compared to 2018. While the content of carotenoids increased in Apatity and Kandalaksha, the FA index also increased. Railway services were advised to pay attention to this. The birches growing in Polyarnye Zori had the lowest variability of the studied parameters, with the exception of morphological parameters.

## 19.4 Conclusion

Monitoring the state of *B. pubescens* for the observation period 2018–2021 showed that the proportion of healthy plants remained almost unchanged at the railway stations of Apatity (33–34%), Olenegorsk (16–20%), and Kandalaksha (7–9%). At the same time, dramatic changes took place in Polyarnye Zori and Murmansk. In Murmansk, an increase in the number of healthy plants (up to 25%) was noted, whereas in Polyarnye Zori, deterioration in the state of this species was revealed (this category is absent), which is associated with sanitary care for plantings (timely

**Table 19.4** Correlation coefficients between studied parameters in leaves of *B. pubescens*

| Parameter | FA | Chlorophill *a* | Chlorophill *b* | Carotenoids |
|---|---|---|---|---|
| *Murmansk* | | | | |
| Leaf length | **−0.97** | 0.02 | **−0.95** | 0.08 |
| Leaf width | −0.43 | −0.76 | −0.37 | −0.72 |
| Leaf area | −0.50 | −0.71 | −0.44 | −0.66 |
| FA index | 1.00 | −0.26 | **1.00** | −0.32 |
| *Olenegorsk* | | | | |
| Leaf length | **−0.98** | 0.33 | **0.97** | −0.29 |
| Leaf width | **−0.82** | 0.68 | **0.99** | 0.12 |
| Leaf area | **−0.88** | 0.61 | **1.00** | 0.02 |
| FA index | 1.00 | −0.15 | **−0.91** | 0.46 |
| *Apatity* | | | | |
| Leaf length | −0.64 | **−0.85** | −0.18 | −1.00 |
| Leaf width | **−0.98** | **−0.99** | −0.75 | −1.00 |
| Leaf area | −0.71 | **−0.90** | −0.28 | −1.00 |
| FA index | 1 | **0.95** | **0.87** | 0.68 |
| *Polyarnye Zori* | | | | |
| Leaf length | **−0.97** | 0.53 | **0.97** | 0.73 |
| Leaf width | **−1.00** | 0.31 | **0.87** | **0.88** |
| Leaf area | **−0.96** | 0.55 | **0.97** | 0.71 |
| FA index | 1.00 | −0.29 | **−0.87** | **−0.88** |
| *Kandalaksha* | | | | |
| Leaf length | **−0.99** | −0.78 | **−0.93** | **−0.70** |
| Leaf width | −0.58 | **−0.99** | −0.40 | **−1.00** |
| Leaf area | −0.29 | **−0.91** | −0.09 | **−0.95** |
| FA index | 1.00 | 0.67 | **0.98** | 0.57 |
| *The tree nursery of PABGI* | | | | |
| Leaf length | −0.36 | **0.96** | **0.83** | **1.00** |
| Leaf width | 0.22 | **0.95** | **1.00** | 0.83 |
| Leaf area | 0.28 | **0.93** | **1.00** | 0.79 |
| FA index | 1.00 | −0.08 | 0.21 | −0.37 |

*Note* Correlation coefficients were shown in bold for the significance level of 0.05

in Murmansk and absent in Polyarnye Zori). The determination of photosynthetic pigments in the leaves of *B. pubescens* in areas near railway stations revealed a predominant increase in photosynthetic activity not associated with the level of the anthropogenic load caused by railway transport activity. An assessment of changes in the stability of plant development using the FA index revealed that the largest deviations from the norm were observed in Olenegorsk in 2018–2019 (FA = 0.058–0.062) and Kandalaksha in 2021 (FA = 0.060). In 2018–2019 in control, this parameter was relatively high, close to critical. It has showed that the FA index in plants is quite variable, with the exception of trees growing in the city of Polyarnye Zori. An analysis of the dynamics of the morphological parameters of the leaf lamina of *B. pubescens* showed that the minimum values were noted in Olenegorsk and Polyarnye Zori in 2018–2019. In 2021, the differences between locations decreased due to an increase in leaf size in the cities. The correlation analysis of all studied parameters of *B. pubescens* leaves showed photosynthetic pigments' significant positive dependence on leaf size in the control; in Olenegorsk and Polyarnye Zori, high correlation coefficients were obtained only for chlorophyll *b*. In Apatity and Kandalaksha, these indicators correlated negatively. A general trend for most cities is the negative dependence of the FA index on the size of morphological features. In general, the native species *B. pubescens* is stable in these environmental conditions of growth and can be recommended as landscaping objects for railway stations.

This work was supported by state task AAAAA 18–118050390076–8 "Collection funds of the Polar-Alpine Botanical Garden-Institute as the basis for biodiversity conservation, development of biotechnologies, optimization of urban environment conditions, phytorehabilitation, and environmental education".

## References

Balandaykin M (2014) Dynamics and distinctions in concentration of the basic photosynthetic pigments of leaves of the birch growing in impure conditions. Chem Plant Raw Mater 1:159–164

Biswal B, Joshi PN, Raval MK, Biswal UC (2011) Photosynthesis, a global sensor of environmental stress in green plants: stress signaling and adaptation. Curr Sci 101(1):47–56

Borda-de-Agua L, Barrientos R, Beja P, Pereira HM (2017) Railway ecology. Springer, Cham (eBook)

Bukharina I, Kuzmin P, Gibadulina I (2013) Analysis of the content of photosynthetic pigments in the leaves of woody plants in urban environment (dy giving an example of Naberezhnye Chelny). Bull Udmurt Univ Ser Biol Earth Sci 1:20–25

Bukharina L, Dvoeglazova A (2010) Bioecological features of herbaceous and woody plants in city. UdmSU Publishing, Izhevsk

Dorsey B, Olsson M, Rew LJ (2015) Ecological effects of railways on wildlife. In van der Ree R, Smith DJ, Grilo C (eds) Handbook of road ecology. Wiley, West Sussex

Gontar O, Zhirov V, Kazakov L, Svyatkovskaya E, Trostenyuk N (2010) Green building in Murmansk region. Kola science center of Russian Academy of Science, Russia, Apatity

Guidelines for the implementation of quality assessment environment as living beings (stability assessment development of living organisms on the level of asymmetry of morphological structures) (2003) M: Ministry of Natural Resources

Kazantsev I (2015) Rail transport as a source of soil contamination with heavy metals. Samara Sci Herald 2(11):94–96

Kozlov M, Wilsey B, Koricheva J, Haukioja E (1996) Fluctuation asymmetry of birch leaves increases under pollution impact. J Appl Ecol 33:1489–1495

Kozlov M, Zverev V, Zvereva E (2022) Leaf size is more sensitive than leaf fluctuating asymmetry as an indicator of plant stress caused by simulated herbivory. J Ecol Ind 140:108970

Kulagin A, Shagieva Y (2005) Woody plants and biological conservation of industrial pollutants. Nauka, Moscow

Lichtenthaler HK, Wellburn AR (1983) Determinations of total carotenoids and chlorophylls a and b of leaf extracts in different solvents. Biochem Soc Trans 11:591–592

Mandra Y, Esaulko A, Klyushin P, Stepanenko E, Zelenskaya T (2019) Environmental assessment of urban areas using plants of different taxa as indicators. J S Russ Ecol Dev 14(4):134–146

Neverova O (2009) Application of phytoindication in the assessment of environmental pollution. J Biosph 1(1):82–92

Nikolaevskiy V, Yakubov H (2008) Environmental monitoring of green plantings in a large city. Methods of research: practical guide. MGUL, Moscow

Sandner T, Matthies D (2017) Fluctuating asymmetry of leaves is a poor indicator of environmental stress and genetic stress by inbreeding in *Silene vulgaris*. J Ecol Ind (79):247–253

Sarsatskaya S (2017) Content of photosynthetic pigments in urban wood species. Bull Kemerovo State Univ 4:9–14

Sokolova G, Tingaeva A (2008) Bioindication of air pollution in Barnaul. Izv Altai State Univ 3(59):19–21

Supuka J (1997) Ecological importance of woody plants in reduction of the reduction on the solid particles impacts in settlements. Folia Dendrologica 1–2:85–95

Transport (2008) Kola encyclopedia. Reference in 5 books. KSC RAS Publishing House, Apatity, vol 1, pp 129–130

Vass I, Aro E (2007) Photoinhibition of photosynthetic electron transport. Primary processes in photosynthesis, basic principles and apparatus. The Royal Society of Chemistry, Cambridge

Viehweger K (2014) How plants cope with heavy metals. Bot Stud 55:35

Voskresenskaya O, Voskresenskii V, Sarbaeva E, Yagdarova O (2014) bInfluence of ultraviolet radiation and microclimate parameters on the pigment count in leaves Betula pendula growing in urban conditions, vol 3, pp 39–45

Yashin D, Zaitsev G (2015) The maintenance of photosynthesis pigments in leaves of European white birch (*Betula pendula*) and English oak (Quercus robur) in the conditions of Ufa industrial center. Acad J Izv Samara Sci Cent Russ Acad Sci 17(6):274–277

Zakharov V (2000) Environmental health: the practice of evaluation. Russian Environmental Policy Center

Zakharov V (2001) Ontogenesis and population (stable development and population variability). Ecology 3:177–191

Zorin A, Korosov A (2007) Characteristics of fluctuating asymmetry of the two species of birch leaf in Karelia. Ecology. Experimental genetics and physiology. Proc Karelian Res Cent 11:28–36

# Chapter 20
# Ecological Revitalization Master Plan of Lipetsk City Based on the HBV Hydrological Modelling of a Small Ungauged Lipovka River (Russia)

**Pavel Terskii, Anatoly Tsyplenkov, Artem Gurinov, Anna Antoniuk, Igor Shchukin, and Aleksey Sayanov**

**Abstract** Water discharge fluctuations in ungauged rivers, especially in the face of climate change and urban sprawl, can be dangerous and destructive. The water discharge dynamics of small urban rivers are highly variable and complex. Understanding these dynamics and their underlying causes is necessary for public awareness. Our work is focused on the small ungauged river Lipovka (Lipetsk, Russia). It has a 50.1 $km^2$ urbanized watershed with an outlet in Komsomolsky Pond in Lipetsk's city center. We used several independent methods to study the hydrological regime of the Lipovka River. Based on in-situ measurements in spring 2022, we estimate the water discharge with a 1% annual exceedance probability according to the regulatory document SP 33-101-2003. Furthermore, we used data from the Lipetsk meteostation to run a Hydrologiska Byråns Vattenbalansavdelning model (HBV) to simulate a daily hydrograph for the 2005–2022 period. We found that the reduced runoff of the Lipovka River was mainly caused by a decline in the effective catchment area, a littered channel, and increased evaporation and filtration into the groundwater. To reestablish the river's natural flow and catchment area, we propose constructing and maintaining a culvert to enable upstream feeding of the Lipovka River by Syrsky Pond. In the context of smart urbanization, our study can be used to design and revitalize small urban rivers that form a crucial component of a city's water-green framework.

---

P. Terskii (✉) · A. Tsyplenkov · A. Antoniuk · A. Sayanov
Faculty of Geography, Lomonosov Moscow State University, Moscow, Russia
e-mail: pavel_tersky@mail.ru

A. Gurinov
Peoples' Friendship University of Russia, Agrarian-Technological Institute, Moscow, Russia

Institute of Geography, Russian Academy of Sciences, Moscow, Russia

I. Shchukin
Perm National Research Polytechnic University, Perm, Russia

P. Terskii
Water Problems Institute, Russian Academy of Sciences, Moscow, Russia

 
M. Korneykova et al. (eds.), *Smart and Sustainable Urban Ecosystems: Challenges and Solutions*, Springer Geography, https://doi.org/10.1007/978-3-031-37216-2_20

**Keywords** Urban river · Ungauged river · Hydrological model · HBV · GLOH2O · Streamflow · Urban planning · Ecological revitalization

## 20.1 Introduction

The biodiversity and quality of water bodies are negatively affected by human-induced landscape changes, nutrient cycling regulation, water purification, and waste treatment (Malhotra et al. 2020; Guo et al. 2022; Pastor et al. 2022). Several techniques, including hydrological stabilization, water filtration, and biodiversity restoration, have been developed to improve the quality of urban rivers (Primo et al. 2022). However, restoration and revitalization cannot be conducted without studying the hydrological regime of a river (Muhar et al. 2018).

The water discharge dynamics of small, urban rivers can be highly variable and complex (Taylor and Owens 2009; Sokolov et al. 2021). Understanding these dynamics and their underlying causes is essential for our awareness of streamflow fluctuations in ungauged basins, especially in the face of climate change and intensified urbanization (Gurnell et al. 2007). Modern Russian regulatory documents (e.g., SP (Handbook) 33-101-2003) on river discharge calculations and their probability analyses are based on regional empirical relationships built for Soviet streamflow data, as presented in SNiP (Construction Rules and Regulations) 2.01.14-83 (Ayzel et al. 2020). Thus, the methodology for estimating streamflow characteristics has not been updated for almost 40 years, raising numerous questions regarding its applicability to contemporary cases (Makarieva et al. 2017; Klimenko and Kharlamova 2019). Therefore, developing and applying alternative methods for streamflow assessment is important, especially for ungauged basins (Nesterova et al. 2019). Although hydrological modelling could tackle this research gap (Pool et al. 2017), streamflow predictions in data-sparse regions remain challenging (Hrachowitz et al. 2013). However, the development of global parameter maps, i.e., hydrologic parameter regionalization (e.g., Beck et al. 2020), could significantly improve the model accuracy by transferring data from data-rich regions (Ma et al. 2021).

In recent decades, the use of hydrological models has grown dramatically, and today, they are standard tools for an increasing number of applications. The Hydrologiska Byråns Vattenbalansavdelning (HBV) model follows this trend (Bergström 1992; Lindström et al. 1997). There are many applications that cover hydrological forecasting, spillway design, studies of the effects of climate change, synoptic water balance mapping, and simulations of groundwater response. (Hundecha and Bárdossy 2004; Shrestha and Solomatine 2008; Thielen et al. 2009; Bergström and Lindström 2015).

The goal of our study was to develop an ecological revitalization strategy for the Lipovka River and Komsomolsky Pond, located at the outlet (Lipetsk, Central Russia). We aimed to determine the flooding stages in Komsomolsky Pond with various annual exceedance probabilities (AEP), considering the anthropogenically disturbed watershed and climate change. The results of the hydrological assessment served as a background for the Lipovka River eco-revitalization master plan. Water budget estimation was used to calculate the water inflow to Komsomolsky Pond during the low-water periods.

## 20.2 Study Area

The small Lipovka River runs through the city of Lipetsk (Fig. 20.1). It is a right tributary of the Voronezh River (a part of the Don River basin). Lipovka is a 3-km-long river, and the total catchment area is 51 km$^2$ in the outlet of Komsomolsky Pond. Downstream, the Komsomolsky Pond Lipovka flows along Skorokhodova Street in an artificial canal created by the Lipetsk ironworks as a bypass along this section of the river at the beginning of the eighteenth century. At present, the Lipetsk springs have dried up, and water intermittently disappears in the source owing to the intensive water removal of the Zadonsk-Yelets horizon. The Lipovka River has a channel of about 6–15 m in width and 0.2–0.3 m in depth (Anichkina et al. 2017).

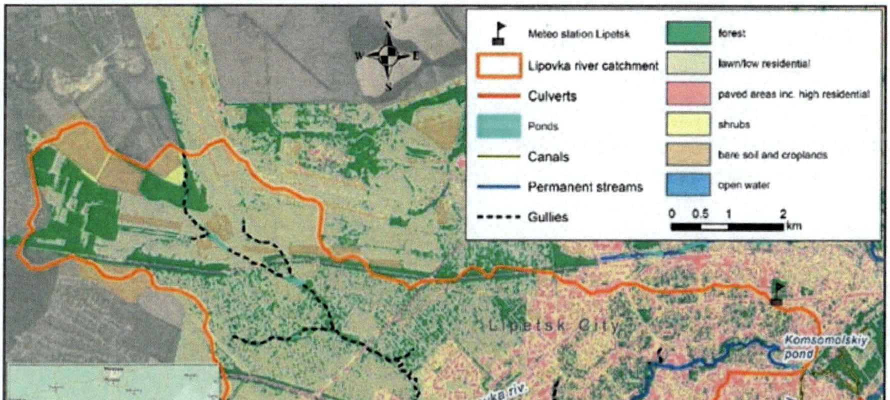


Fig. 20.1 The land use/land cover (LULC) map of the Lipovka River basin based on Sentinel-2 satellite image classification

### 20.2.1 Landscape Setting

The central part of Lipetsk is located on the right bank of the Voronezh River at the confluence of the Studenovka and Lipovka Rivers. The small valleys of the Voronezh tributaries and gullies are deeply incised into the interfluves, thus forming the natural boundaries of the city's historic center. The city lies at the edge of two major geomorphological domains: the Oksko-Donskaya Lowland and the Srednerusskaya Upland. The two main complexes of forms represent the topography of the city territory. The first is the glacial and fluvioglacial landforms of the interfluves formed during the Middle Pleistocene glaciation, which were further transformed by erosion, mass movement, and karst processes. The second is the fluvial landforms presented by the Voronezh River valley, its tributary valleys (Lipovka and Studenovka Rivers), and small erosional landforms (gullies, arroyos, and dry washes).

On the Lipovka River valley slopes, there are massive rock outcrops of Devonian limestone up to 15 m in height, which are of natural and cultural heritage value. However, locals often use gullies and small valleys as unauthorized storage sites for construction and domestic waste. Thus, small erosional landform slopes and bottoms are heavily littered, significantly affecting the hydrological connectivity of the Lipovka basin.

The land cover map (see Fig. 20.1) was constructed using Sentinel-2 maximum likelihood classification. The Lipetsk city land cover is mainly residential. Most paved lands are situated in factory territories and modern residential districts. There are some areas with destroyed and sliced soil layers in the northern part of the city. The forested areas are situated in the upper third of the Lipovka catchment and city center.

### 20.2.2 Hydrometeorological Setting

The climate of Lipetsk is temperate continental with warm summers and moderately cold winters. The mean January (the coldest month) air temperature is −8.6 °C. The snow–water equivalent can exceed 100 mm. Annual precipitation sums area 480–500 mm, evapotranspiration ranges from 350 to 370 mm, and the potential evapotranspiration is 570–580 mm. Precipitation varies from season to season, with the lowest amounts in winter, with a minimum in February of 20–25 mm. Monthly and daily maxima occur in June and July, and can exceed 50–100 mm, caused by heavy rainfall.

Rivers in the Lipetsk region are mainly snow-fed, have a pronounced spring freshet, and have an Eastern-European hydrological regime. Such hydrological regimes are characterized by pronounced spring snowmelt freshets, summer–winter low-water periods, and rain floods during summer and autumn. The period of the frozen rivers is approximately 120–130 days in winter. Ice cover usually cracks at the end of March. Small rivers have rapid snow-fed freshets because of compact catchments, developed gullies, and relatively narrow floodplains.

The Lipetsk city water bodies include rivers, streams, gullies (with temporary runoff), and artificial water bodies, such as ponds, canals, culverts, and reservoirs. Natural surface runoff in city areas has been transformed substantially through the construction of buildings, road infrastructure, and land reclamation. The catchment area was strongly modified by deforestation; almost all forests were removed, except for several parks upstream and within the floodplain along the streams. Several streams were buried underground in pipes to serve as drainage arteries in the city's sewer system, whereas others were filled by ground. In the lower reach, after Komsomolsky Pond, the Lipovka River is a channelized stream discharging into the Voronezh River 1.5 km downstream from its natural mouth.

In this study, we examined the Lipovka River along its entire watercourse, with the most downstream point in Komsomolsky Pond. It was built in the middle of the eighteenth century to meet factory needs. Historically, ponds have been filled with water from the Voronezh River through a system of pumps and channels. Reconstruction work has now taken place at Komsomolsky Pond: treatment facilities have been rebuilt, new locks have been installed to regulate the water level, new fountains have been installed, and a pipeline has been laid along the bottom of the reservoir.

## 20.3 Materials and Methods

### *20.3.1 Field Surveys*

The field survey was conducted in April 2022. During the field survey in April 2022, we examined the entire river valley and channel, particularly the ecological conditions in the control cross-sections. Topographic surveys were conducted by using a DGPS receiver (EFT M2). Accurate measurements of the reference points of the water surface and valley landforms (shoulder and footslope breakpoints of the channel, floodplain, terraces, etc.) were performed to obtain a longitudinal water surface profile and cross-sectional elevation profiles. Water discharge measurements were performed using the floating method (John et al. 1978). The water temperature and conductivity were measured using a YSI Pro1030 portable probe at 32 control points.

### *20.3.2 Water Discharge Assessment (Regulatory Documents)*

The freshet water discharge ($Q_{1\%}^{\text{freshet}}$) with a 1% AEP was estimated using the so-called reduction equation according to SP 33-101-2003. The maximum water discharge was calculated as follows:

$$Q_{1\%}^{\text{freshet}} = \frac{K_0 h_{1\%} \mu_{1\%} A}{(A+c)^n} \delta\delta_1\delta_2, \tag{20.1}$$

where $h_{1\%}$ is a specific water discharge with a 1% AEP, mm; $A$ is a catchment area, km$^2$; $K_0$ is freshet coefficient; $n$ is a regional reduction coefficient set to 0.25; $c$ is a regional declining reduction coefficient describing the reduction decline in a small sub-basin, set to 2; $\delta$, $\delta_1$, $\delta_2$—area empirical coefficients describing the basin lake cover, forest cover, and wetland cover.

We used the so-called equation of peak intensity from SP 33-101-2003 to estimate rain flood water discharge with a 1% AEP ($Q_{1\%}^{\text{rain}}$). This equation also uses several empirical coefficients when the overall formula is:

$$Q_{1\%}^{\text{rain}} = q_{1\%} \varphi H_{1\%} \delta A, \tag{20.2}$$

where $q_{1\%}$ is a relative maximum specific water discharge; $\varphi$ is the empirical streamflow coefficient; $H_{1\%}$ is a maximum daily rainfall amount, mm; $A$ is a catchment area, km$^2$. The $q_{1\%}$ and $H_{1\%}$ were taken from eponymous maps (Kopilov 1986).

### *20.3.3 Water Discharge Modelling*

The HBV is a lumped conceptual catchment model (Lindström et al. 1997) that has relatively few model parameters (see Table 20.1) and minimal forcing input requirements—daily air temperature (T), precipitation (P), and potential evapotranspiration (PET). A daily time step was used in the model, which had an unsaturated zone store, two groundwater stores, and 15 parameters that could be calibrated (see Table 20.1). We selected the HBV model because it is low in complexity, agile, and computationally efficient (Beck et al. 2020).

Meteorological data (T and P) for the 2005–2022 period was taken from the Lipetsk meteostation (WMO code 27,930) located 9 km north of Komsomolsky Pond (i.e., study outlet). The daily air temperature, precipitation, and snow depth data were obtained from 2005 to 2022 (https://www.rp5.ru). Daily PET was calculated using the Oudin equation (Oudin et al. 2005) for a latitude of 52.5°, based on the daily station air temperature (T). Because the Lipovka River is ungauged, we could not use water discharge data to calibrate the free parameters. Therefore, we used the HBV global parameter maps created by Beck et al. (2020). In their dataset, ten alternative sets (folds) of parameters exist for every 0.05° cell. We ran all scenarios to study the water discharge variability (Fig. 20.2).

The final parameter set (Table 20.1) was manually chosen based on measured water discharges in 2022 and by comparing simulated and published water yield regional means (Kopilov 1986). Snowmelt runoff was calibrated by comparing the simulated snow water equivalent (SWE) accumulation and measured snow depth at Lipetsk meteostation (cf. Fig. 20.1 for location). Model performance was measured

**Table 20.1** HBV model parameters were calibrated for the Lipovka River (2005–2022)

| Parameter | Description | Process | Unit | Adjusted value |
|---|---|---|---|---|
| TT | Temperature threshold determining snowfall or rainfall, and melt or refreezing | Precipitation/ melt/refreeze | °C | 0.1 |
| CFMAX | Degree-day factor of snow melt and refreezing | Melt/refreeze | mm $°C^{-1}$ $day^{-1}$ | 6.4 |
| SFCF | Snowfall correction factor | Snowfall | — | 0.55 |
| SP | Seasonal variability of degree-day factor | Melt/refreeze | — | 0.9 |
| CFR | Coefficient of refreezing of melted snow | Refreeze | — | 0.05 |
| CWH | Maximum water-holding content of the snowpack | Excess flow | — | 0.1 |
| FC | Maximum soil moisture storage | Infiltration/ recharge | mm | 360 |
| LP | Soil moisture value above which Ea reaches Ep (wilting point) | Evaporation | — | 0.9 |
| BETA | Nonlinearity coefficient of upper zone recharge | Infiltration/ recharge | — | 3.7 |
| $K_0$ | Runoff coefficient for quick flow (from the upper zone) | Quick flow | 1 $day^{-1}$ | 0.1 |
| $K_1$ | Runoff coefficient for normal flow (from the upper zone) | Normal flow | 1 $day^{-1}$ | 0.1 |
| $K_2$ | Runoff coefficient for base flow (from the lower zone) | Base flow | 1 $day^{-1}$ | 0.25 |
| UZL | The threshold for quick flow | Quick flow | mm | 60.9 |
| PERC | The maximum rate of percolation to the lower zone | Percolation | mm $day^{-1}$ | 0.5 |
| MAXBAS | Flow routing delay | Routing | day | 2 |

using the Nash–Sutcliffe coefficient (NSE; Nash and Sutcliffe 1970) and coefficient of determination ($R^2$).

### 20.3.4 *Flood Stage Estimation*

The highest water stage of a certain AEP for the free-of-ice conditions for each cross-section was calculated based on the maximum discharge of the relevant AEP and stage–discharge curve, which was constructed considering the hydraulic and morphometric properties of the channel and floodplain. The maximum capacity of the channel in each design segment is computed using the curve.

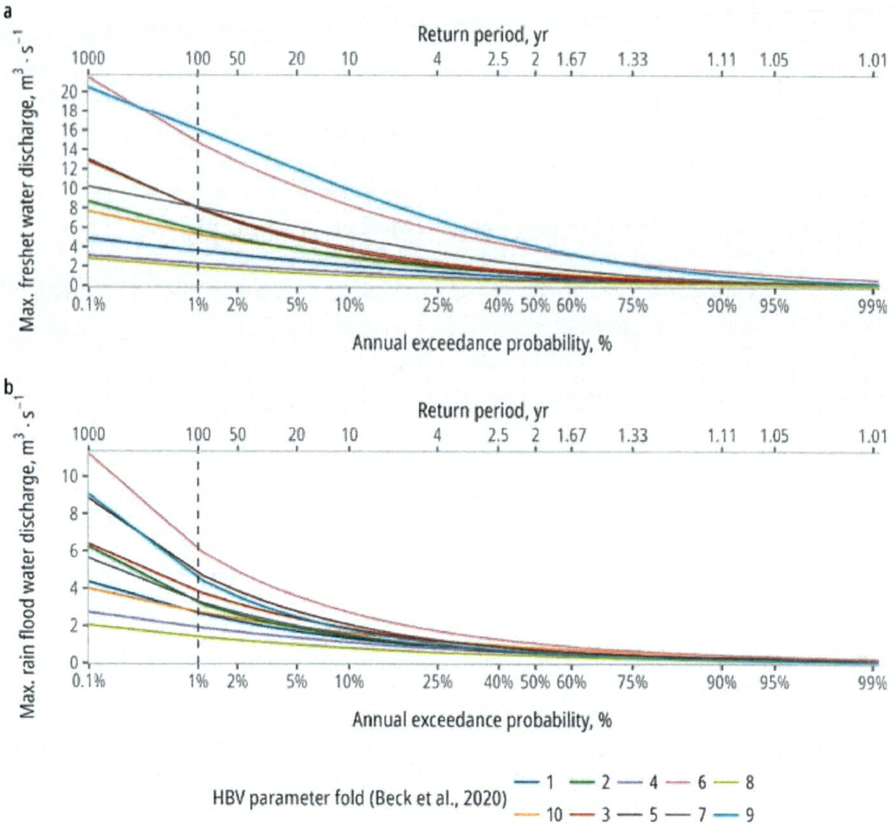


**Fig. 20.2** Water discharge variability in HBV modelling according to various parameter sets for freshet (**a**) and rain (**b**) floods

Hydraulic radius, width, and cross-sectional area were calculated based on the water level. Subsequently, the Chézy–Manning formula (Chanson 2004) was used to obtain the average velocity in the section for each piece of the channel cross-section. As a result, the stage–discharge curve was estimated for each cross-section.

## 20.4 Results

### *20.4.1 Field Surveys*

The Lipovka River watershed area is 50.1 km$^2$ (with an outlet in Komsomolsky Pond). It has only one tributary on the left, with a watershed area of 21.4 km$^2$. The catchment of the unnamed river is fully developed, and streamflow is regulated. During the field survey in April 2022, no streamflow was observed. Before the conflux, the streamflow of Lipovka was also regulated by Syrsky Pond. Therefore, we hypothesized that the effective catchment area begins only after confluence (A = 22.6 $km^2$).

The measured discharges in April 2022 were approximately 5–10 l $s^{-1}$ at the beginning of the river directly downstream of Syrsky Pond, 80–100 l $s^{-1}$ in the middle, 180–200 in the Kamenny Log, 250–270 l $s^{-1}$ through the Komsomolsky Pond, and 320–350 l $s^{-1}$ in the outlet of the Lipovka Canal running into the Voronezh River.

### *20.4.2 Regulatory Documents*

The study area's rivers belong to the watercourses group with an indicator of the expected annual specific water discharge of 3.8 l $s^{-1}$ $km^2$. The regional mean annual water discharge of Lipovka watershed (Kopilov 1986) should be 190 l $s^{-1}$ for the entire catchment and *ca.* 90 l $s^{-1}$ for the lower part of the catchment.

According to the reduction equation (Eq. 20.1), the maximum freshet water discharges with a 1% AEP is 520 l $s^{-1}$ for the effective catchment (A = 22.6 $km^2$) and 1311 l $s^{-1}$ for the whole area (A = 50.1 $km^2$). Equation 20.2 suggests that the maximum rain flood discharges with a 1% AEP should be 570 l $s^{-1}$ and 910 l $s^{-1}$, respectively.

### *20.4.3 Hydrological Modelling*

The run of tenfold parameter sets from the 2020th study (Beck et al. 2020) for the whole catchment area suggested that the mean annual freshet discharge with a 1% AEP varied from 1939 to 15 095 l $s^{-1}$ with a mean of 6927 l $s^{-1}$ (see Fig. 20.2). The rain flood discharges were significantly lower, ranging from 1514 to 5648 l $s^{-1}$ with a mean of 3379 l $s^{-1}$ (1% AEP). It should be noted that these results correspond to an uncalibrated model and serve only descriptive needs.

The HBV model was calibrated based on the regional surface/ground runoff relationship and snow water equivalent (NSE = 0.86, $R^2$ = 0.90). This allowed us to adapt the 8th CV-fold (from Beck et al. 2020) for local conditions. Therefore, in the case of the gauging station's absence, we decided that the HBV model represents snow melting runoff satisfactorily and generally corresponds to measured discharges (see Sect. 4.1). The simulated streamflow and snow cover time series (compared to the observed snow cover values) are shown in Fig. 20.3; Table 20.2.

Simulated mean annual hydrograph is shown on the Fig. 20.4. The mean annual specific water discharge for the inlet to Komsomolsky Pond was 104 mm during 2005–2021, whereas the mean annual precipitation was 540 mm. The average inflow into Pond is 6200 $m^3$ from mid-March to early April. The annual inflow into Pond is 22,000 $m^3$. Calculated using the Oudin method, 10.9 $m^3$ of water evaporates annually. The computation of the maximum levels of the pond was based on fieldwork of the

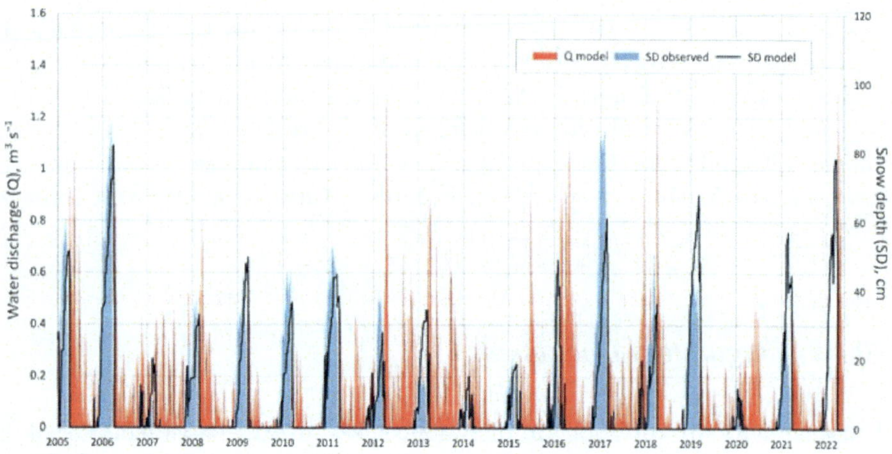


**Fig. 20.3** Simulated water discharge (Q, $m^3$ $s^{-1}$) and snow depth (SD, cm) time series for the Lipovka River for 2005–2022. The observed snow depth is indicated by blue bars

**Table 20.2** Long-term water budget components of the Lipovka River catchment for 2005–2022

| Runoff characteristic | Units | Value |
|---|---|---|
| Annual total runoff | mm | 105 |
| Annual precipitation | mm | 483 |
| Annual Evapotranspiration | mm | 365 |
| Annual Potential Evapotranspiration | mm | 581 |
| Snowmelt (freshet) runoff | mm | 16 |
| Rainwater runoff | mm | 89 |
| Surface runoff | % | 39 |
| Lateral, soil, and groundwater runoff | % | 61 |
| Mean maximum freshet water discharge | $m^3$ $s^{-1}$ | 0.91 |
| Mean maximum rain flood water discharge | $m^3$ $s^{-1}$ | 0.63 |
| Maximum freshet water discharge | $m^3$ $s^{-1}$ | 1.50 |
| Maximum rain flood water discharge | $m^3$ $s^{-1}$ | 0.97 |

locks located in the southern part of the pond. The water level at which overflow occurred through the locks was taken as 110.12 m a.s.l. The discharge overflowing through the walls of the locks will be 0.3 $m^3$ $s^{-1}$, provided that the water stage increases by 0.1 m.

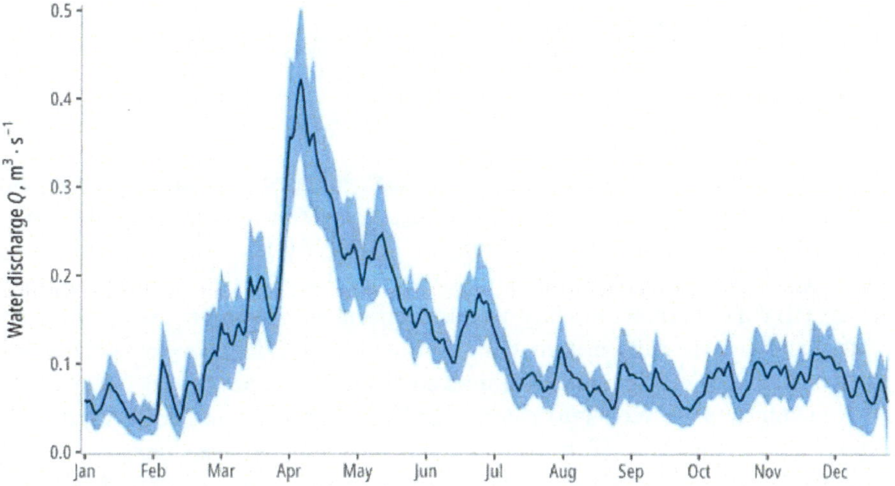


**Fig. 20.4** The mean daily water discharge (Q, $m^3\ s^{-1}$) at the inlet at Komsomolsky Pond. The colored shaded areas represent the range corresponding to ± SE. The mean values and standard errors were estimated using the HBV model for 2005–2022

### *20.4.4 Komsomolsky Pond Water Stages*

At an average water stage of 110.12 m, the pond had a volume of approximately 22 000 $m^3$. This volume is filled in 4–6 days during low-water periods with a water discharge *of ca.* 50 l $s^{-1}$. The daily inflow volume reaches six pond volumes during extreme floods, with a daily discharge of approximately 1.5 $m^3\ s^{-1}$, as opposed to the ordinary flood of 500 l $s^{-1}$. In any case, floods completely pass through Komsomolsky Pond. Approximately 8 $m^3\ s^{-1}$ is the gate opening's maximum capacity (*ca.* 700,000 $m^3\ day^{-1}$).

According to the calculated water discharge curves, the Lipovka Canal could pass a short-term flood with a discharge of up to 9 $m^3\ s^{-1}$. The hazard of water spilling over the Lipovka Canal edges arises in the event of a lengthy rain flood or spring flood owing to the filling of the channel network and inflow from the surrounding areas below Komsomolsky Pond. It is advisable to create a pond water-level management system to avoid floodplain inundation.

## 20.5 Discussion

### *20.5.1 Water Balance*

The long-term maximum water discharge calculated based on regulatory documents (SP) and the HBV model are as follows: Spring freshet (mainly snowmelt) discharges with a 1% annual exceedance probability are 1.5 (HBV) and 1.31 (SP) $m^3\ s^{-1}$, while rain flood discharges are 0.97 (HBV) and 0.91 (SP) $m^3\ s^{-1}$. Global parameter maps for the HBV model showed highly diverse results in both hydrograph estimation and water balance, especially in low AEP discharges (Fig. 20.3). However, calibrating the model with SWE significantly improved its prediction power.

Applying the HBV model to several small Swedish catchments, Seibert and Beven (2009) showed that even a few field measurements can effectively constrain prediction uncertainties. In the case of the Lipovka River, hydrological modelling was the only method used for hydrograph reconstruction. Even with several discharge measurements, it was possible to set up the HBV model and obtain relatively reliable water balance results.

At the same time, there are many unaccounted local factors that reduce reliability and bring uncertainties to the discharge calculation results. These factors are mainly culverts, small dams across the riverbed, sewers, and clutter of the river valley. The maximum spring freshet discharges and rainwater discharges of the Lipovka River are of the same order. However, daily snowmelt discharge is usually higher owing to the daily time step in the modelling. Heavy rain causes short-term floods that might be much higher than daily discharges because high-intensity rains usually have one–two hours peak amount, and the actual maximum short runoff peak cannot be modelled with daily time step.

The regulatory document approach (SP 33-101-2003) was based on regional and outdated relationships (Ayzel et al. 2020). Therefore, without gauged analog rivers, the approach cannot account for both local synoptic situations (rainfall and temperature) and particular catchment features, such as ground and surface water interaction and anthropogenic disturbance of the river valley. Modelling, on the contrary, may address these features by adjusting the parameters.

### *20.5.2 Ecological Revitalization Master Plan*

Komsomolsky Pond is one of the critical components of the hydrological balance of the Lipovka River. It is situated in the city center and acts as a barrier between surface runoff from the city's central area and water from the upstream Lipovka. The pond is continuously filled in the summer with water from the Voronezh River. Maintaining an average water level in the pond, releasing the minimum sanitary discharge into the Lipovka Canal, removing sediments and pollutants delivered by the river from the upstream, and flood protection for the downstream surrounding areas are the requirements of the pond water balance management system.

Even though the pond only has a small capacity and primarily serves as a transit container during floods, it is nevertheless crucial to be ready for them to avoid the lengthy discharge of water into the Lipovka Canal. Shield gates control the pond's water level and the amount of water that drains into the Lipovka Canal, preventing floods. On the other hand, efficient congestion management can optimize the water consumption from the Voronezh River and reduce the electricity costs of pumping.

It is advisable to develop and install an automatic system to regulate streamflow to control the water stage in the Komsomolsky Pond and to provide consistent runoff into the Lipovka River's artificial channel (canal) downstream of the Komsomolsky Pond. Figure 20.5 illustrates an example of the calculated water levels in Komsomolsky pond during the operation of the automatic level maintenance system.

These levels are maintained by managing the water pumps at the water intake station of the Voronezh River. The shield gates are located at the outlet of the pond. Shield gates are remotely and automatically controlled in response to signals from water-level sensors. The automatic level maintenance system uses the following algorithm. Existing pumps fill the pond with water from the Voronezh River until it reaches its highest water level (Lv. 1), at which point they are shut off. Through the shield gate, water runs continuously from the pond downstream into the canal.

**Fig. 20.5** Water stages of Komsomolsky pond while the automatic level maintenance system operates. Lv. 1 represents the highest calculated water level in the pond (turning off the pump), Lv. 2 represents the lower calculated water level in the pond (turning on the pump), Lv. 3 illustrates the level at which the shield gate opens, Lv. 4 represents the water level with an unfavorable precipitation forecast, and Lv. 5 represents the stable water level with the mean annual discharge

The pumps are turned on when the water level reaches a lower calculated level (Lv. 2), and remain on until Lv. 1 is reached. When a lot of rain and water from the Lipovka River entered the pond, the water level increased to Lv. 3. The shield gate was opened slightly to allow the water to flow smoothly into the canal. The shield gate closes when Lv.1 is reached.

To avoid inundation downstream of Komsomolsky pond, it is advisable to perform a preliminary lowering of the water level in the pond in case of considerable precipitation. To do this, the shield gate is gently opened, and the water level is lowered to Lv.4 when warnings about the probability of extremely heavy precipitation are received from meteorological services or the Ministry of Emergency Situations. As a result, the Komsomolsky Pond water level is automatically controlled, and the overflow into the Lipovka Canal is managed to prevent flooding of the city's residential districts. To perform water purification through the Lipovka River downstream and Komsomolsky Pond, it is recommended to settle aquatic plants in gabion structures for biofiltration against organic and mineral pollutants, as well as for decorative purposes. With the help of gabion structures filled with substrate, plants can be settled at certain artificial heights (cf. Figure 20.6). Aquatic planting zones in Komsomolsky Pond should be close to the stormwater sewer points. Another local biofiltration treatment facility should be constructed directly opposite to the stormwater sewers.

Gabion structures similar to aquatic plants can be placed at the lower Lipovka Canal, where a series of complex landscaping and shoreline reconstruction projects are planned. Biofiltration zones installed along the canal are frequently at a stable minimum permanent water level. Recreational zones along the river are designed to not be inundated at a water stage with a 10% AEP and are constructed of water-invulnerable materials (stones, gabions, and composite boards).

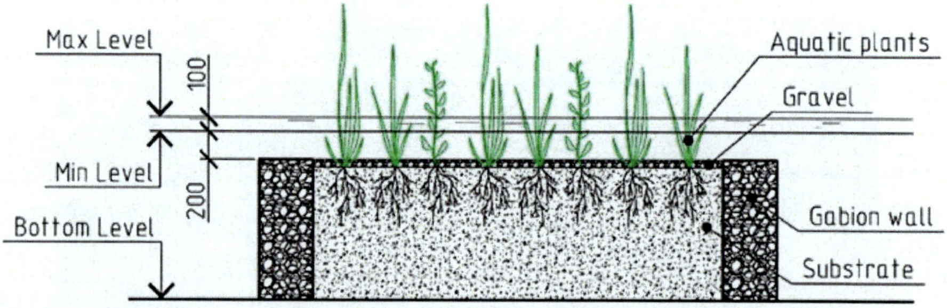


**Fig. 20.6** An illustration of how gabion construction is used for sowing aquatic plants in Komsomolsky Pond

## 20.6 Conclusion

Water discharge dynamics of small urban rivers are highly variable and complex. Understanding these dynamics and their underlying causes is essential for our awareness of water discharge fluctuations in ungauged basins, especially in the face of climate change and urban sprawl. To study the hydrological regime of the small urban Lipovka River, we used two methods: empirical equations recommended by regulatory documents, and hydrological modelling. The anticipated freshet discharges with a 1% AEP are 0.69 (HBV) and 0.52 (SP 33-101-2003) $m^3 s^{-1}$. In contrast, rain flood discharges are 0.45 (HBV) and 0.57 (SP 33-101-2003) $m^3 s^{-1}$. However, we believe that HBV is a better predictor of maximum water discharge. However, more precise modelling is still required for extreme event forecasting and accurate assessment of their effects on the urban environment.

The reduced runoff of the Lipovka River was mainly caused by a decline in the effective catchment area, a littered channel, and increased evaporation and filtration into the groundwater. To reestablish the river's natural flow and catchment area, we propose constructing and maintaining a culvert to enable upstream feeding of the Lipovka River by Syrsky Pond.

The following are some recommendations for urban planning and water management based on the suggested approach to modelling the runoff of a small urban river:

- To install a system of automatic water level control in Komsomolsky Pond to manage the discharge downstream and prevent flooding in downstream areas.
- Use the pond's estimated water balance to determine the amount of pumping water running through the pumping station, which provides the required water level during the warm season.

Our study can serve as the basis for designing and revitalizing small urban rivers, as they are a crucial component of the city's water-green framework, providing a variety of ecosystem services in the context of smart urbanization.

**Acknowledgements** The authors sincerely thank the Guild of Landscape Engineers (LAEN) for providing the funding. This study contributes to the State Task No. FMGE-2019-0005 (Institute of Geography, Russian Academy of Science); State Task No. 121051100166-4 (Laboratory of Soil Erosion and Fluvial Processes, Faculty of Geography, Lomonosov Moscow State University); and RUDN University Strategic Academic Leadership Program. Anatoly Tsyplenkov and Pavel Terskii performed this research according to the Development program of the Interdisciplinary Scientific and Educational School of M.V. Lomonosov Moscow State University "Future Planet and Global Environmental Change".

## References

Anichkina NV, Belyaeva LN, Povkh TV, Schetinkin AG (2017) Rivers of the upper basin of the Don river in the Lipetsk region. Adv Curr Nat Sci 60–65

Ayzel GV, Belozerov EV, Kurochkina LS (2020) Determination of estimated hydrological characteristics in the absence of hydrometric observations: potential for the use of regional gridded river runoff reanalysis. Water Sect Russ 83–101. https://doi.org/10.35567/1999-4508-2020-2-6

Beck HE, Pan M, Lin P, Seibert J, van Dijk AIJM, Wood EF (2020) Global fully distributed parameter regionalization based on observed streamflow from 4,229 headwater catchments. J Geophys Res Atmos 125:e2019JD031485. https://doi.org/10.1029/2019JD031485

Bergström S (1992) The HBV model-its structure and applications. SMHI Rep Hydrol 32

Bergström S, Lindström G (2015) Interpretation of runoff processes in hydrological modelling—experience from the HBV approach. Hydrol Process 29:3535–3545. https://doi.org/10.1002/hyp.10510

Chanson H (2004) The hydraulics of open channel flow: an introduction; basic principles, sediment motion, hydraulic modelling, design of hydraulic structures, 2nd edn. Elsevier Butterworth Heinemann, Amsterdam

Guo H, Han J, Qian L, Long X, Sun X (2022) Assessing the potential impacts of urban expansion on hydrological ecosystem services in a rapidly urbanizing Lake Basin in China. Sustainability 14:4424. https://doi.org/10.3390/su14084424

Gurnell A, Lee M, Souch C (2007) Urban rivers: hydrology, geomorphology, ecology and opportunities for change. Geogr Compass 1:1118–1137. https://doi.org/10.1111/j.1749-8198.2007.00058.x

Hrachowitz M, Savenije HHG, Blöschl G, McDonnell JJ, Sivapalan M, Pomeroy JW, Arheimer B, Blume T, Clark MP, Ehret U, Fenicia F, Freer JE, Gelfan A, Gupta HV, Hughes DA, Hut RW, Montanari A, Pande S, Tetzlaff D, Troch PA, Uhlenbrook S, Wagener T, Winsemius HC, Woods RA, Zehe E, Cudennec C (2013) A decade of predictions in ungauged basins (PUB)—a review. Hydrol Sci J 58:1198–1255. https://doi.org/10.1080/02626667.2013.803183

Hundecha Y, Bárdossy A (2004) Modeling of the effect of land use changes on the runoff generation of a river basin through parameter regionalization of a watershed model. J Hydrol 292:281–295. https://doi.org/10.1016/j.jhydrol.2004.01.002

John P, Johnson F, Sutcliffe P (1978) An integrating float method of discharge measurement. Proc Inst Civ Eng 65:569–588. https://doi.org/10.1680/iicep.1978.2806

Klimenko DE, Kharlamova EO (2019) Engineering methods for calculating the higher water levels of the unexplored Trans-Ural lakes. Inž izysk 12:28–36. https://doi.org/10.25296/1997-8650-2018-12-9-10-28-36

Kopilov AP (ed) (1986) Atlas of hydrological calculation maps and nomograms. Gidrometeoizdat, Leningrad

Lindström G, Johansson B, Persson M, Gardelin M, Bergström S (1997) Development and test of the distributed HBV-96 hydrological model. J Hydrol 201:272–288. https://doi.org/10.1016/S0022-1694(97)00041-3

Ma K, Feng D, Lawson K, Tsai W-P, Liang C, Huang X, Sharma A, Shen C (2021) Transferring hydrologic data across continents—leveraging data-rich regions to improve hydrologic prediction in data-sparse regions. Water Resour Res 57:e2020WR028600. https://doi.org/10.1029/2020WR028600

Makarieva OM, Beldiman IN, Lebedeva LS, Vinogradova TA, Nesterova NV (2017) On the issue of validity of recommendations from SP 33-101-2003 for assessment of maximum flow characteristics of small rivers in the permafrost zone. Inž izysk 50–63. https://doi.org/10.25296/1997-8650-2017-6-7-50-63

Malhotra K, Lamba J, Shepherd S (2020) Sources of stream bed sediment in an urbanized watershed. CATENA 184:104228. https://doi.org/10.1016/j.catena.2019.104228

Muhar S, Sendzimir J, Jungwirth M, Hohensinner S (2018) Restoration in integrated river basin management. In: Schmutz S, Sendzimir J (eds) Riverine ecosystem management: science for governing towards a sustainable future. Springer International Publishing, Cham, pp 273–299
Nash JE, Sutcliffe JV (1970) River flow forecasting through conceptual models part I—a discussion of principles. J Hydrol 10:282–290. https://doi.org/10.1016/0022-1694(70)90255-6
Nesterova NV, Makarieva OM, Post DA (2019) Methods of mathematical modelling for calculating flow characteristics of ungauged rivers in engineering design tasks (by the example of the Khemchik River, Tyva Republic, Russia). IOP Conf Ser Earth Environ Sci 381:012068. https://doi.org/10.1088/1755-1315/381/1/012068
Oudin L, Hervieu F, Michel C, Perrin C, Andréassian V, Anctil F, Loumagne C (2005) Which potential evapotranspiration input for a lumped rainfall–runoff model?: part 2—Towards a simple and efficient potential evapotranspiration model for rainfall–runoff modelling. J Hydrol 303:290–306. https://doi.org/10.1016/j.jhydrol.2004.08.026
Pastor AV, Tzoraki O, Bruno D, Kaletová T, Mendoza-Lera C, Alamanos A, Brummer M, Datry T, De Girolamo AM, Jakubínský J, Logar I, Loures L, Ilhéu M, Koundouri P, Nunes JP, Quintas-Soriano C, Sykes T, Truchy A, Tsani S, Jorda-Capdevila D (2022) Rethinking ecosystem service indicators for their application to intermittent rivers. Ecol Ind 137:108693. https://doi.org/10.1016/j.ecolind.2022.108693
Pool S, Viviroli D, Seibert J (2017) Prediction of hydrographs and flow-duration curves in almost ungauged catchments: Which runoff measurements are most informative for model calibration? J Hydrol 554:613–622. https://doi.org/10.1016/j.jhydrol.2017.09.037
Primo KR, Silva RA e., Olimpio BC, Silva GB, Silva APM, Schulz HE, da Silva AM (2022) Assessing an ecological revitalization project of urban streams in Brazilian Southeastern region by analytic hierarchy process. Water Air Soil Pollut 233:225. https://doi.org/10.1007/s11270-022-05695-6
Seibert J, Beven KJ (2009) Gauging the ungauged basin: how many discharge measurements are needed? Hydrol Earth Syst Sci 13:883–892. https://doi.org/10.5194/hess-13-883-2009
Shrestha DL, Solomatine DP (2008) Data-driven approaches for estimating uncertainty in rainfall-runoff modelling. Int J River Basin Manage 6:109–122. https://doi.org/10.1080/15715124.2008.9635341
Sokolov D, Chalov S, Tereshina M, Erina O, Shinkareva G (2021) Hydrological regime of the urban Setun River. IOP Conf Ser Earth Environ Sci 834:012024. https://doi.org/10.1088/1755-1315/834/1/012024
Taylor KG, Owens PN (2009) Sediments in urban river basins: a review of sediment–contaminant dynamics in an environmental system conditioned by human activities. J Soils Sed 9:281–303. https://doi.org/10.1007/s11368-009-0103-z
Thielen J, Bartholmes J, Ramos M-H, de Roo A (2009) The European flood alert system—Part 1: concept and development. Hydrol Earth Syst Sci 13:125–140. https://doi.org/10.5194/hess-13-125-2009

# Chapter 21
# Development of Environmentally Safe Concentration of Silver Content in Soils of Different Resistance to Chemical Pollution

**Natalia Tsepina, Sergey Kolesnikov, and Tatiana Minnikova**

**Abstract** To manage the quality of urban soils effectively, it is necessary to develop maximum permissible concentrations of various pollutants in the soil. This list includes pollution with heavy metals, such as silver (Ag). At the moment, neither in Russia nor in the world, maximum permissible concentrations of silver in the soil have been developed. Therefore, it is relevant to establish environmentally safe concentrations of silver in soils with different resistance to chemical pollution, which can later be used in soil pollution normalizing. The work objective is to develop environmentally safe concentrations of silver in soils with different resistance to chemical pollution. The following are selected as research objects for the development of environmentally safe concentrations of silver Haplic Chernozems Calcic, Haplic Arenosols Eutric, and Haplic Cambisols Eutric. Environmentally safe concentrations of Ag in soils were developed based on the degree of violation of ecosystem functions of the soil under the influence of silver pollution. Quantitative guidelines for the development of these standards are proposed. The environmentally safe concentration of Ag in Haplic Chernozems Calcic, Haplic Arenosols Eutric, and Haplic Cambisols Eutric is 4.4, 0.9, and 0.8 mg of Ag/kg of soil. The developed environmentally safe concentrations of Ag can also be used to regulate the quality of similar soils in other regions of the world.

**Keywords** Normalizing · Haplic chernozems calcic · Haplic Arenosols Eutric · Haplic Cambisols Eutric · Length of roots · Enzymatic activity · Ecosystem functions

---

N. Tsepina (✉) · S. Kolesnikov · T. Minnikova
Southern Federal University, 344090 Rostov-on-Don, Russia
e-mail: cepina@sfedu.ru

S. Kolesnikov
e-mail: kolesnikov@sfedu.ru

T. Minnikova
e-mail: tminnikova@sfedu.ru



M. Korneykova et al. (eds.), *Smart and Sustainable Urban Ecosystems: Challenges and Solutions*, Springer Geography, https://doi.org/10.1007/978-3-031-37216-2_21

## 21.1 Introduction

In recent years, silver has increasingly become a polluting element as a result of anthropogenic activity. Silver (Ag) enters the environment from industrial sources, which include coal burning at thermal power plants, metal smelting at non-ferrous metallurgy enterprises, cement production, as well as synthesis, processing, and disposal of products based on silver nanoparticles (Künniger et al. 2014; Michels et al. 2017; Xing et al. 2004). The silver content in polluted soils can reach up to 35.9 mg/kg (Yildirim and Sasmaz 2017). The negative effect of Ag is manifested in a decrease in the length of roots and biomass of plants, suppression of the growth and reproduction of earthworms, a decrease in the number of soil bacteria, and inhibition of the activity of soil enzymes (Rahmatpour et al. 2017; Samarajeewa et al. 2017; Shoults-Wilson et al. 2011). As a result of the interaction of silver with humans, toxicity to the respiratory and gastrointestinal systems of humans, skin, liver, and brain cells was noted (Dziendzikowska et al. 2012; Hussain et al. 2001; Oberdorster et al. 2005). To manage the quality of soils, including urban soils, effectively, it is necessary to develop maximum permissible concentrations (MPCs) of various pollutants in the soil. This list includes pollution with heavy metals, such as silver. Now, neither in Russia nor in the world, MPCs of silver in the soil have been developed. Therefore, it is relevant to establish an environmentally safe concentration of silver in soils with different resistance to chemical pollution. The developed environmentally safe concentrations of silver can later be used for soil standardization with the help of biological indicators (Kolesnikov et al. 2000, 2013). The presented set of indicators provides an informative picture of the biological processes occurring in the soil and its ecological state.

The work objective is to assess environmentally safe concentrations of silver in soils with different resistance to chemical pollution.

## 21.2 Materials and Methods

Soils with different resistance to chemical pollution were selected as research objects for the development of environmentally safe silver concentrations. Soils differed in their texture, soil acidity, and organic matter content: Haplic Chernozems Calcic, Haplic Arenosols Eutric, and Haplic Cambisols Eutric (World Reference Base for Soil Resources. International Soil Classification System for Naming Soils and Creating Legends for Soil Maps 2022). The concentrations recorded by the authors in uncontaminated soil for Ag were 0.01–1 mg/kg (Alekseenko and Alekseenko 2013; Jones et al. 1986). The soil samples were selected from the top layer of 0–10 cm since silver is usually accumulated in the surface layers of the soil. In polluted soils, silver content ranged from 8 to 35.9 mg/kg according to Kabata-Pendias (2010); Yildirim and Sasmaz (2017), up to 7000 mg/kg in the soils of ore deposits (Druzhinin and Karelina 2008). The study was conducted under laboratory conditions. The doses were calculated based on the background concentrations of silver. Since the MPC of silver has not been established, its content can be expressed as a derived concentration, which for heaviest metals is about three to four background concentrations in the soil (Kolesnikov et al. 2019). The background silver content in Haplic Chernozems Calcic, Haplic Cambisols Eutric, and Haplic Arenosols Eutric is 0.303, 0.282, and 0.215 mg/kg, respectively. The soil was incubated in vegetative vessels in 3 replications in the climatic chamber Binder for 30 days. The mass of soil in each vessel is 200 g. Favorable conditions for soil biota were set up in the climatic chamber: 25–30% humidity, temperature 24–25 °C, change of lighting (day and night). The silver background content in soils was determined by ICP MS method at the All-Russian Scientific Research Geological Institute named after A.P. Karpinsky (St. Petersburg, Russia). Soil samples were dried and sieved through a 0.25-mm sieve, packed in thick paper bags and transported to the laboratory.

The biological properties of the soil were investigated first, because they were the first to react to external influences and are much more sensitive and informative compared to other soil properties. The change in biological parameters of soils was evaluated 30 days after contamination. Silver was introduced into the soil in the amount 1, 10, and 100 mg/kg, respectively. In this experiment, silver nitrate, which is quite soluble in water, was used to achieve maximum silver toxicity. It is also worth noting the high ecotoxicity of silver nitrate in studies conducted earlier (Cvjetko et al. 2017; Tripathi et al. 2017).

Silver ecotoxicity was assessed using biological methods (Kazeev et al. 2016). The total number of bacteria, catalase and dehydrogenases activity, and phytotoxic properties of the soil (length of root) were determined. To assess the soil state according to biological indicators, the result of calculating the integral indicator of the biological state (IIBS) of the soil was used (Kolesnikov et al. 2000). The calculation of IIBS allows comparing different biological indicators with different units of measurement and making a conclusion about the state of soils after pollution. To calculate environmentally safe concentrations of silver, regression equations were used describing the dependence of the decrease in IIBS values on silver concentration in the soil.

To verify the obtained data for significance, variance analysis was carried out, followed by the determination of the least significant difference (LSD). Variation statistics (mean values, dispersion) was determined, and reliability of different samples was established by using dispersion analysis (Student's t test).

## 21.3 Results and Discussion

The integral indicator of biological state of soils (IIBS) is used just for a holistic assessment of the degree of violation of ecosystem functions. As can be seen from Fig. 21.1, with an increase in silver concentration, the degree of the IIBS decrease for Haplic Chernozems Calcic, Haplic Cambisols Eutric, and Haplic Arenosols Eutric increases.

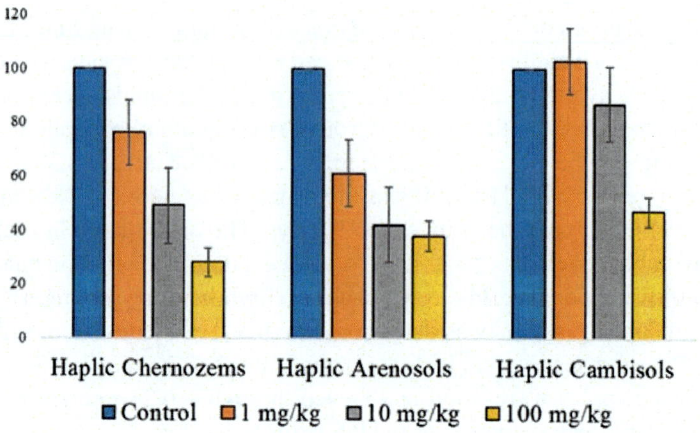


**Fig. 21.1** Integral indicator of the biological state of soils with different resistance to silver pollution, % of control

At a dose of 1 mg/kg of silver, the decrease in IIBS of Haplic Chernozems Calcic by 3%, at 10 mg/kg—by 10%, and at 100 mg/kg—by 27% relative to the control values, respectively, was observed. A dose of 1 and 10 mg/kg of silver caused an IIBS decrease in Haplic Cambisols Eutric by 13 and 27% relative to the control, respectively. At a dose of 100 mg/kg, IIBS of Haplic Cambisols Eutric decreased by 39% relative to the control. IIBS of Haplic Arenosols Eutric decreased at a dose of 1 mg/kg of silver by 15%, at 10 mg/kg—by 30%, and at 100 mg/kg—by 42%, relative to the control, respectively.

To calculate environmentally safe concentrations of silver, regression equations were calculated describing the dependence of the decrease in IIBS values on silver content in each type of soil (Table 21.1). Functional violation is directly related to the silver concentration in the soil. First, a violation of information, then—biochemical, physicochemical, chemical, and holistic, and then physical functions was observed (Kolesnikov et al. 2019, 2002).

Ecological functions of the soil are not violated when the IIBS of the soil is reduced by 5%, and the silver concentration is less than 0.5 mg/kg. The function of soil

**Table 21.1** Regression equations for soils with different tolerance to silver pollution and ecologically safe silver concentrations

| Soil type | Regression equation | Ecologically safe concentration, mg/kg |
|---|---|---|
| Haplic Chernozems Calcic | $y = -4.704 \ln(x) + 96.955$, $R^2 = 0.99$ | 4.4 |
| Haplic Arenosols Eutric | $y = -7.138 \ln(x) + 88.803$, $R^2 = 0.98$ | 0.8 |
| Haplic Cambisols Eutric | $y = -6.664 \ln(x) + 90.101$, $R^2 = 0.98$ | 0.9 |

information is violated when the IIBS is reduced by 5–10%, and silver concentration is in the range from 1.5 to 4.4 mg/kg. The biochemical, physicochemical, chemical, and the complete function of the soil are validated if a decrease in IIBS occurs by more than 25%, while the silver concentration in the soil is more than 106 mg/kg. A decrease in IIBS by more than 10% indicates serious violations in soil functioning. The dose of silver, which causes a violation of the holistic functions responsible for soil fertility, can be considered an environmentally safe concentration of silver in this soil, the excess of which is unacceptable.

The authors have proposed the quantitative guidelines for the development of these norms for three soils that differ greatly in resistance to chemical pollution in Haplic Chernozems Calcic, Haplic Cambisols Eutric, and Haplic Arenosols Eutric. An environmentally safe concentration of silver in Haplic Chernozems Calcic is 4.4 mg/kg of silver in the soil, in Haplic Cambisols Eutric—0.9 mg/kg, and Haplic Arenosols Eutric—0.8 mg/kg.

The higher the silver concentration in the soil, the more effective and efficient the method of sanitation should be. When the silver concentration in Haplic Chernozems Calcic is less than 0.5 mg/kg, there is no violation of environmental functions, and soil sanitation is not required. When the silver concentration is in the range of 4.4–106 mg/kg, the application of organic and mineral fertilizers (phosphorous fertilizers, lime, etc.) and adsorbents (ion exchange resins, zeolites, etc.) is already required as part of chemical recultivation. The developed environmentally safe concentrations of silver can also be used in the normalizing of soils and other geographical areas.

The series of informativeness (averaged for all types of soils) of biological indicators is the following sequence: length of radish roots (−0.86) > dehydrogenases activity (−0.81) > catalase activity (−0.76) > total number of bacteria (−0.71).

Of the four biological indicators, the most informative was the length of roots, and the least informative was the total number of bacteria. But at the same time, all indicators have high correlation coefficients with a silver content of more than (−0.70).

When studying the effect of silver pollution on the biological properties of soils, it was found that all bioindicators had high sensitivity and informative value. This indicates the expediency of their use in monitoring, diagnostics, and normalizing of silver-contaminated soils. Environmentally safe concentrations of silver in soils were developed based on the degree of violation of ecosystem functions of the soil under the influence of silver. When soils are polluted with silver biogeocenotical functions of the soil are violated.

## 21.4 Conclusion

The increasing trend of soil pollution with silver, as well as the absence of the MPC of silver in soils, determines the necessity for its normalizing. The environmentally safe concentration of silver in soils was developed based on the degree of violation of ecosystem functions of the soil under the influence of silver. The IIBS was used for a holistic assessment of the degree of violation of ecosystem functions. The dose of soil-polluting silver, which causes a violation of the holistic functions responsible for soil fertility, can be considered an environmentally safe concentration of silver in this soil, the excess of which is unacceptable. The authors have proposed the quantitative guidelines for the development of these norms for three soils that differ greatly in resistance to chemical pollution. The high resistance to silver contamination is due the following parameters: the high content of organic matter which is about (3.7%), the neutral reaction of the soil medium pH (7.8), and heavy-loamy granulometric composition of Haplic Chernozems Calcic. The high in both mobility and ecotoxicity of silver in the following soils is due to the light granulometric composition (sandy loam) of Haplic Cambisols Eutric and the acidic reaction of the medium (pH = 5.8) in Haplic Arenosols Eutric furthermore the low humus content (1.8 and 2.3%, respectively). It can be concluded that each type of soil, has its own environmentally safe concentration of silver. The environmentally safe concentration of silver in Haplic Chernozems Calcic, Haplic Cambisols Eutric, and Haplic Arenosols Eutric is 4.4, 0.9, and 0.8 mg of silver/kg of soil, respectively.

**Acknowledgements** The study was financially supported by a grant from the President (MK-1168.2022.5) and the Ministry of Science and Higher Education of the Russian Federation (no. FENW-2023-0008).

## References

Alekseenko VA, Alekseenko AV (2013) Chemical elements in geochemical systems. Clarks of soils of residential landscapes: monograph Rostov n/A. Southern Federal University Press, p 380

Cvjetko P, Milošić A, Domijan AM, Vinković VI, Tolić S, Peharec ŠP, Letofsky-Papst I, Tkalec M, Balen B (2017) Toxicity of silver ions and differently coated silver nanoparticles in *Allium cepa* roots. Ecotoxicol Environ Saf 137:8–28. https://doi.org/10.1016/j.ecoenv.2016.11.009

Druzhinin AV, Karelina EV (2008) The main types of industrial silver deposits. Bull RUDN Eng Res Ser 1:35–41

Dziendzikowska K, Gromadzka-Ostrowska J, Lankoff A, Oczkowski M, Krawczyńska A, Chwastowska J et al (2012) Time-dependent biodistribution and excretion of silver nanoparticles in male Wistar rats. J Appl Toxicol 32:920–928. https://doi.org/10.1002/jat.2758

Hussain N, Jaitley V, Florence AT (2001) Recent advances in the understanding of uptake of microparticulates across the gastrointestinal lymphatics. Adv Drug Deliv Rev 50:107–142. https://doi.org/10.1016/S0169-409X(01)00152-1

Jones KC, Davies BE, Peterson PJ (1986) Silver in Welsh soils: physical and chemical distribution studies. Geoderma 37:157–174. https://doi.org/10.1016/0016-7061(86)90028-5

Kabata-Pendias A (2010) Trace elements in soils and plants. 4th Edition. Crc Press, Boca Raton, FL, pp 548

Kazeev KS, Kolesnikov SI, Akimenko YV, Dadenko EV (2016) Methods of biodiagnostics of terrestrial ecosystems. Rostov–on–Don: SFU Publishing House, p 356

Kolesnikov SI, Kazeev KS, Val'kov VF (2000) Effects of heavy metal pollution on the ecological and biological characteristics of common chernozem. Russ J Ecol 31:174–181

Kolesnikov SI, Kazeev KS, Valkov VF (2002) Ecological functions of soils and the effect of contamination with heavy metals. Eurasian Soil Sci 12:1335–1340

Kolesnikov SI, Yaroslavtsev MV, Spivakova NA, Kazeev KS (2013) Comparative assessment of the biological tolerance of chernozems in the South of Russia towards contamination with Cr, Cu, Ni, and Pb in a model experiment. Eurasian Soil Sci 46:176–181

Kolesnikov SI, Kazeev KS, Akimenko YV (2019) Development of regional standards for pollutants in the soil using biological parameters. Environ Monit Assess 191:544. https://doi.org/10.1007/s10661-019-7718-3

Künniger T, Gerecke AC, Ulrich A, Huch A, Vonbank R, Heeb M, Wichser A, Haag R, Kunz P, Faller M (2014) Release and environmental impact of silver nanoparticles and conventional organic biocides from coated woodenfaçades. Environ Pollut 184:464–471. https://doi.org/10.1016/j.envpol.2013.09.030

Michels C, Perazzoli S, Soares M (2017) Inhibition of the enriched culture of ammonium-oxidizing bacteria by two different nanoparticles: silver and magnetite. Common Environ Sci 586:995–1002. https://doi.org/10.1016/j.scitotenv.2017.02.080

Oberdorster G, Oberdorster E, Oberdorster J (2005) Nanotoxicology: an emerging discipline evolving from studies of ultrafine particles. Environ Health Perspect 113:823–839. https://doi.org/10.1289/ehp.7339

Rahmatpour S, Shirvani M, Mosaddeghi MR, Farshid N, Bazarganipour M (2017) Dose–response effects of silver nanoparticles and silver nitrate on microbial and enzyme activities in calcareous soils. Geoderma 285:313–322. https://doi.org/10.1016/j.geoderma.2016.10.006

Samarajeewa AD, Velicogna JR, Princz JI, Subasinghe RM, Scroggins RP, Beaudette LA (2017) Effect of silver nano-particles on soil microbial growth, activity and community diversity in a sandy loam soil. Environ Pollut 220:504–513. https://doi.org/10.1016/j.envpol.2016.09.094

Shoults-Wilson WA, Reinsh BB, Tsyusko OV, Bertsh PM, Lowry GV, Unrin JM (2011) Role of particle size and soil type in the toxicity of silver nanoparticles to worms. Soil Sci Soc Am J 75:365–377. https://doi.org/10.2136/sssaj2010.0127nps

Tripathi A, Liu S, Kumar SP, Kumar N, Chandra PA, Tripathi DK, Chauhan KD, Sahi S (2017) Differential phytotoxic responses of silver nitrate ($AgNO_3$) and silver nanoparticle (AgNPs) in *Cucumis sativus* L. Plant Genet Resour 11:255–264. https://doi.org/10.1016/j.plgene.2017.07.005

World Reference Base for Soil Resources. International Soil Classification System for Naming Soils and Creating Legends for Soil Maps (2022) 4th ed.; International Union of Soil Sciences (IUSS): Vienna, Austria, 234

Xing G, Zhu J, Xiong Z (2004) Ag, Ta, Ru, and Ir enrichment in surface soil: Evidence for land pollution of heavy metal from atmospheric deposition. Glob Biogeochem Cycles 18:1–5. https://doi.org/10.1029/2003GB002123

Yildirim D, Sasmaz A (2017) Phytoremediation of As, Ag, and Pb in contaminated soils using terrestrial plants grown on Gumuskoy mining area (Kutahya Turkey). J Geochem Explor 182:228–234. https://doi.org/10.1016/j.gexplo.2016.11.005

# Chapter 22
# Cost–Benefit Assessment for Maintenance of Urban Green Infrastructure at the University Campus in Moscow: Application of GreenSpaces and TreeTalker Technologies to Regulating Ecosystem Services

**V. Matasov, V. Grigoreva, K. Makhinya, M. Kozyreva, O. Romzaikina, O. Maximova, A. Pakina, A. Filyushkina, and V. Vasenev**

**Abstract** Trees are a key element of urban green infrastructure (UGI). They supply a wide range of benefits—ecosystem services—from air purification to places for recreation. UGI is also an expenditure item in the budget due to the investments required for its establishment and maintenance. Accounting for ecosystem services in monetary terms allows considering direct and indirect benefits of green spaces together with costs, has not only implications for decision-making but also could be instrumental in changing landowners' perceptions of UGI and its importance. In this paper, we used the data from advanced tree monitoring technologies to compare monetary values of regulating ecosystem services to maintenance costs for the case of campus of the People's Friendship University of Russia (RUDN) in Moscow, Russia. Inventories of UGI elements were conducted by means of a field survey and remote sensing, that resulted in GIS project in GreenSpaces software. The same program was used to keep track of costs of the different types of maintenance work based on information from Greening Department of RUDN. Biophysical parameters of tree functioning were obtained from more than 60 TreeTalker sensors installed on the major species during the two years. The monetary value of four major regulating

V. Matasov (✉) · V. Grigoreva · K. Makhinya · M. Kozyreva · O. Romzaikina
RUDN University, Miklukho-Maklaya St., 6, 117198 Moscow, Russia
e-mail: matasov_vm@pfur.ru

A. Pakina
Lomonosov Moscow State University, Leninskie Gory-1, 119991 Moscow, Russia

O. Maximova · A. Filyushkina
Association Landscape Engineers Guild, Menzhinskogo St., 21, 129327 Moscow, Russia

V. Vasenev
Soil Geogrpaphy and Landscape Department, Wageningen University, Wageningen, The Netherlands



M. Korneykova et al. (eds.), *Smart and Sustainable Urban Ecosystems: Challenges and Solutions*, Springer Geography, https://doi.org/10.1007/978-3-031-37216-2_22

ecosystem services (air purification, climate regulation, water transpiration, carbon sequestration) of the trees was then assessed, allowing a comparison of the costs for maintenance and some of the benefits derived. The results show that even when considering just monetary value from regulating services, it outweighs the costs for maintenance of the UGI in the amount of 1.5 million rubles or about a quarter of the costs.

**Keywords** Ecosystem services · Value · TreeTalker · R3GIS · Green area maintenance

## 22.1 Introduction

Urban green infrastructure (UGI) plays a crucial role in ensuring good quality of life and human well-being, as well as in the sustainable development and functioning of urban environment (Gómez-Baggethun and Barton 2013; Haase et al. 2014; Kondo et al. 2018), especially at the times of crisis (Fagerholm et al. 2022). One of the most substantial components of UGI are trees, growing on roadsides, in backyards, as well as in the parks, gardens and urban forests and are even included in vertical greening. Urban trees provide a wide range of benefits, referred to as ecosystem services (ES), such as supporting, regulating, provisioning and cultural (Burkhard et al. 2018; Klimanova and Illarionova 2020). Carbon sequestration as part of climate mitigation, provision of shade and cooling effect, aesthetic and recreational values, habitat for birds and other species is the most widely studied ecosystem services provided by urban trees. Many of these services fall under the category of so-called non-market services (e.g., recreation in a park), i.e., not traded on markets and often they are also not easy to quantify / determine value both in biophysical and socio-economic terms (Small et al. 2017). This is also partly linked to the challenge of stakeholders often not realizing the full range of services and values provided by the UGI and its components.

At the same time, having trees in cities coincide with costs that include not only those for planting and maintenance, but also a long list of potential indirect costs, such as damage to buildings and pavements by tree roots, damage and injury from falling trees, disruption to traffic during maintenance, carbon emissions through operating machinery, blockage of drains by leaf litter and air pollution by volatile organic compounds emitted by foliage, to name but a few (Vogt et al. 2015). Calculation of costs is also not a straight-forward exercise, as it involves specific rates for labor and other components, technologies and materials to be used, all set by the official documents.

This complex balance of understanding and valuation of UGI's benefits and costs is the reason why planting trees in the cities can be and often is a controversial issue, involving many stakeholders with different interests and preferences, as well as more importantly with differences in awareness of the entirety of the costs and benefits associated with urban trees. Therefore, in order to support decision-making, it is paramount to investigate benefits and costs and explore ways to better quantify them. In recent years, a number of studies devoted to the question have emerged (Bolund and Hunhammar 1999; Russo et al. 2017), including reviews (e.g., Song et al. 2018). Many of them have quantified the benefits, while the costs have received less attention. Suggesting that understanding of the balance between costs and benefits (namely, which way it lies) remains poor.

At the same time, advanced GIS-based technologies for monitoring tree health and greenery maintenance allow considering these costs and aim to minimize them. One of such systems is GreenSpaces, developed by R3GIS company (https://www.r3gis.com/greenspaces). Its main tasks include inventory of current state and changes of main elements of UGI, management and control of maintenance activities. The company is also developing features for valuation of ES integrated in the app along with TreeTalker integration (Valentini et al. 2019; Matasov et al. 2020). This company entered the Russian market relatively recently and thus the app is still undergoing the process of adaptation to the institutional, biophysical, climatic and economic conditions of Russian cities.

Aim of this study was to determine whether the costs of maintaining urban trees outweigh the benefits they provide in order to support decision- and policy-makers. Benefits were represented through four main regulating ecosystem services measured in real time with TreeTalkers. Data on costs for their establishment and maintenance of trees have been collected from the authorities managing the site—the campus of the Peoples' Friendship University of Russian (RUDN) in Moscow, Russian Federation.

## 22.2 Materials and Methods

### *22.2.1 Study Site*

Founded in 1960, the RUDN University campus (No 1 in Russia and No 42 in World University Green Metrics Ranking) (GreenMetric 2023) is situated in southwestern Moscow, in the Konkovo and Obruchevsky districts. The southwest of Moscow is located within the Teplostan upland, making the territory much higher than the rest of the city (Fig. 22.1). That, together with the prevailing western winds, an abundance of greenery, and a lack of big industrial areas, makes the territory one of the most favorable places to live in the capital. The campus occupies 0.34 km$^2$ of land and includes 20 teaching blocks and 14 dormitories.

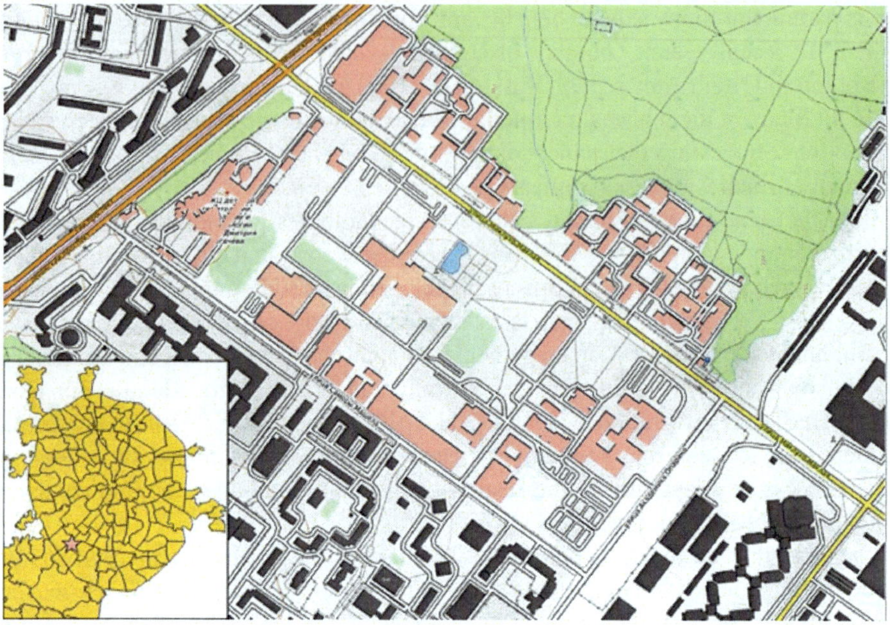

**Fig. 22.1** Location of the study site (in Moscow)

## *22.2.2 Assessment of Costs and Benefits*

Our study was conducted in three steps:

(1) We carried out inventory on campus and identified the structure of green infrastructure as well as the number of trees and their parameters.
(2) We gathered data on costs for establishment and maintenance of the trees using several calculation methods and primal information from RUDN Greenery Department, which were then compared to the benefits.
(3) Based on findings from the previous steps and using TreeTalker data, we have calculated the value of regulating ecosystem services provided by the trees on campus.

### 22.2.2.1 Inventory of Green Infrastructure

We used basic classification of green spaces in R3GIS, which is represented by four groups: trees, shrubs, lawns, flower beds. Trees can be a linear object or a point object. Shrubs can be designated as a single object, a line (hedge) if work is estimated per meter in length, and a polygon (shaped/unshaped) if work calculations are estimated per square meter. Lawns are polygonal objects and are represented by several types (by grass composition and type of care). Flowerbeds represent different variations of mixborders, flowerbeds and modular flowerbeds. Differences in classification are due to the longevity of plants: perennials and annuals. There are also ampelous plants that decorate the front entrances of buildings.

To describe the structure of green spaces, we used field observations with georeferencing using the GPS receiver GarmineTrex 32 and visual description of elements of green infrastructure with fixation of attributes. For trees, we described: species, trunk diameter, canopy diameter and tree height. For shrubs—species, height, length for hedges in addition. Green infrastructure elements with complex shapes were digitized in QGIS using DJI Mavic 2 PRO images (Hasselblad and Mapir DJI Inspire 2 Survey 3 near-infrared (NIR) and visible spectrum (RGB) camera on board, flying altitude 50 m). The images were taken in September 2019. The resulting images (with a resolution of 3–4 cm per pixel) were combined into a single georeferenced image and DEM. We used ForestTools package algorithms in R to obtain a shape file with the exact canopy shapes and georeferencing of each tree (R Core Team 2014).

#### 22.2.2.2 Assessment of Regulating Services and Their Monetary Value

In order to conduct assessment of regulating services and their value, we have chosen four biophysical indicators, which can be measured in real time with TreeTalker (Matasov et al. 2020): carbon sequestration, water transpiration, $PM_{10}$ absorption and energy use for climate regulation. We installed about 60 devices on major species of trees to collect their biophysical functioning data during 3 vegetation seasons in 2019–2021 to make extrapolation possible for the whole number of trees on campus.

To value biophysical flows in terms of money, we made such suggestions. To estimate, carbon deposition was used world prices for $CO_2$ emissions (40$ per ton—(World Bank 2021). Transpiration was estimated through the cost of drainage in prices of GUP Mosvodostok (https://мосводосток.рф/subscriber-service/rates/—14 rubles/m$^3$). The energy spent on transpiration was estimated as the work of a climate-conditioning system in rubles at prices of GUP Mosenergosbyt (https://www.mosenergosbyt.ru/individuals/services/pricelist.php—5.5 rubles/kWh). Literature data were used to estimate $PM_{10}$ air purification—$4500/ton (Nowak et al. 2018).

#### 22.2.2.3 Costs for Green Infrastructure Maintenance

The documents of RUDN Greenery Department for 2019 and 2020, job descriptions of the gardener and florist, the calendar of work on the maintenance of green areas were used to determine the costs of maintaining green areas. The data on the costs of planting material, its quantity and types of plants, garden tools and consumables were also obtained from the documents and interviews.

To compare real costs of the maintenance in RUDN campus with norms of maintenance for the Moscow City, we used Government Regulation No. 743-PP dated September 10, 2002 "On Approval of the Rules of Creation, Maintenance and Protection of Green Plantations and Natural Communities of the City of Moscow" (Decree of the Moscow City Government No 2002). We used information on the number of necessary «man working hours» and «machine working hours» for different types of work dedicated to trees and lawns care processes. Thus, the total amount of working hours per year was calculated in accordance with the regulations. These values were recalculated to cost equivalents based on average market prices and salaries.

## 22.3 Results

### *22.3.1 Green Infrastructure Composition of the Study Site*

As a result of the inventory, a map of RUDN campus green infrastructure was created (Fig. 22.2), the areas of all existing objects of green infrastructure presented in Table 22.1. Open lawns (33%) and lawns under crowns (13%) dominate the structure of green areas. The share of sealed areas on campus is a bit over 50%. These include parking lots and roadways, 33%, and buildings, 18%.

The RUDN campus has 1707 trees. The four most common species here are *Tilia cordata, Betula pendula, Salix caprea* and *Populus tremula.* They account for 64.7%

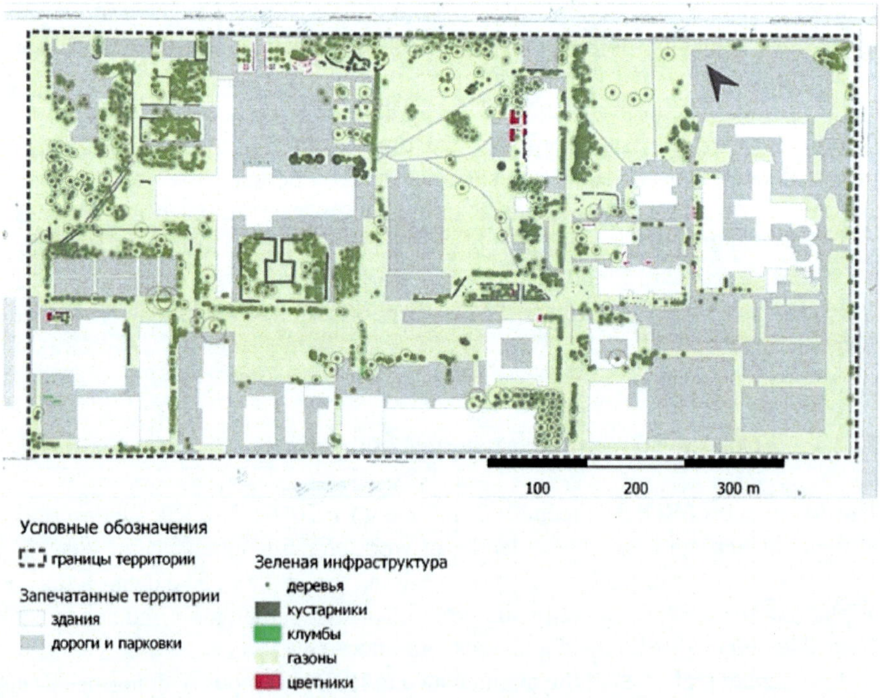


**Fig. 22.2** RUDN campus green infrastructure

**Table 22.1** Share of land cover types and green infrastructure elements

| Green elements | Area, sq. m | Share of area, % | Number, pcs |
|---|---|---|---|
| The entire territory | 347,314 | 100 | |
| Sealed: | 179,502 | 51 | |
| Roads and parking slots | 116,099 | 33 | |
| Buildings | 63,402 | 18 | |
| Green area: | 167,812 | 48 | |
| Lawns: | 163,235 | 47 | |
| Open lawns | 115,465 | 33 | |
| Lawns under the canopy | 47,770 | 13 | |
| Bushes: | 4100 | 0.6 | 260 |
| Artificial shapes | 3522 | 1 | |
| Natural shapes | 578 | 0.16 | |
| Flower beds: | 402 | 0.12 | |
| Annuals | 234 | 0.07 | |
| Perennial | 168 | 0.05 | |
| Trees: | | | 1707 |
| Trunk area | 74 | 0.02 | |

of all trees. The distribution of tree diameters, heights and canopy areas by species is shown in Annex 1. Rare individuals exceed the height of 20 m, the most typical height is about 10–13 m. Trunk diameters in most cases range from 10 to 30 cm.

### *22.3.2 Costs for Maintenance*

Actual spending consists of three main categories—employee salaries, supplies for planting, and equipment purchases or repairs (Table 22.2). In 2020, lawn seeds, seedlings for flower beds, soil and fertilizers were purchased for a total of about 1 million rubles, which amounted to 18% of the total cost. Among the equipment used are all kinds of soil drills, lawn mowers, long pruners and other stuff. There were spent on them about 10% of the total amount, i.e., about 0.5 million. Other 72% went on salaries of employees according to their employment by months and types of work (see Annex 2). Thus, the total costs amounted to 5.5 million rubles.

Compared to regulation norms (Decree No. 743-PP) on the man-hours for different types of work care for 1700 trees and an area of 34 hectares, we get the total number of man-hours about 21,035. Based on average salaries and the number of working days in 2020, we get an estimated amount of 213 rubles per working hour of a gardener which in total gives an amount of about 3.5 million rubles.

**Table 22.2** Amount and types of the costs

| Type | Category | Units | Quantity | Costs, rub | Costs, % |
|---|---|---|---|---|---|
| Flower seedling | Consumables | pieces | 36,000 | 829,056.00 | 17.90 |
| Soil and fertilizers | | cubic meter | 100 | 142,800.00 | |
| Grass seeds | | kg | 60 | 24,000.00 | |
| Equipment repair | Equipment | - | - | 36,000.00 | 10.74 |
| Petrol brush | | pieces | 3 | 192,000.00 | |
| Lawn mower | | pieces | 2 | 144,000.00 | |
| Gas shears | | pieces | 1 | 67,200.00 | |
| Pole cutter | | pieces | 1 | 60,000.00 | |
| Office equipment | | pieces | 4 | 14,400.00 | |
| Soil drill echo $d = 12$ cm | | pieces | 1 | 10,022.84 | |
| Soil drill echo $d = 20$ cm | | pieces | 1 | 10,051.37 | |
| Soil drill echo $d = 25$см | | pieces | 1 | 10,517.42 | |
| Soil drill echo EA-410 | | pieces | 1 | 53,273.42 | |
| Personnel of service jobs | Personnel | person | 7 | 3,823,012.05 | 71.37 |
| Personnel of planning jobs | | person | 1 | 95,333.33 | |
| Personnel of planting jobs | | person | 1 | 52,962.96 | |
| | | | Total | 5,564,629.41 | 100 |

### 22.3.3 *Benefits Obtained from the Trees: Regulating Ecosystem Services*

According to the results of the analysis, we can conclude that the RUDN campus trees produces regulating ecosystem services worth about 7.2 million rubles per year (Table 22.3), with most of it being climate regulation through transpiration (1144 MWh/yr). The second most valuable is air quality regulation (2263 kg/yr), slightly less valuable is rainwater removal through transpiration (11,185 m$^3$/yr) and carbon deposition (14.8 t/yr) is valued the least.

To sum up, with this estimate, we get that the balance of benefits to the territory due to regulating ecosystem services at the RUDN campus is more than 1.5 million rubles.

**Table 22.3** Biophysical volume of ES and their cost estimation

| Ecosystem service | Biophysical flow, units | Cash flow, thousand rubles |
|---|---|---|
| Tree carbon accumulation | 14.8 t/yr | 6.2 |
| Energy absorbed | 1144 MWh/yr | 6292.2 |
| Total transpiration | 11,185 m$^3$/yr | 156.6 |
| Total PM absorbed | 2263 kg/yr | 712.8 |

## 22.4 Discussion

Our assessment of costs and benefits associated with the maintenance of trees as part of the UGI concludes that the latter even based on regulating services value only slightly outweighs the former. Similar findings have been reached by Song et al. (2018): these estimates are often within the same range if not the same. Regulating services are the most common to be used in such cost–benefit assessments (Song et al. 2018). Of the differences, it can be noted that several studies have concluded water retention to be more valuable than air purification, which is probably due to the large number of studies conducted in the subtropical climate zone, where there is significantly more precipitation than in Moscow. Carbon sequestration has the least effect most of the time. And microclimate regulation through shading and transpiration turns out to be the most significant of all regulatory services.

At the same time, trees provide other benefits than regulating services, such as cultural services. (Song et al. 2018) In their review demonstrated that benefits from the latter (for example, aesthetics) are often larger than those of water purification and carbon sequestration. Authors of another review of cultural services provided by the UGI have concluded that while parks, gardens and forests have been examined in studies often, more comprehensive assessments of services from other components of UGI (such as trees, tree lines) as well as network as a whole are needed (Cheng et al. 2021). In line with the call to assess various components of UGI is the fact that lawns, while being one of the most common components of open green spaces and UGI, their importance for the human well-being and urban sustainability is still underestimated in many parts of the world, including Russia (Ignatieva et al. 2020). Their transpiration and PM10 removal rates, especially microclimate regulation, could be similar or even higher than trees due to the larger area they occupy. Moreover, as with trees, cultural services of lawns have not been represented in our valuation. Other studies have demonstrated their importance for people as well as a whole range of preferences toward their composition (Ignatieva et al. 2017; Yang et al. 2019).

Methodologically, we tested a combination of hardware and software: TreeTalkers and GreenSpaces. We discovered that while they have potential for application in such assessments, first a few limitations need fixing. For example, R3GIS didn't contain some of the management functions inherent in Moscow climate (like snow drilling in spring) and some of the elements of UGI that are present in Moscow, such as stone flowerbeds. Moreover, it became evident that there is a need for development of similar sensors for lawns (LawnTalker) which will be able to assess the biophysical processes in such important green infrastructure elements and translate them into ecosystem services numbers to be used in the decision support.

## 22.5 Conclusion

The results show that even when considering just monetary value from regulating services, it outweighs the costs for maintenance of the UGI by about a quarter of the former. Findings from such studies could play a major role in transforming a decision-maker's perspective of an urban tree, as the most typical element of UGI, from a liability on the landowner' balance sheet into an asset. This outcome could add to the current understanding of the value of a tree among decision-makers, that is largely determined by the value of a felling ticket or the value of a cubic meter of its timber, and the necessary costs of its valuation are forced by the need to avoid the risks of damage to someone else's property in the event of a possible fall of a tree. This study presents the foundation for future research, which needs to conduct more comprehensive assessments of UGI and its components. Moreover, as suggested by previous studies, decision-making could benefit largely from findings and process of the transdisciplinary research involving stakeholders (Ignatieva et al. 2017; Willcock et al. 2016). Such close engagement with among other decision-makers would be beneficial for the researchers and their studies in terms of knowledge and navigation of the currently applied procedures for calculation of costs.

## 22.6 Conflicts of Interest

The authors declare no conflict of interest.

**Acknowledgements** Data collection was done with support of "RUDN University Strategic Academic Leadership Program". Data analysis was done under the Russian Scientific Foundation Project # 19-77-300-12.

## Annex 1

See Fig. 22.3.

## Annex 2

See Table 22.4.

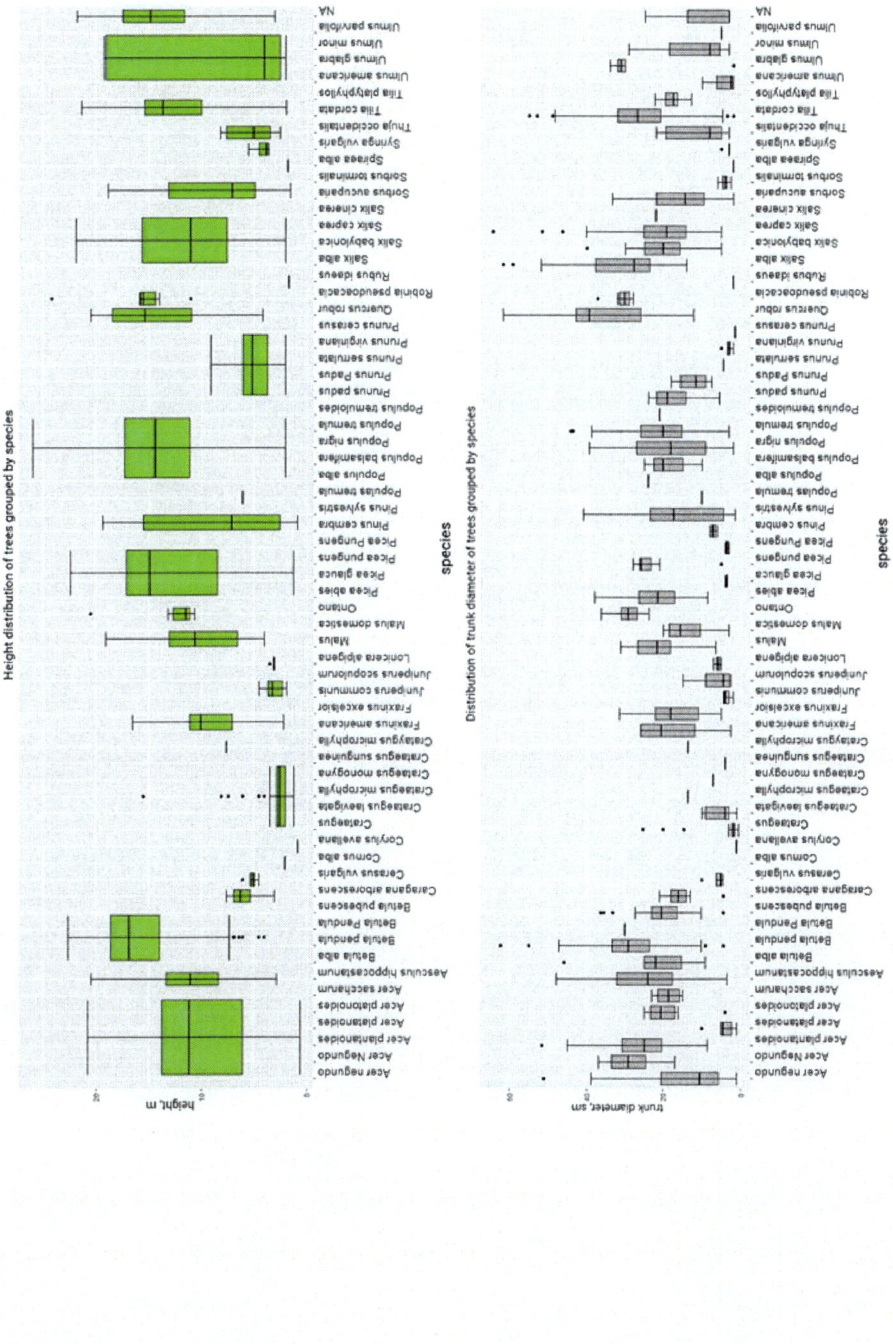


**Fig. 22.3** Distribution of tree characteristics by species: height (**a**), diameter at breast height (**b**), canopy area (**c**)

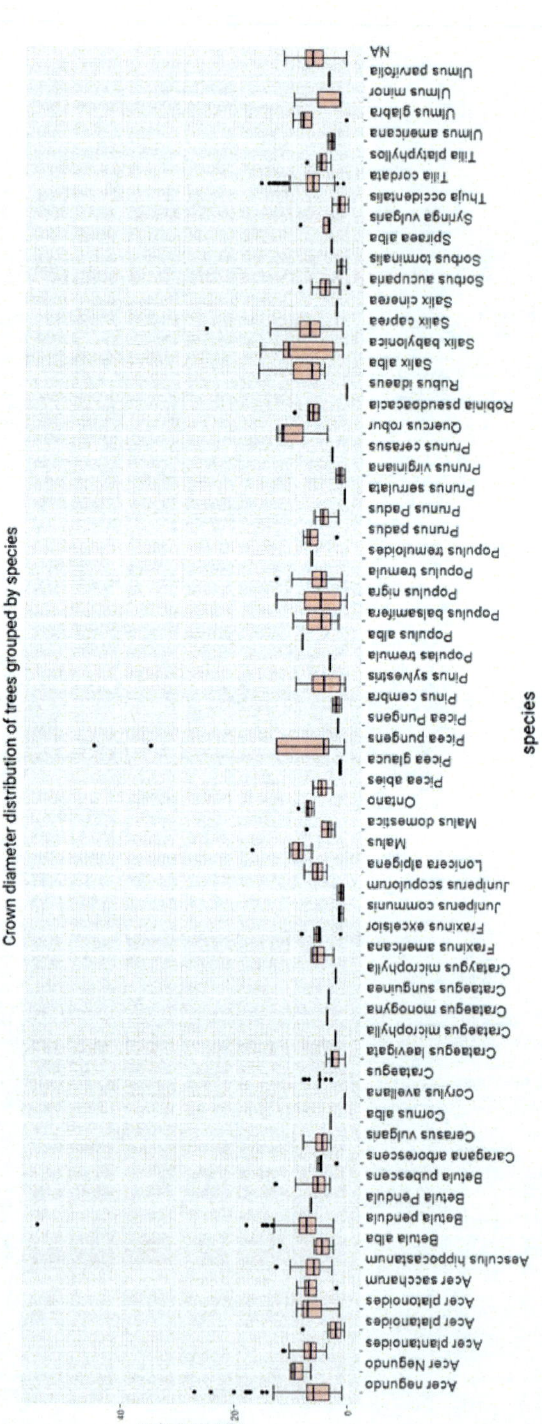


**Fig. 22.3** (continued)

**Table 22.4** Distribution of different work types across seasons and their working hours assessment

| Type of jobs | Job category | Man-hours/ year | January | February | March | April | May | June | July | August | September | October | November | December |
|---|---|---|---|---|---|---|---|---|---|---|---|---|---|---|
| Snow removal | Service | 2800 | 700 | 700 | 350 | | | | | | | | 350 | 700 |
| Equipment repair | Service | 9 | | | 9 | | | | | | | | | |
| Pruning trees and shrubs | Service | 280 | | | 140 | 140 | | | | | | | | |
| Cleaning after winter | Service | 240 | | | | 240 | | | | | | | | |
| Addition of mineral fertilizers | Service | 200 | | | | 40 | 40 | 40 | 40 | 40 | | | | |
| Chemical protection of trees from pests and diseases | Service | 120 | | | 20 | 20 | 20 | 20 | 20 | 20 | | | | |
| Chemical protection of shrubs from pests and diseases | Service | 720 | | | 120 | 120 | 120 | 120 | 120 | 120 | | | | |
| Hedge trimming | Service | 420 | | | | | | 140 | 140 | 140 | | | | |

(continued)

**Table 22.4** (continued)

| Type of jobs | Job category | Man-hours/year | January | February | March | April | May | June | July | August | September | October | November | December |
|---|---|---|---|---|---|---|---|---|---|---|---|---|---|---|
| Removing overgrowth on trees | Service | 50 | | | | | 10 | 10 | 10 | 10 | 10 | | | |
| Overseeding of lawn grasses | Service | 120 | | | | 60 | 60 | | | | | | | |
| Flower garden care: watering | Service | 3000 | | | | | 600 | 600 | 600 | 600 | 600 | | | |
| Flower garden care: weeding | Service | 600 | | | | | 120 | 120 | 120 | 120 | 120 | | | |
| Lawn mowing | Service | 3750 | | | | | 750 | 750 | 750 | 750 | 750 | | | |
| Cleaning of flower beds and flower beds before winter | Service | 175 | | | | | | | | | | 175 | | |
| Removal of the territory from fallen leaves | Service | 385 | | | | | | | | | 192.5 | 192.5 | | |

(continued)

**Table 22.4** (continued)

| Type of jobs | Job category | Man-hours/year | January | February | March | April | May | June | July | August | September | October | November | December |
|---|---|---|---|---|---|---|---|---|---|---|---|---|---|---|
| Lawn reclamation (locally) | Service | 0 | | | 1день/ 100 кв.м - 2 человека | | | | | | | | | |
| personnel management of service work | Service | 1582 | 131.833 | 169.5 | 169.5 | 169.5 | 169.5 | 169.5 | 169.5 | 169.5 | 169.5 | 169.5 | 169.5 | 169.5 |
| conclusion of contracts | Planning | 9 | | | 9 | | | | | | | | | |
| Planting design development | Planning | 63 | | | 31.5 | 31.5 | | | | | | | | |
| Planning new lawns | Planning | 45 | | | | 9 | 9 | 9 | 9 | 9 | | | | |
| Purchase of soil | Planning | 9 | | | 9 | | | | | | | | | |
| Purchase of perennial and annual plants | Planning | 54 | | | | | 54 | | | | | | | |
| Planting annual flower seedlings | Planting | 125 | | | | | 125 | | | | | | | |

(continued)

**Table 22.4** (continued)

| Type of jobs | Job category | Man-hours/ year | January | February | March | April | May | June | July | August | September | October | November | December |
|---|---|---|---|---|---|---|---|---|---|---|---|---|---|---|
| Planting annual perennial flowers | Planting | 25 | | | | | 25 | | | | | | | |
| Planting shrubs and trees | Planting | 50 | | | | | 50 | | | | | | | |
| | | Total man-hours/ year | January | February | March | April | May | June | July | August | September | October | November | December |
| Total | | 14,756 | 831.833 | 869.5 | 858 | 830 | 2152.5 | 1978.5 | 1978.5 | 1978.5 | 1842 | 537 | 519.5 | 869.5 |

## References


Bolund P, Hunhammar S (1999) Ecosystem services in urban areas. Ecol Econ 29:293–301. https://doi.org/10.1016/S0921-8009(99)00013-0

Burkhard B, Maes J, Potschin-Young M, Santos-Martín F, Geneletti D, Stoev P, Kopperoinen L, Adamescu C, Adem Esmail B, Arany I, Arnell A, Balzan M, Barton DN, van Beukering P, Bicking S, Borges P, Borisova B, Braat L, M Brander L, Bratanova-Doncheva S, Broekx S, Brown C, Cazacu C, Crossman N, Czúcz B, Daněk J, de Groot R, Depellegrin D, Dimopoulos P, Elvinger N, Erhard M, Fagerholm N, Frélichová J, Grêt-Regamey A, Grudova M, Haines-Young R, Inghe O, Kallay T, Kirin T, Klug H, Kokkoris I, Konovska I, Kruse M, Kuzmova I, Lange M, Liekens I, Lotan A, Lowicki D, Luque S, Marta-Pedroso C, Mizgajski A, Mononen L, Mulder S, Müller F, Nedkov S, Nikolova M, Östergård H, Penev L, Pereira P, Pitkänen K, Plieninger T, Rabe S-E, Reichel S, Roche P, Rusch G, Ruskule A, Sapundzhieva A, Sepp K, Sieber I, Šmid Hribar M, Stašová S, Steinhoff-Knopp B, Stępniewska M, Teller A, Vackar D, van Weelden M, Veidemane K, Vejre H, Vihervaara P, Viinikka A, Villoslada M, Weibel B, Zulian G (2018) Mapping and assessing ecosystem services in the EU—lessons learned from the ESMERALDA approach of integration. One Ecosyst 3. https://doi.org/10.3897/oneeco.3.e29153

Cheng X, Van Damme S, Uyttenhove P (2021) A review of empirical studies of cultural ecosystem services in urban green infrastructure. J Environ Manage 293:112895. https://doi.org/10.1016/j.jenvman.2021.112895

Decree of the Moscow City Government No. 743-PP (2002) On approving the rules for creation, maintenance and protection of green areas and natural communities in the City of Moscow. Dated September 10, 2002

Fagerholm N, Samuelsson K, Eilola S, Giusti M, Hasanzadeh K, Kajosaari A, Koch D, Korpilo S, Kyttä M, Legeby A, Liu Y, Præstholm S, Raymond C, Rinne T, Stahl Olafsson A, Barthel S (2022) Analysis of pandemic outdoor recreation and green infrastructure in Nordic cities to enhance urban resilience. NPJ Urban Sustain 2:25. https://doi.org/10.1038/s42949-022-00068-8

Gómez-Baggethun E, Barton DN (2013) Classifying and valuing ecosystem services for urban planning. Ecol Econ 86:235–245. https://doi.org/10.1016/j.ecolecon.2012.08.019

UI GreenMetric (2023) https://greenmetric.ui.ac.id/profile/rudn.ru. Accessed 16 Jan 2023

Haase D, Larondelle N, Andersson E, Artmann M, Borgström S, Breuste J, Gomez-Baggethun E, Gren Å, Hamstead Z, Hansen R, Kabisch N, Kremer P, Langemeyer J, Rall EL, McPhearson T, Pauleit S, Qureshi S, Schwarz N, Voigt A, Wurster D, Elmqvist T (2014) A quantitative review of urban ecosystem service assessments: concepts, models, and implementation. Ambio 43:413–433. https://doi.org/10.1007/s13280-014-0504-0

Ignatieva M, Eriksson F, Eriksson T, Berg P, Hedblom M (2017) The lawn as a social and cultural phenomenon in Sweden. Urban for Urban Greening 21:213–223. https://doi.org/10.1016/j.ufug.2016.12.006

Ignatieva M, Haase D, Dushkova D, Haase A (2020) Lawns in cities: from a globalised urban green space phenomenon to sustainable nature-based solutions. Land 9:73. https://doi.org/10.3390/land9030073

Klimanova OA, Illarionova OI (2020) Green infrastructure indicators for urban planning: applying the integrated approach for Russian largest cities. GES 13:251–259. https://doi.org/10.24057/2071-9388-2019-123

Kondo M, Fluehr J, McKeon T, Branas C (2018) Urban green space and its impact on human health. IJERPH 15:445. https://doi.org/10.3390/ijerph15030445

Matasov V, Belelli Marchesini L, Yaroslavtsev A, Sala G, Fareeva O, Seregin I, Castaldi S, Vasenev V, Valentini R (2020) IoT monitoring of urban tree ecosystem services: possibilities and challenges. Forests 11:775. https://doi.org/10.3390/f11070775

Nowak DJ, Hirabayashi S, Doyle M, McGovern M, Pasher J (2018) Air pollution removal by urban forests in Canada and its effect on air quality and human health. Urban for Urban Greening 29:40–48. https://doi.org/10.1016/j.ufug.2017.10.019

R Core Team (2014) R: a language and environment for statistical computing. R Foundation for Statistical Computing, Vienna, Austria

Russo A, Escobedo FJ, Cirella GT, Zerbe S (2017) Edible green infrastructure: an approach and review of provisioning ecosystem services and disservices in urban environments. Agric Ecosyst Environ 242:53–66. https://doi.org/10.1016/j.agee.2017.03.026

Small N, Munday M, Durance I (2017) The challenge of valuing ecosystem services that have no material benefits. Glob Environ Chang 44:57–67. https://doi.org/10.1016/j.gloenvcha.2017.03.005

Song XP, Tan PY, Edwards P, Richards D (2018) The economic benefits and costs of trees in urban forest stewardship: a systematic review. Urban for Urban Green 29:162–170. https://doi.org/10.1016/j.ufug.2017.11.017

Valentini R, Belelli Marchesini L, Gianelle D, Sala G, Yarovslavtsev A, Vasenev V, Castaldi S (2019) New tree monitoring systems: from Industry 4.0 to Nature 4.0. Ann Silvicult Res 43. https://doi.org/10.12899/asr-1847

Vogt J, Hauer RJ, Fischer BC (2015) The costs of maintaining and not maintaining the urban forest: a review of the urban forestry and arboriculture literature. 41(6): 293–323

Willcock S, Hooftman D, Sitas N, O'Farrell P, Hudson MD, Reyers B, Eigenbrod F, Bullock JM (2016) Do ecosystem service maps and models meet stakeholders' needs? A preliminary survey across sub-Saharan Africa. Ecosyst Serv 18:110–117. https://doi.org/10.1016/j.ecoser.2016.02.038

World Bank (2021) State and trends of carbon pricing 2021. The World Bank

Yang F, Ignatieva M, Larsson A, Zhang S, Ni N (2019) Public perceptions and preferences regarding lawns and their alternatives in China: a case study of Xi'an. Urban for Urban Greening 46:126478. https://doi.org/10.1016/j.ufug.2019.126478

# Chapter 23
# Urban Livability: A Place of Environmental Compounds in a Mindset

**Vitaly A. Kryukov and Elena I. Golubeva**

**Abstract** Urban livability depends on a lot of compounds, including the environmental ones, most of them dependent on green infrastructure (GI) and its distribution. Two interviews of the public in Moscow and Saint-Petersburg ($N = 352$) and experts from Belgrade in the field of urban studies ($N = 27$) were conducted to assess the importance of 23 compounds of urban livability. Public respondents are more interested in social and economic issues (income rise, job creation, improvement of healthcare, education, housing quality, safety, etc.) than in environmental ones. The improvement of parks and public spaces (8th place), the design of new ones (12th) and pollution reduction (9th) are the most significant environmental compounds, while the conservation of secluded silent shelters (20th) and expansion of protected areas (18th) are the least demanded. The local environmental issues are not properly expressed in public answers. The most significant differences between respondents are related to the place of residence, not to demographic, education level or access to the GI. Saint-Petersburg citizens are more interested in the design of new parks and public spaces compared to Moscow citizens that are keener on pollution reduction and nature shelters for relaxation. Belgrade experts consider environmental compounds of livability as significantly more important. Local pressing issues of the urban environment are expressed in the opinion of experts from Belgrade (air pollution, public transportation, wastewater treatment and economic problems). Considerable gap between the professional community and public in terms of environmental compounds of livability is yet to be explored.

**Keywords** Livability · Well-being · Urban ecology · Sustainable cities · Mindset · GIS

V. A. Kryukov (✉)
Aston University, Birmingham, UK
e-mail: vitkryukov@gmail.com

E. I. Golubeva
Lomonosov Moscow State University, Moscow, Russian Federation


M. Korneykova et al. (eds.), *Smart and Sustainable Urban Ecosystems: Challenges and Solutions*, Springer Geography, https://doi.org/10.1007/978-3-031-37216-2_23

## 23.1 Introduction

Livability can be described as the ability to be and dwell in some physical spaces with the appropriate environmental and cultural prerequisites (Onnom 2018; Yassin 2019). At the same time, most researchers imply their definitions of livability depending on the aims and scope of surveys. One of the most popular and comprehensive methods of livability assessment is subjective well-being (SWB), which may be characterized as "people's evaluations of their lives, which include both cognitive judgements of one's life satisfaction in addition to affective evaluations of mood and emotions" (Diener et al. 1999), p. 213. The positive relation between livability and SWB has been revealed earlier (Florida 2008; Okulicz-Kozaryn and Valente 2019). SWB mostly depends on basic needs, according to Maslow's pyramid (Maslow 1987) like physiological and security ones, but the highest state of SWB is only possible by satisfying psychological and self-development needs. Urban livability may be defined as the unique type of livability at large (Kashef 2016). Besides architectural characteristics of human-made grey infrastructure (buildings, roads, utilities and other sealed surfaces), determining important dimensions of urban livability, a liveable environment also heavily depends on the ecosystems' state – the rate of transformation, fragmentation, and connectivity. One of the most significant compounds of urban livability is green infrastructure (GI), according to the World Health Organization (WHO) (2004), which can be characterized as "an interconnected network of natural areas and other open spaces that conserves natural ecosystem values and functions, sustains clean air and water, and provides a wide array of benefits to people and wildlife" (Benedict and McMahon 2006, p. 1). GI properly designed with the main planning principles is capable of providing a livable environment: multifunctionality (useful for the environment, economics, and community), connectivity of landscape systems, habitability (for people and other living organisms), resiliency (ability to revive after some disturbances and changes), identity (defining a visual image of a place) and return to investments (reducing costs for citizens, businesses, government entities) (Rouse and Bunster-Ossa 2013). GI ecosystem services are varying widely and comprise most issues of urban SWB. Most SWB environmental compounds are strongly dependent on GI elements.

Meanwhile, GI is generally examined by case studies, covering parts of regions, cities, their districts or even residential blocks. From this perspective, it becomes evident that urban spaces provide a specific set of services related to geographical zonal features and peculiarities of environmental policy, spatial planning and conflicts of various stakeholders (Bush et al. 2021; Haase and Wolff 2022).

Therefore, urban GI protection fully corresponds with three UN Sustainable Development Goals – № 11 (sustainable cities and communities), № 14 (life below water) and № 15 (life on land) (United Nations 2017). The ecosystem services of urban GI are yet to be evaluated properly, especially cultural ones (Maes et al. 2016), but their effect usually might be comparable to huge financial costs to provide a healthy, safe and comfortable place of residence (Birtles et al. 2013). The emphasis is also placed on the blue infrastructure as evaluations of its services, protection of natural wetland systems and adequate design of water supply networks and runoff collectors will significantly mitigate the adverse effects of global climate change (Wong and Brown 2009). Since the understanding of unique green–blue infrastructure contribution to life-enabling, sustaining and fulfilling (Sydney Metropolitan Catchment Management Authority (SMCMA) 2012) is rising amongst the academic community, government and public, robust surveys of urban ecosystem services supplied by GIS comparative assessments and financial evaluations are emerging consistently but they are predominantly dedicated to the most developed cities (Boehnke et al. 2022; Haase and Wolff 2022; Kabisch et al. 2014; Klimanova et al. 2021). Such studies are accelerating the process of green transition to sustainable urban policies which should be based on the implementation of an ecosystem services approach (Burkhard et al. 2014; Hamel et al. 2021; McPhearson et al. 2015).

The urbanization rate is expected to increase mostly by an ambiguous expansion of cities in less developed countries (United Nations, Department of Economic and Social Affairs 2019) complicated by the great constellation of gaps in urban planning, which will require organizing a comfortable environment with a strong support of multidisciplinary teams. Besides, megapolises should be in the focus of livability surveys as their share of the world population is estimated to reach about 39% by 2035 (United Nations, Department of Economic and Social Affairs 2019).

This survey is aimed at revealing the ratio between various livability compounds in the mindset of urban residents. A huge variety of questions arise from this goal, but this study is focused on the place of ecological compounds determined by the natural peculiarities of study areas and human-caused impact on the environment. Moreover, the perceptions of these issues by various population groups in specific cities to organize a livable environment, according to the equity principle of sustainable development (Emas 2015), are yet to be explored.

A wide variety of various methods and techniques are used to contemplate public perception of livability issues: focus groups (Raymond et al. 2014), expert scores, narrative methods (Burkhard and Maes 2017), design of hypothetical scenarios of willingness to pay (West et al. 2006), photo-elicitation surveys (Burkhard and Maes 2017) and content analysis of surveys conducted earlier (Dennis and Bower 2008). Some approaches are considered to be integrating a few techniques and providing the most accurate and feasible imprint of the mindsets, e.g., Perceived Residential Environmental Quality – PREQ (Debek and Janda-Dębek 2015) or various international comparative ratings and indexes: The Global Liveability Index,[1] Mercer's Quality of Living Ranking,[2] Arcadis Sustainable Cities Index,[3] etc.

At the same time, objective quantitative indicators alone are insufficient to complete a livability assessment due to the need to consider subjective feelings and

[1] Global Liveability Report of The Economist Intelligence Unit (EIU) (2022). https://www.eiu.com/n/campaigns/global-liveability-index-2022/.

[2] Mercer's Quality of Living Ranking (2019). https://mobilityexchange.mercer.com/insights/quality-of-living-rankings.

[3] Arcadis Sustainable Cities Index (2018). https://www.arcadis.com/en/global/our-perspectives/sustainable-cities-index-2018/citizencentric-cities/.

emotions (Levent and Nijkamp 2006; Marans and Stimson 2011). The conjugation of public questionnaires and interviewing experts is considered to be the most reliable way of livability assessment (Debek and Janda-Dębek 2015). However, most researchers have applied their specific conceptual frameworks and techniques to gauge the quality of urban life (e.g. Marans and Stimson 2011), residential satisfaction (e.g. Ramkissoon et al. 2013), sense of place (Campelo et al. 2014) and other similar or almost identical aspects of urban livability. The essential problems caused by this variety are applicable to the specific projects and case studies only, insufficient efforts to compare to other surveys and extremely different sets of assessed indicators (Debek and Janda-Dębek 2015).

## 23.2 Data and Methods

### *23.2.1 Study Areas*

The citizens of the two largest Russian megapolises, Moscow and Saint-Petersburg (13.0 and 5.6 million people, respectively), and respondents from the surrounding Moscow region (federal subject in Russian administrative division) and Leningrad region (8.5 and 1.9 million people, respectively)[4] have been interviewed. Both megapolises may be characterized as extremely polarized cities with highly dense urban cores and remote outskirts of New Moscow, towns within Saint-Petersburg or even pristine wetlands within Yuntolovsky or Pike Lake protected areas.[5] The choice of case study areas is explained by the fact that citizens of these regions have quite high rates of the Internet use (the share of active online users from Moscow is 87.7% and from Saint-Petersburg is 80.9%),[6] which makes it easier to carry out online interviews. Nearby regions beyond megapolises are strongly oriented towards these cities as the conjugations of jobs, social and transport infrastructure, and public life. Besides, these megapolises and nearby subjects are highly developed Russian regions, economically and socially,[7] which has made us suggest that these residents are strongly aware of urban environment quality and ecological compounds of livability, according to Maslow's pyramid.

Moreover, it has turned out that some citizens of nearby cities have been identifying themselves as Moscow or Saint-Petersburg residents while snowballing samples.

---

[4] Mosstat, demographics (In Russian). https://mosstat.gks.ru/folder/64634.

Petrostat, demographics (In Russian). https://petrostat.gks.ru/storage/mediabank/Возраст-пол%20нас%20СПб%202019.pdf.

[5] The Directorate of Saint-Petersburg protected areas. https://oopt.spb.ru/.

[6] FinExpertiza. Survey on the activity of Internet users across Russian regions. https://finexpertiza.ru/press-service/researches/2022/kolich-inter-polz-vyrosl/.

[7] The Federal State Statistics Service (Rosstat). Social and economic indicators of Russian regions. https://rosstat.gov.ru/folder/210/document/13204.

Experts in the fields of urban planning, environmental management and architecture, residing in Belgrade (Serbia) with a population of around 1.7 million people[8] and its suburbs, have been interviewed during the scientific and professional conference[9] to reveal the professional perception of urban livability. Moscow and Saint-Petersburg professionals interviewed during such conferences had no appropriate experience in urban planning, which makes us incorporate Serbian results in this study and plan the further extension of the public interviews to Belgrade and expert interviews to Moscow and Saint-Petersburg, respectively.

### *23.2.2 Public Survey*

Text individual mass social survey reflected in online questionnaire entitled "Comfort urban environment – what it is?" has been conducted in January and February 2022 via Google Forms (the number of respondents $N = 352$) using freelance workers to distribute the link. It has been decided not to involve respondents who have personal connections with authors to exclude additional bias in the results. Four classic questions represent the introduction: gender, age, education and city. Moreover, each respondent must specify his postal code to identify his approximate place of residence.

The core formalized interview question comprised 23 items representing the main compounds of urban livability (Table 23.1). This compounds' set did not constellate all of the possible particular features, but such a bunch may be applied to a wide range of cities across developing countries. Each respondent specified the importance of each compound through a 5-point Likert scale: from 0 («not important at all») to 4 («extremely important»). Moreover, the additional optional unformalized open question, suggesting to type their proposals to improve the urban environment, has been included in the interview.

### *23.2.3 Spatial Analysis*

The second stage of results processing implied the use of postal codes obtained from respondents through the second open question of the interview. This approach has been already successfully implemented while accessing GI services (Dushkova et al. 2021; Rozzi 2021; Rushton et al. 2007). Accessibility of large (more than 100 and 500 ha) green cores and GI share of total area have been estimated according to barriers, such as highways and water objects through QGIS tools and OpenStreetMap

[8] RZS Srbije. Statistical office of the Republic of Serbia. https://www.stat.gov.rs/en-US/oblasti/stanovnistvo/procene-stanovnistva.

[9] "Lokalna samouprava u planiranju i uredjenju prostora i naselja" ("Local self-governing in planning and organisation of space and settlement" in Veliko Gradište, Serbia, 16–18 June 2022).

**Table 23.1** Urban livability compounds (orange—economic, blue—social, green—environmental)

| № | Livability factor |
|---|---|
| 1 | Jobs creation |
| 2 | Income rise |
| 3 | Improvement of healthcare and education quality |
| 4 | Design of healthcare and education facilities |
| 5 | Improvement of sport facilities |
| 6 | Design of sport facilities |
| 7 | Design of new parks and public spaces |
| 8 | Improvement of parks and public spaces |
| 9 | Quality of housing |
| 10 | Design of religious facilities |
| 11 | Security improvement |
| 12 | Digitalization |
| 13 | Reducing air and water pollution |
| 14 | Expansion of protected areas |
| 15 | Control of environmental violations |
| 16 | Improvement of public transportation |
| 17 | Increasing convenience for private transportation |
| 18 | Expansion of leisure objects |
| 19 | Expansion of education programmes |
| 20 | Architectural design and conservation of historical sites |
| 21 | Modernization of utilities network |
| 22 | Conservation of secluded silent places |
| 23 | Reducing noise pollution |

data. Openrouteservice API[10] has been harnessed to estimate the actual proximity areas of each respondent via QGIS ORS Tools plugin. Hence, 352 accessibility areas in the form of 1-h isochrone are represented by variously distorted circles. Such procedures of isochrones correction, significantly improving the accuracy of results, are already known from various surveys (Efentakis et al. 2013; Kaszczyszyn and Sypion-Dutkowska 2019; Lantseva and Ivanov 2016).

Landsat 8 imageries (2021) were used to assess the GI share of the total area occupied by 1-h proximity isochrone. Normalized difference vegetation index (NDVI) has been estimated to identify GI areas:

$$\text{NDVI} = (\text{NIR} - \text{RED})/(\text{NIR} + \text{RED}) \tag{23.1}$$

where NIR – reflectance value of near-infrared light, RED – reflectance value of red visible light (Tarpley et al. 1984). NDMI has not been used for such analysis due

[10] Openrouteservice API. https://openrouteservice.org/.

to great variations during different weather conditions (Gao 1996), as well as EVI which is mostly used for natural areas with great leaf area and may be considerably distorted by some buildings, especially the multistorey ones (Huete et al. 2002). At the same time, deep-routed NDVI implementation in the field of urban ecology is quite successful for the demarcation of green and grey infrastructure (Aryal et al. 2022; Liu et al. 2015).

The threshold value NDVI = 0.18 was used for the delimitation of green and barren areas in QGIS, following to comparison of different values to Google Satellite images in natural colours. Further, raster images were transformed into vector data to gain intersections between GI polygons and 1-h isochrones. GI shares within 1-h proximity areas of each respondent have been estimated through QGIS spatial statistics.

### 23.2.4 *Expert Interviews*

Respondents to these interviews who attended the conference mentioned above have expressed willingness to participate in the social survey for 3 days. Most of the Serbian experts live in Belgrade and nearby towns ($N = 27$). This questionnaire has included the same items as the public survey but has been advanced by questions about the field of professional activity, affiliated institution, position and completion of doctoral degree.

The results of the public and experts' interviews have been checked via IBM SPSS v.27 using nonparametric tests (Mann–Whitney $U$ and Kruskal Wallis) for ordinal data, which are widely common in Likert-scaled surveys (Dushkova et al. 2021).

## 23.3 Results

### 23.3.1 *Public Survey*

The age and gender sample structure is quite close to those published by Moscow and Saint-Petersburg statistical authority branches.[11] The near-zero prevalence of respondents who identified themselves as female has been gained through this interview. The citizens aged 40–49 prevail, while the share of respondents above 40 years is compatible with the younger respondents – 48.7% versus 51.3% (Fig. 23.1).

[11] Mosstat, demographics (In Russian). https://mosstat.gks.ru/folder/64634.

Petrostat, demographics (In Russian). https://petrostat.gks.ru/storage/mediabank/Возраст-пол%20нас%20СПб%202019.pdf.

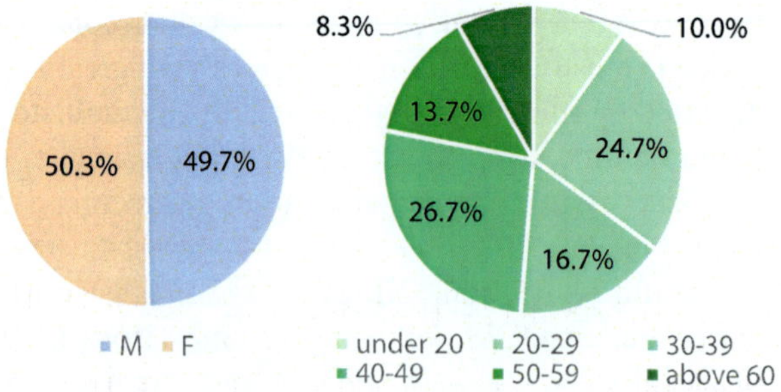


**Fig. 23.1** Gender and age of respondents (public survey)

The citizens of Moscow and Moscow region amount to 63.3% of the total respondents' number. Nonetheless, two samples of respondents comprising (1) Moscow and Moscow region residents and (2) Saint-Petersburg and Leningrad region residents are comparable because of the sufficient respondents' number (Fox et al. 2007).

Only 14.3% of respondents have not completed either secondary professional education or a higher degree of education. Most of them (65.6%) are falling into the category of respondents aged below 20, which is making us able to assume the higher expected education rate. The share of respondents who completed higher education is quite similar to the opposite group (49.0% vs. 51.0%).

The public mindset of Moscow and Saint-Petersburg regions tends to be more economic- and social-oriented rather than ecologically inclined (Fig. 23.2). The leaders of public perception are income rise (№ 2), improvement of education and healthcare quality (№ 3), jobs creation (№ 1), housing quality (№ 9) and safety (№ 11), while conservation of secluded spaces for relaxation (№ 22), programmes of additional education (№ 19), design of sport (№ 6) and religious facilities (№ 10) are the least important.

The most important environmental issues are related to the design of new parks and public spaces (№ 7), their improvement (№ 8) and reduction of air and water pollution (№ 13). Meanwhile, the extension of protected areas took 18th place only, while the conservation of secluded silent places (№ 22) has become the least demanded compound. Besides this compound, some social ones make up the list of outsiders: programmes of additional education (№ 19), design of sport (№ 6) and religious facilities (№ 10).

Several patterns of public perception have been identified through the analysis of responses (Table 23.2):

- men are keener on the design of religious (№ 10) and sport facilities (№ 6), while women are more interested in the improvement of healthcare and education quality (№ 3).
- citizens above the age of 40 are more interested in religious facilities (№ 10).
- the large set of factors is considered by dwellers with higher education to be more important: improvement of education and healthcare quality (№ 3), development of public transport (№ 16), conservation of architectural and historic peculiarities

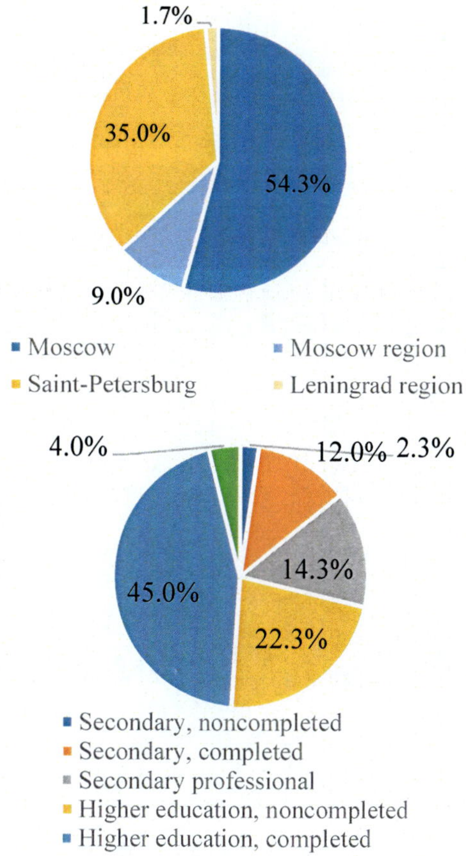


**Fig. 23.2** The place of residence and education degree (public survey)

(№ 20) and upgrades of utilities (№ 21); while respondents without such education degree prefer religious facilities (№ 10).

- the citizens of Moscow and Moscow region are significantly more interested in three environmental compounds – pollution reduction (№ 13), nature shelters for relaxation (№ 22), noise pollution (№ 23) – as well as in the design and improvement of sport facilities (№ 5–6), upgrades of utilities (№ 21) and jobs creation (№ 1), whereas the citizens of Saint-Petersburg and Leningrad region are chiefly keen on the design of parks and public spaces (№ 7) and their improvement (№ 8). Such disparity may be related to the higher incomes of Moscow citizens and their psychological possibility to concentrate on the more comprehensive social and environmental issues. In the meantime, Saint-Petersburg does not possess such green spaces in the city centre as Moscow does.
- there are no significant disparities of the importance of protected areas expansion (№ 14) and control of environmental violations (№ 15) in the sample.

**Table 23.2** Importance values of 1–23 urban livability compounds, their differences (*D*) in the various samples and *p* of null hypotheses (any differences between the samples of respondents are insignificant), according to the Mann–Whitney U test. Low (<0.05) *p* refuting null hypotheses are in bold

| No. of compound | Gender | | | | Age | | | |
|---|---|---|---|---|---|---|---|---|
| | M | F | *D* | $p_U$ | Below 40 | Above 40 | *D* | $p_U$ |
| 1 | 3.39 | 3.26 | 0.13 | 0.056 | 3.34 | 3.31 | 0.03 | 0.312 |
| 2 | 3.40 | 3.54 | – 0.14 | 0.099 | 3.49 | 3.46 | 0.03 | 0.373 |
| 3 | 3.30 | 3.48 | **– 0.19** | **0.017** | 3.38 | 3.40 | – 0.03 | 0.399 |
| 4 | 3.05 | 3.21 | – 0.15 | 0.398 | 3.12 | 3.14 | – 0.01 | 0.395 |
| 5 | 2.83 | 2.62 | 0.21 | 0.364 | 2.75 | 2.69 | 0.05 | 0.394 |
| 6 | 2.79 | 2.43 | **0.35** | **0.006** | 2.60 | 2.61 | – 0.01 | 0.399 |
| 7 | 3.14 | 3.03 | 0.11 | 0.333 | 3.02 | 3.16 | – 0.14 | 0.200 |
| 8 | 3.13 | 3.17 | – 0.04 | 0.281 | 3.14 | 3.16 | – 0.03 | 0.391 |
| 9 | 3.28 | 3.32 | – 0.05 | 0.291 | 3.32 | 3.28 | 0.04 | 0.300 |
| 10 | 2.11 | 1.58 | **0.53** | **0.004** | 1.56 | 2.14 | **– 0.58** | **0.001** |
| 11 | 3.12 | 3.28 | – 0.16 | 0.082 | 3.23 | 3.17 | 0.06 | 0.282 |
| 12 | 3.06 | 3.05 | 0.01 | 0.399 | 3.11 | 2.99 | 0.12 | 0.215 |
| 13 | 3.17 | 3.13 | 0.05 | 0.394 | 3.25 | 3.05 | 0.20 | 0.084 |
| 14 | 2.79 | 2.75 | 0.04 | 0.395 | 2.77 | 2.78 | – 0.01 | 0.390 |
| 15 | 3.03 | 2.95 | 0.08 | 0.372 | 2.99 | 2.98 | 0.01 | 0.397 |
| 16 | 3.22 | 3.09 | 0.14 | 0.283 | 3.12 | 3.19 | – 0.07 | 0.278 |
| 17 | 3.19 | 3.13 | 0.07 | 0.389 | 3.13 | 3.19 | – 0.06 | 0.267 |
| 18 | 2.94 | 2.80 | 0.14 | 0.255 | 2.99 | 2.74 | 0.25 | 0.073 |
| 19 | 2.66 | 2.56 | 0.09 | 0.260 | 2.49 | 2.74 | – 0.25 | 0.120 |
| 20 | 2.94 | 2.85 | 0.09 | 0.332 | 2.80 | 2.99 | – 0.19 | 0.116 |

(continued)

**Table 23.2** (continued)

| No. of compound | Gender | | | | Age | | | |
|---|---|---|---|---|---|---|---|---|
| | M | F | *D* | $p_U$ | Below 40 | Above 40 | *D* | $p_U$ |
| 21 | 3.17 | 3.03 | 0.13 | 0.325 | 3.09 | 3.11 | − 0.02 | 0.399 |
| 22 | 2.65 | 2.58 | 0.07 | 0.346 | 2.73 | 2.50 | 0.23 | 0.110 |
| 23 | 2.97 | 2.83 | 0.13 | 0.237 | 2.90 | 2.64 | 0.26 | 0.397 |
| Average | 3.01 | 2.94 | 0.07 | | 2.97 | 2.97 | − 0.01 | |
| No. of compound | Education | | | | Place of residence | | | |
| | Without higher completed | With higher completed | *D* | $p_U$ | Moscow and Moscow region | Saint-Petersburg and Leningradsky region | *D* | $p_U$ |
| 1 | 3.41 | 3.24 | 0.17 | 0.122 | 3.40 | 3.19 | **0.21** | **0.048** |
| 2 | 3.42 | 3.53 | − 0.11 | 0.164 | 3.42 | 3.57 | − 0.16 | 0.091 |
| 3 | 3.27 | 3.52 | **− 0.25** | **0.030** | 3.42 | 3.35 | 0.07 | 0.281 |
| 4 | 3.07 | 3.20 | − 0.13 | 0.393 | 3.20 | 3.01 | 0.19 | 0.389 |
| 5 | 2.75 | 2.69 | 0.05 | 0.224 | 2.85 | 2.50 | 0.35 | 0.396 |
| 6 | 2.65 | 2.56 | 0.10 | 0.272 | 2.74 | 2.37 | **0.37** | **0.014** |
| 7 | 3.07 | 3.10 | − 0.03 | 0.342 | 2.97 | 3.28 | **− 0.31** | **0.025** |
| 8 | 3.10 | 3.20 | − 0.09 | 0.177 | 3.05 | 3.33 | **− 0.28** | **0.044** |
| 9 | 3.19 | 3.41 | − 0.23 | 0.070 | 3.35 | 3.22 | 0.13 | 0.171 |
| 10 | 2.01 | 1.66 | **0.35** | **0.048** | 1.94 | 1.66 | 0.28 | 0.122 |
| 11 | 3.17 | 3.23 | − 0.06 | 0.350 | 3.20 | 3.20 | 0.00 | 0.388 |
| 12 | 3.11 | 2.99 | 0.12 | 0.377 | 3.08 | 3.01 | 0.07 | 0.241 |

(continued)

**Table 23.2** (continued)

| No. of compound | Gender | | | | Age | | | |
|---|---|---|---|---|---|---|---|---|
| | M | F | *D* | $p_U$ | Below 40 | Above 40 | *D* | $p_U$ |
| 13 | 3.03 | 3.27 | – 0.24 | 0.077 | 3.32 | 2.85 | **0.47** | **0.000** |
| 14 | 2.78 | 2.76 | 0.02 | 0.388 | 2.77 | 2.77 | 0.00 | 0.384 |
| 15 | 2.95 | 3.02 | – 0.07 | 0.239 | 2.97 | 3.01 | – 0.04 | 0.393 |
| 16 | 2.99 | 3.33 | **– 0.34** | **0.002** | 3.16 | 3.14 | 0.03 | 0.371 |
| 17 | 3.16 | 3.16 | – 0.01 | 0.355 | 3.12 | 3.24 | – 0.12 | 0.209 |
| 18 | 2.99 | 2.74 | 0.25 | 0.218 | 2.89 | 2.83 | 0.07 | 0.220 |
| 19 | 2.58 | 2.65 | – 0.07 | 0.346 | 2.67 | 2.50 | 0.17 | 0.193 |
| 20 | 2.78 | 3.01 | – 0.22 | 0.038 | 2.99 | 2.73 | 0.26 | 0.145 |
| 21 | 2.99 | 3.21 | – 0.22 | 0.034 | 3.20 | 2.93 | 0.27 | 0.028 |
| 22 | 2.65 | 2.58 | 0.08 | 0.378 | 2.77 | 2.35 | 0.41 | 0.021 |
| 23 | 2.82 | 2.99 | – 0.17 | 0.146 | 3.04 | 2.65 | 0.39 | 0.028 |
| Average | 2.95 | 3.00 | – 0.05 | | 3.02 | 2.90 | 0.12 | |

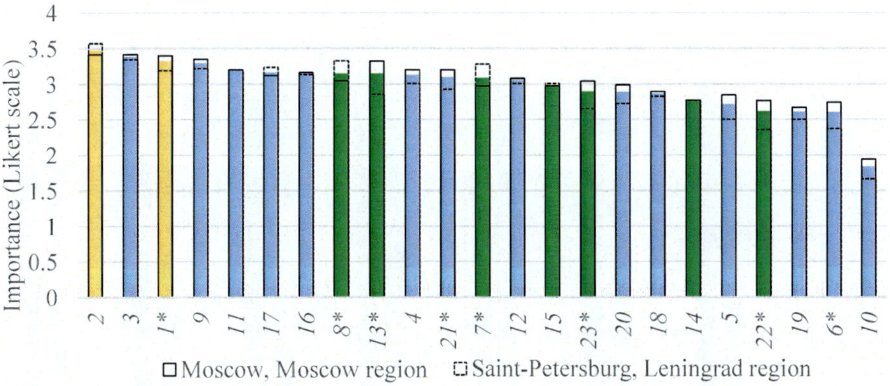


**Fig. 23.3** Urban livability compounds, descending according to their importance: orange – economic, blue – social, green – environmental (public survey of Moscow, Moscow region, Saint-Petersburg, Leningrad region citizens). *Statistically significant differences

Therefore, the essential ecological issues (air pollution predominantly by mobile sources; stormwater retention; protection of wildlife habitats, coupled with reducing of free space due to ongoing urbanization of Moscow megapolis[12]) are not properly reflected in the residents' answers.

According to our results (Table 23.2), the distinctions of environmental compounds related to the place of residence (5 out of 7) are significantly more important than ones related to gender, age and education level ($p$ of all compounds > 0.05). At the same time, no significant relation between the respondents' batches and the perception of economic issues (№ 1–2) has been revealed.

The hypothetical assumption about the less importance of 13–15 livability compounds for the respondents possessing GI most might be made (Fig. 23.3). However, most of the livability compounds have no significant relation to the GI share within 1-h proximity isochrones ($r_{s\,max} = 0.20$, negligible relationship) as well as to the access to the large GI cores ($r_{s\,max} = 0.17$, negligible relationship) (Fig. 23.4).

### 23.3.2 Expert Interviews

The respondents aged 20–29 make up most of the sample (Fig. 23.5). Any comparisons between the respondents aged below 40 and above 40 are unreliable due to the small sample. Therefore, two groups (below 30 and above 30) have been compared. Twelve experts received PhD degrees. Only three experts specialize in environmental

[12] Moscow Department of land use and nature protection (2016–2020) Reports on the environmental condition of Moscow city (In Russian). https://www.mos.ru/eco/documents/doklady/.

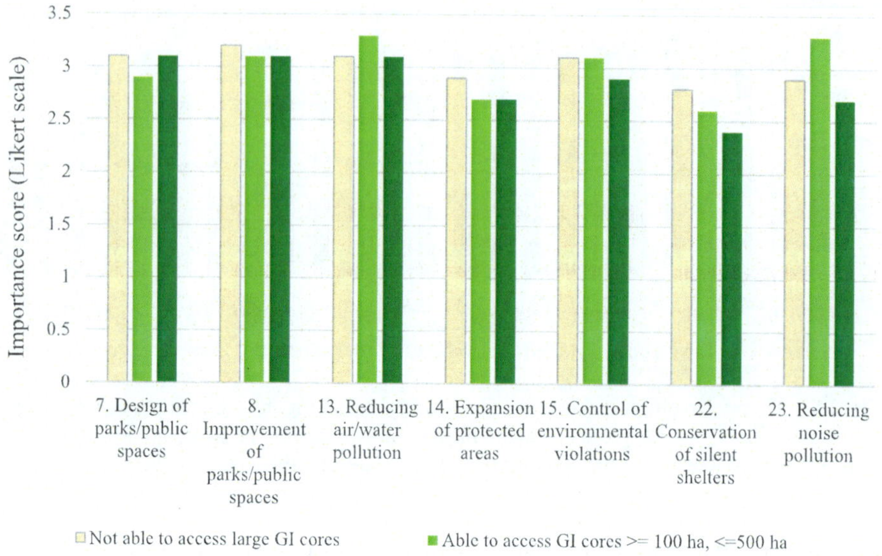


**Fig. 23.4** The mean importance values of environmental compounds depending on GI access

protection and biogeography, while most of the respondents are focused on professional occupations in the field of spatial planning and architecture or social studies (regional, social, cultural geography and tourism).

The leaders of expert choices in Belgrade are: reducing pollution (№ 13), job creation (№ 1), improvement of public transportation (№ 16), income rise (№ 2), and design of new parks and public spaces (№ 7). Serbian experts are quite interested in the improvement of the ecological state (Fig. 23.6). The key differences between Moscow and Belgrade's perceptions are related to the local geographical peculiarities: common unfavourable weather conditions including high pressure, temperature inversions, smog in Belgrade (Đurić and Vujović 2020), even resulted in public protests[13]; lack of high capacity rail system, heavy reliance on buses, frequent traffic congestions due to motorization increase (Jovanovic 2013); inadequate, but improving waste management and wastewater treatment,[14] etc.

However, no significant relation between the perception of environmental factors and groups of respondents has been revealed ($p_U$ of all compounds > 0.05). Jobs creation, improvement of healthcare and education quality (№ 3), and digitalization (№ 12) are more important for women experts, and expansion of leisure objects (№ 18) – for men experts. The experts aged below 30 are more interested in the

[13] N1 news. https://rs.n1info.com/english/news/hundreds-of-belgrade-citizens-in-protest-against-air-pollution/.

[14] Coalition 27 (2018) Chapter 27 in Serbia: No-Progress Report. https://rs.boell.org/sites/default/files/uploads/2018/06/izvestaj_k27_2018_eng_web.pdf.

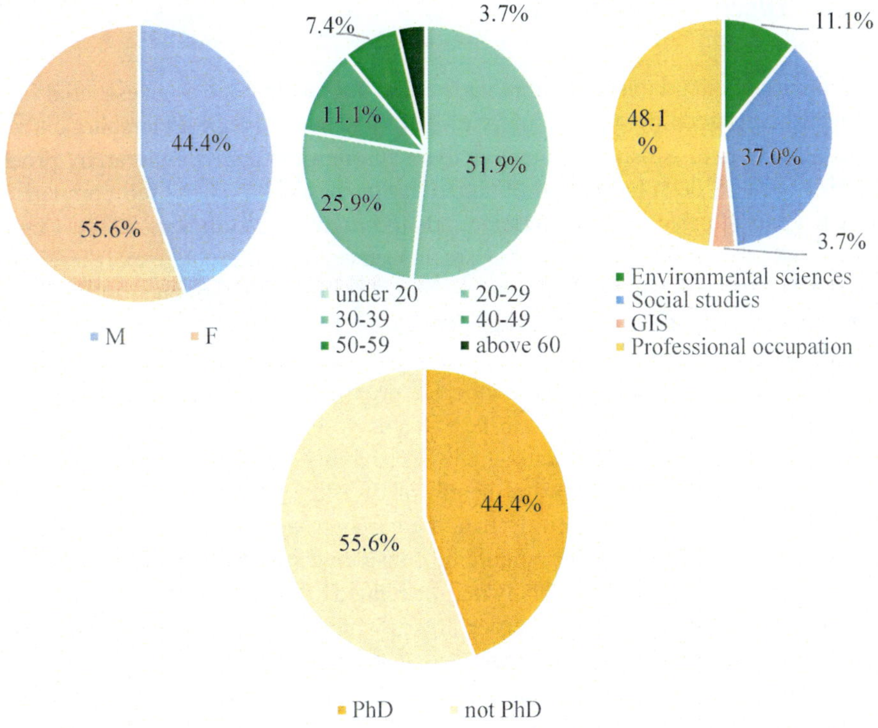


**Fig. 23.5** From left to right: gender, age, field of the professional activity, position/academic degree of Serbian experts

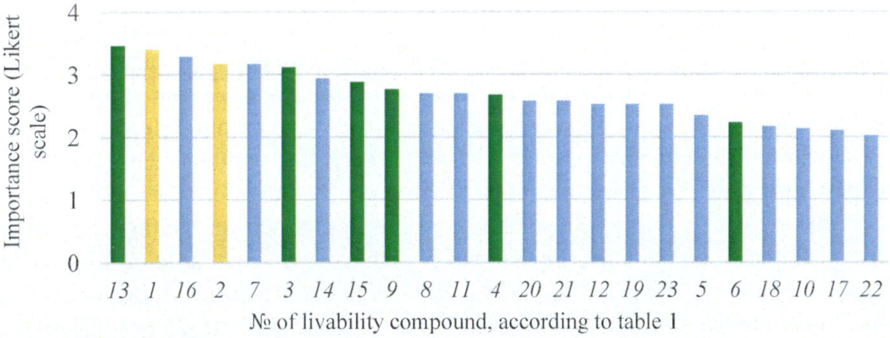


**Fig. 23.6** Urban livability compounds, descending according to their importance: orange – economic, blue – social, green – environmental, interviews of Belgrade experts

development of sport facilities (№ 5). The experts without an academic degree tend to be keener on the problem of job creation (№ 1).

## 23.4 Discussion of Results

The environmental bunch of compounds could not be regarded as an essential factor of well-being, according to the interviews of Moscow and Saint-Petersburg citizens. Similar results of Moscow content analysis through the analytic hierarchy process (AHP) (Saaty 2004) have been obtained earlier based on pair-wised comparisons of each pair of urban livability compounds (Kryukov and Golubeva 2020). Content analysis of scientific and realty assessments surveyed earlier and expert interviews have been applied. The sum of environmental factors was estimated to be 2 times less important than the sum of social factors. Moreover, the economic elements of urban livability, such as income level, have not been assessed at that time.

No significant relation between the GI proximity within the specific city and reported SWB has been found earlier (Dushkova et al. 2021). Regardless of this fact, it has been repeatedly elicited that GI proximity, size and availability strongly influence the physical and mental health of dwellers (Ridder et al. 2004; Dennis et al. 2020; Hunter et al. 2019). Hence, the gap between objective GI value and its subjective perception by public should be considered significant and even alarming.

Possible limitations of this survey may be related to the inaccurate locations of post service areas, even though such procedure is widely used. Interviews within focus groups distinguished by demographic, education or answers to the main question may result in more reliable outcomes, but such a study seems to be less feasible as larger financial costs are required and rare respondents are ready to provide specific locations of their residence places. The spatial resolution of Landsat 7 imageries (30 m) has not brought significant inaccuracy compared to the maximum distances from post office locations to residents more than 2 km. Moreover, some preferences in the importance of specific livability compounds might have been determined by the place of interview attending, mostly at home. However, the role of environmental compounds seems to be not overstated as interviews have not been carried out within urban forests, parks or public gardens. At the same time, the demographic of residents is quite close to the official statistical data of Moscow and Saint-Petersburg.

However, it is crucial to implement an approach that combines both public and expert interviews in urban planning of green infrastructure, since the livability concept includes subjective evaluations of comfort based on feelings and objectively estimated compounds (Levent and Nijkamp 2006), such as harmful pollutant emissions, low proximity of education and healthcare facilities, high unemployment rate, outdated utilities, emergency housing conditions, etc. Hence, the experts related to urban studies and possessing a high level of awareness and consciousness have expressed their strong demand for the solving of environmental problems, even though most experts do not specialize in ecology and environmental studies. Such disparity between the grassroots and professional perceptions may be an important argument for changes in the system of spatial planning and environmental education.

The psychological linkage between the place of residence and green infrastructure is quite complicated, but it is considered to be more important than factors of gender and age, which corresponds to similar surveys conducted before (Lamond and Everett 2019). Besides, the essential problem of the low involvement of citizens in environmental design exists. The current demand for the preservation of the most valuable urban green infrastructure parts – protected areas – is rather distinct, but at the same time, it involves merely a small number of activists. The detected gap between livability compounds will likely become more significant as global and regional challenges are translating into great economic and social problems. Nonetheless, environmental issues and in particular the design of green infrastructure must remain at the core part of urban management. Most of the green and blue spaces are almost indispensable, and the recovery costs of commensurate ecosystem functions would be enormous, given actual economic transformations.

## 23.5 Conclusion

The inconsistency between public and expert mindsets in terms of environmental compounds' role should be surveyed further. Despite current international and regional economic and social challenges, environmental factors are still important to provide a comfortable urban space. This blueprint of expert opinions should be implemented not only in urban environmental policy but also unveiled to the public and embedded into environmental education and propaganda.

Besides questionnaires dissemination amongst public and experts, this survey may be continued in the following areas:

- comparing the results of online and on-site text individual mass interviews;
- dynamic interviews over several years to reveal responses to social, economic and environmental issues by different population groups;
- gathering focus groups of old timers/new settlers and residents with a standard 5-day on-site work week/residents working mostly online or attached by freelance contracts;
- narrative analysis of chosen respondents to identify the main points of interest in the areas of their residence;
- AHP (analytic hierarchy process) implementation into experts' interviews;
- design hypothetical scenarios of willingness to pay for the protection of the nearby environment – how much the different groups of residents are going to pay for it?

These issues are strongly linked to the applications of environmental policy and should be analysed to develop the new eco-oriented mindset of urban citizens. Nonetheless, it would seem that the ecological dimension may be neglected as basic needs satisfaction will decline over years due to economic and social shocks, according to Maslow's pyramid. But the place of environmental factors in livability should not be overstated, from our standpoint. On the contrary, the delicate programmes of environmental education should become one of the prior goals of urban environmental policy around the world, especially in developing countries substantially vulnerable to economic issues.

## References

Aryal J, Sitaula C, Aryal S (2022) NDVI Threshold-based urban green space mapping from Sentinel-2A at the Local Governmental Area (LGA) level of Victoria, Australia. Land 11(3). https://doi.org/10.3390/land11030351

Benedict MA, McMahon ET (2006) Green infrastructure: linking landscapes and communities. Island Press, Washington, DC

Birtles P, Hore J, Dean M, Hamilton R, Dahlenburg J, Moore J, Bailey M (2013) Creating a liveable city—the role of ecosystem services. https://doi.org/10.13140/2.1.4643.6167

Boehnke D, Krehl A, Mörmann K, Volk R, Lützkendorf T, Naber E, Becker R, Norra S (2022) Mapping urban green and its ecosystem services at microscale—a methodological approach for climate adaptation and biodiversity. Sustainability 14:9029. https://doi.org/10.3390/su14159029

Bonnes M, Bonaiuto M, Aiello A, Perugini M, Ercolani AP (1997) A transactional perspective on residential satisfaction. In: Despres C, Piché D (eds) Housing surveys. Advances in theory and methods. Crad, Quebec, Canada

Burkhard B, Kandziora M, Hou Y, Müller F (2014) Ecosystem service potentials, flows and demands-concepts for spatial localisation, indication and quantification. Landscape Online 34. https://doi.org/10.3097/LO.201434

Burkhard B, Maes J (2017) Mapping ecosystem services. Pensoft Publishers, Sofia

Bush J, Ashley G, Foster B, Hall G (2021) Integrating green infrastructure into urban planning: developing Melbourne's green factor tool. Urban Plan 6:20–31. https://doi.org/10.17645/up.v6i1.3515

Campelo A, Aitken R, Thyne M, Gnoth J (2014) Sense of place: the importance for destination branding. J Travel Res 53(2):154–166. https://doi.org/10.1177/0047287513496474

De Ridder K, Adamec V, Bañuelos A, Bruse M, Bürger M, Damsgaard O, Dufek J, Hirsch J, Lefebre F, Pérez-Lacorzana JM, Thierry A, Weber C (2004) An integrated methodology to assess the benefits of urban green space. Sci Total Environ 334–335:489–497. https://doi.org/10.1016/j.scitotenv.2004.04.054

Debek M, Janda-Dębek B (2015) Perceived Residential Environment Quality and Neighborhood Attachment (PREQ & NA) indicators by Marino Bonaiuto, Ferdinando Fornara, and Mirilia Bonnes—Polish adaptation. Pol J Appl Psychol 13:111–162. https://doi.org/10.1515/pjap-2015-0032

Dennis B, Bower T (2008) Using content analysis software to analyze survey comments. portal: Libr Acad 8:423–437. https://doi.org/10.1353/pla.0.0015

Dennis M, Cook PA, James P et al (2020) Relationships between health outcomes in older populations and urban green infrastructure size, quality and proximity. BMC Public Health 20:626. https://doi.org/10.1186/s12889-020-08762-x

Diener E, Lucas RR (1999) Personality and subjective well-being. In: Kahneman E, Diener E, Schwarz N (eds) Well-being: the foundations of hedonic psychology. Russell Sage Foundation, New York, pp 213–229

Đurić M, Vujović D (2020) Short-term forecasting of air pollution index in Belgrade, Serbia. Meteorol Appl 27. https://doi.org/10.1002/met.1946

Dushkova D, Ignatieva M, Hughes M, Konstantinova A, Vasenev VI, Dovletyarova EA (2021) Human dimensions of urban blue and green infrastructure during a pandemic. Case study of Moscow (Russia) and Perth (Australia). Sustainability 13:4148. https://doi.org/10.3390/su13084148

Efentakis A, Grivas N, Lamprianidis G, Magenschab G, Pfoser D (2013) Isochrones, traffic and DEMOgraphics. In: Proceedings of the 21st ACM SIGSPATIAL international conference on advances in geographic information systems, 2013, pp 548–551. https://doi.org/10.1145/2525314.2525325

Emas R (2015) The concept of sustainable development: definition and defining principles. Brief GSDR 10:13140. https://doi.org/10.13140/RG.2.2.34980.22404

Florida R (2008) Who's your city? Basic Books, New York

Fox N, Hunn A, Mathers N (2007) Sampling and sample size calculation. Yorkshire & the Humber: The NIHR RDS for the East Midlands

Gao BC (1996) NDWI—a normalized difference water index for remote sensing of vegetation liquid water from space. Remote Sens Environ 58:257–266. https://doi.org/10.1016/S0034-4257(96)00067-3

Haase D, Larondelle N, Andersson E, Artmann M, Borgstrom S, Breuste J, Elmqvist T (2014) A quantitative review of urban ecosystem service assessments: concepts, models, and implementation. Ambio 43:413. https://doi.org/10.1007/s13280-014-0504-0

Haase D, Wolff M (2022) Enabling ecosystem services at the neighborhood scale while allowing for urban regrowth: the case of Halle, Germany. Ecol Soc 27(1):22

Hamel P, Guerry AD, Polasky S et al (2021) Mapping the benefits of nature in cities with the InVEST software. npj Urban Sustain 1:25. https://doi.org/10.1038/s42949-021-00027-9

Huete A, Didan K, Miura T, Rodriguez EP, Gao X, Ferreira LG (2002) Overview of the radiometric and biophysical performance of the MODIS vegetation indices. Remote Sens Environ 83(1):195–213. https://doi.org/10.1016/S0034-4257(02)00096-2

Hunter RF, Cleland C, Cleary A, Droomers M, Wheeler BW et al (2019) Environmental, health, wellbeing, social and equity effects of urban green space interventions: a meta-narrative evidence synthesis. Environ Int 130:104923. https://doi.org/10.1186/s12889-020-08762-x

Jovanovic M (2013) The challenges of the urban transport strategy of Belgrade. Glasnik Srpskog Geografskog Drustva 93:35–44. https://doi.org/10.2298/GSGD1303035J

Kabisch N, Larondelle N, Artmann M (2014) Urban ecosystem services in Berlin, Germany and Salzburg, Austria: climate regulation and recreation function. Human-Environmental Interactions in Cities: Challenges and Opportunities of Urban Land Use Planning and Green Infrastructure, Cambridge, pp 66–80

Kashef M (2016) Urban livability across disciplinary and professional boundaries. Front Archit Res 5(2):239–253. https://doi.org/10.1016/j.foar.2016.03.003

Kaszczyszyn P, Sypion-Dutkowska N (2019) Walking access to public transportation stops for city residents. A comparison of methods. Sustainability 11:3758. https://doi.org/10.3390/su11143758

Klimanova O, Illarionova O, Grunewald K, Bukvareva E (2021) Green infrastructure, urbanization, and ecosystem services: the main challenges for Russia's largest cities. Land 10(12):1292. https://doi.org/10.3390/land10121292

Kryukov VA, Golubeva EI (2020) Assessment of the contribution of environmental and social factors to liveability in Moscow. Bull Moscow Univ Ser 5 Geogr 4:32–41. https://doi.org/10.6084/m9.figshare.19518148.v1

Lamond J, Everett G (2019) Sustainable Blue-Green Infrastructure: a social practice approach to understanding community preferences and stewardship. Landsc Urban Plan 191:103639. https://doi.org/10.1016/j.landurbplan.2019.103639

Lantseva AA, Ivanov SV (2016) Modeling transport accessibility with open data: case study of St. Petersburg. Procedia Comput Sci 101:196–206. https://doi.org/10.1016/j.procs.2016.11.024

Levent TB, Nijkamp P (2006) Quality of urban life: a taxonomic perspective. J Stud Reg Sci 36(2):269–281. https://doi.org/10.2457/srs.36.269

Liu Y, Wang Y, Peng J, Du Y, Liu X, Li S, Zhang D (2015) Correlations between urbanization and vegetation degradation across the world's metropolises using DMSP/OLS nighttime light data. Remote Sens 7(2):2067–2088. https://doi.org/10.3390/rs70202067

Maes J, Zulian G, Thijssen M, Castell C, Baró F, Ferreira AM, Melo J et al (2016) Mapping and assessment of ecosystems and their services. Urban ecosystems. Publications Office of the European Union, Luxembourg. https://ec.europa.eu/environment/nature/knowledge/ecosystem_assessment/pdf/102.pdf

Marans RW, Stimson RJ (2011) Investigating quality of urban life: theory, methods and empirical research. Springer, Dordrecht

Maslow A (1987) Motivation and personality, 3rd edn. Longman, White Plains

McPhearson T, Andersson E, Elmqvist T, Frantzeskaki N (2015) Resilience of and through urban ecosystem services. Ecosyst Serv 12:152–156. https://doi.org/10.1016/j.ecoser.2014.07.012

Okulicz-Kozaryn A, Valente RR (2019) Livability and subjective well-being across European cities. Appl Res Qual Life 14:197–220. https://doi.org/10.1007/s11482-017-9587-7

Onnom W (2018) Development of a liveable city index (LCI) using multi criteria geospatial modelling for medium class cities in developing countries. Sustainability 10(2). https://doi.org/10.3390/su10020520

Ramkissoon H, Graham S, Liam D, Weiler B (2013) Testing the dimensionality of place attachment and its relationships with place satisfaction and pro-environmental behaviours: a structural equation modelling approach. Tour Manag 36:552–566. https://doi.org/10.1016/j.tourman.2012.09.003

Raymond CM, Kenter JO, Plieninger T, Turner NJ, Alexander KA (2014) Comparing instrumental and deliberative paradigms underpinning the assessment of social values for cultural ecosystem services. Ecol Econ 107:145–156. https://doi.org/10.1016/j.ecolecon.2014.07.033

Rouse DC, Bunster-Ossa IF (2013) Green infrastructure: a landscape approach. APA Planning Advisory Service Report No. 571, American Planning Association, Chicago

Rozzi GC (2021) ZipcodeR: advancing the analysis of spatial data at the ZIP code level in R. Softw Impacts 9:100099. https://doi.org/10.1016/j.simpa.2021.100099

Rushton G, Armstrong M, Gittler J, Greene BR, Pavlik CE, West M, Zimmerman D (2007) Geocoding health data: the use of geographic codes in cancer prevention and control, research and practice. CRC Press, Boca Raton. https://doi.org/10.1201/9780849384332

Saaty TL (2004) Decision making with the analytic hierarchy process. Int J Serv Sci 1(1):83–98. https://doi.org/10.1504/IJSSCI.2008.017590

Steel P, Schmidt J, Shultz J (2008) Refining the relationship between personality and subjective wellbeing. Psychol Bull 134:138–161

Sydney Metropolitan Catchment Management Authority (SMCMA) (2012) Sydney metropolitan CMA 2012 catchment action plan—a plan for Sydney's liveability. SMCMA, Parramatta

Tarpley JD, Schneider SR, Money RL (1984) Global vegetation indices from the NOAA-7 meteorological satellite. J Climate Appl Meteorol 23:491–494. https://doi.org/10.1175/1520-0450(1984)023%3c0491:GVIFTN%3e2.0.CO;2

United Nations (2017) Resolution adopted by the General Assembly on 6 July 2017. https://ec.europa.eu/environment/nature/knowledge/ecosystem_assessment/pdf/102.pdf. https://documents-dds-ny.un.org/doc/UNDOC/GEN/N17/207/63/PDF/N1720763.pdf?OpenElement

United Nations, Department of Economic and Social Affairs, Population Division (2019) World urbanization prospects: the 2018 revision (ST/ESA/SER.A/420). United Nations, New York

West P, Igoe J, Brockington D (2006) Parks and peoples: the social impact of protected areas. Annu Rev Anthropol 35:251–277. https://doi.org/10.1146/annurev.anthro.35.081705.123308

Wong TH, Brown RR (2009) The water sensitive city: principles for practice. Water Sci Technol 60(3):673–682. https://doi.org/10.2166/wst.2009.436

World Health Organization (WHO) (2004) Promoting mental health: concepts, emerging evidence, practice: summary report. World Health Organization, Geneva. https://doi.org/10.1111/j.1466-7657.2004.00268.x

Yassin HH (2019) Liveable city: an approach to pedestrianization through tactical urbanism. Alexandria Eng J 58(1):251–259. https://doi.org/10.1016/j.aej.2019.02.005

# Index


**B**
Baikal EM-1, 225, 227, 228, 230–234
*Betula pubescens*, 237–239, 241–248
Bioclimatic gradient, 199, 200
Biodiversity, 131, 132, 139, 140

**C**
Calcic Chernozems, 78, 80–85, 181, 182
Carbon balance, 103, 107, 159
Carbon loss, 176
Carbon neutrality, 103, 160
Chechenia Republic, 101, 103, 159, 161, 166
Community level physiological profile, 198, 200, 201
Complex mineral fertilizer, 225, 227, 228, 230–234

**D**
Dissolved organic carbon, 169–171, 173–178

**E**
Ecoacoustics, 131, 132
Ecological features, 238
Ecological revitalization, 252, 262
Ecological soil state, 111, 115, 116, 118
Ecosystem services, 111–115, 117–119, 277–280, 284, 285
Environmental pollution, 208
Enzymatic activity, 2, 225, 228, 233, 270

**F**
Forest ecosystem, 111, 114, 117
Functional diversity of soil microbial community, 48, 54, 55

**G**
GIS, 297
GLOH2O, 252
Grazing intensity, 101, 102, 106, 107
Green area maintenance, 281
Green infrastructure, 196
Greening, 237, 238

**H**
Haplic Arenosols Eutric, 269, 271–274
Haplic Cambisols Eutric, 269, 271–274
Haplic chernozem, 225, 227, 230–234
Haplic Chernozems Calcic, 269, 271–274
Heavy metals, 28, 33–35, 38, 39, 43–47, 49, 51, 52, 56, 67, 69–73
Human-made soil constructions, 196
Humus, 181, 185, 187, 189, 190, 192
Hydrocarbons, 1, 2, 4–8
Hydrological model, 252, 259, 262, 265
Hydrologiska Byråns Vattenbalansavdelning (HBV), 251, 252, 256–259, 261, 262, 265

**I**
Industrial barren, 207–209, 214, 215
Industrial waste, 11, 13, 15, 16, 18–20, 23




M. Korneykova et al. (eds.), *Smart and Sustainable Urban Ecosystems: Challenges and Solutions*, Springer Geography, https://doi.org/10.1007/978-3-031-37216-2

L

Lawn, 11, 12, 15, 18, 24
Length of roots, 270, 271, 273
Livability, 295–300, 304, 307, 309–311

M

Marigold, 11–13, 15, 19, 20, 23, 24
Microbial diversity, 144, 154
Microbial respiration, 104, 106
Microbial specific respiration, 195, 198, 200, 204
Microelements, 78, 80, 81, 83–85
MicroResp, 147
Mindset, 297, 302, 311
Mining industry, 70
Mobile phosphorus, 87–89, 91, 93, 98
Mobile potassium, 87, 91–93, 98
Murmansk region, 238, 239

N

Natural grasslands, 102
Natural soils, 182
Normalizing, 269, 273

O

Organic matter, 181, 182, 185, 190

P

Park and recreational areas of the city, 78
Peatland, 169–171, 174–177
Phytoremediation, 207–209, 214, 218–220
Phytotoxicity, 11–13, 15, 18, 19, 21, 23, 24, 225, 226, 228
Picea abies, 123, 125–129
Plant diversity, 47, 53, 54
Podzol soils, 87, 89, 90, 92–94, 96
Pollution, 1, 2, 5–8, 225, 226, 230, 232

R

R3GIS, 279, 280, 285
Rail station areas, 242
Risk elements, 30, 34
Rostov-on-Don, 225–227

S

Season, 87, 89, 91, 93, 98
Seasonal dynamic, 176
Soil erosion assessment, 208
Soil organic matter, 143, 148, 149, 151, 153, 154
Soil pollution, 27, 29, 30, 32, 34, 35, 37, 39
Soil stock, 87, 89, 93
Song Meter, 131, 133
Streamflow, 252, 256, 258, 259, 263
Subarctic, 12, 13, 15, 24
Subarctic zone, 1, 2
Sustainable cities, 296, 297

T

Technosols, 143–146, 148, 151, 154, 161, 163
Temperate forest, 131, 132, 139
Temporal dynamics, 131, 134–140
Total number of bacteria, 229, 230
Trans-Urals of the Republic of Bashkortostan, 67
TreeTalker, 277, 279–281, 285
TreeTalker+, 123–125, 127–129

U

Ungauged river, 251
Urban ecology, 301
Urban forest parks, 182
Urban green infrastructures, 143–145, 159–162, 166
Urban landscape, 113, 115
Urban landscaping, 124, 125
Urban park, 45, 49, 50, 53, 57
Urban planning, 265
Urban river, 251, 252, 265
Urban soils, 1, 2, 4–8, 13, 20–24

V

Value, 277, 278, 280–282, 285, 286

W

Well-being, 296, 310
West Siberia, 170, 171
Wildfires, 111–113, 116, 117

Z

Zc index, 35, 36